The ECONOMICS of INTERNATIONAL TRADE and the ENVIRONMENT

The
ECONOMICS
of
INTERNATIONAL
TRADE and the
ENVIRONMENT

Edited by
Amitrajeet A. Batabyal
Hamid Beladi

LEWIS PUBLISHERS
Boca Raton London New York Washington, D.C.

Library of Congress Cataloging-in-Publication Data

Batabyal, Amitrajeet A., 1965-
    The economics of international trade and the environment / edited by Amitrajeet
Batabyal and Hamid Beladi.
      p. cm.
    A collection of 20 papers by the editors and other scholars.
    Includes bibliographical references and index.
    ISBN 1-56670-530-4
     1. International trade. 2. Environmental protection. 3. Environmental policy. I. Beladi,
Hamid. II. Title.

HF1379 .B38 2000
363.73—dc21   00-046048

Acknowledgment

For discussions and comments on this book, we thank Swapna Batabyal, Ravi Batra, Fathali Firoozi, Chuck Ingene, Ron Jones, Jim Kahn, and the contributors of the individual chapters.

We thank the Utah Agricultural Experiment Station at Utah State University, the Arthur J. Gosnell endowment at the Rochester Institute of Technology, and Keith Criddle for financial support.

Finally, a big thank you to Ruby Vazquez, who spent an enormous amount of time typing, and typing, and typing …, and to Qing Xu for expert research assistance.

Batabyal would like to dedicate this book to the memory of Sutapa Batabyal.

Amitrajeet A. Batabyal
Hamid Beladi

About the Editors

Amitrajeet A. Batabyal, Ph.D. is Arthur J. Gosnell Professor of Economics at the Rochester Institute of Technology. He earned his B.S. degree in Agricultural Economics with Honors and with Distinction at Cornell University in 1987, his M.S. degree in Agricultural and Applied Economics at the University of Minnesota in 1990, and his Ph.D. in Agricultural and Resource Economics at the University of California, Berkeley, in 1994. Professor Batabyal has taught undergraduate courses in international economics for business, international trade theory, and microeconomic theory, and has taught graduate courses in environmental economics, microeconomic theory, and operations research.

Professor Batabyal has published over 200 books, book chapters, journal articles, and book reviews. As well, he has received many awards and honors including the Robert W. Purcell Scholarship for Research and Economics at the American Enterprise Institute for Public Policy Research, the James E. and Velva L. Rose Prize in International Development at Cornell University, and the College of Business Research Publication Award at Utah State University. He currently is book review editor of the *American Journal of Agricultural Economics*, associate editor of the *Journal of Regional Science*, and editorial council member of the *Review of Development Economics*. His research interests lie in environmental economics, natural resource economics, development economics, international trade theory, and the interface of economics with biology, philosophy, and political science.

Hamid Beladi, Ph.D. is Professor of Economics and holds the Niehaus Chair in Business Administration at the University of Dayton. He earned his Ph.D. at the Utah State University in 1983. From 1992 to 1997, Professor Beladi was the William J. Hoben Research Scholar in International Business at the University of Dayton. He has taught graduate and undergraduate courses in international trade, research methods, mathematical economics, microeconomic theory, and development economics.

Professor Beladi is currently editor of *International Review of Economics and Finance* and associate editor of *Review of International Economics*. He serves on the editorial boards for the *Journal of International Trade and Economics Development* and the *Review of Economic Development*, and is a founding member of the International Economics and Finance Society. He has received the Faculty Scholarship Award and School of Business Scholarship Award from the University of Dayton. Professor Beladi's research interests include international trade and finance, international technology transfer, strategic and trade theoretic issues, foreign investment in LDCs and designing of incentive-compatible contracts, migration models of economic development, decision-making under uncertainty, and natural resource and environmental economics.

Contributors

Edward B. Barbier
Department of Economics
University of Wyoming
Laramie, Wyoming

Scott Barrett
School of Advanced International Studies
Johns Hopkins University
Washington, D.C.

Amitrajeet A. Batabyal
Department of Economics
Rochester Institute of Technology
Rochester, New York

Hamid Beladi
Department of Economics and Finance
University of Dayton
Dayton, Ohio

Christoph Böhringer
Institut für Energiewirtschaft und Rationelle
 Energieeanwendung
Universität Stuttgart
Germany

Graciela Chichilnisky
Department of Economics
Columbia University
New York, New York

Brian R. Copeland
Department of Economics
University of British Columbia
Vancouver, British Columbia

J. Andres Espinosa
American Express
Phoenix, Arizona

Stefan Felder
University of Berne
Berne, Switzerland

Lewis R. Gale
Department of Economics and Finance
University of Southwestern Louisiana
Lafayette, Louisiana

Bryan W. Husted
Instituto de Empresa
Madrid, Spain

Hiro Lee
Research Institute for Economics
Kobe University
Kobe, Japan

Jeanne M. Logsdon
Anderson School of Management
University of New Mexico
Albuquerque, New Mexico

James R. Markusen
Department of Economics
University of Colorado
Boulder, Colorado

Jose A. Mendez
Department of Economics
Arizona State University
Tempe, Arizona

John D. Merrifield
Division of Economics and Finance
University of Texas
San Antonio, Texas

Edward R. Morey
Department of Economics
University of Colorado
Boulder, Colorado

Nancy D. Olewiler
Department of Economics
Simon Fraser University Burnaby
Vancouver, British Columbia

Carlo Perroni
University of Western Ontario
Ontario, Canada

Michael Rauscher
Institute of Economics
University of Rostock
Rostock, Germany

H. David Robison
Department of Economics
LaSalle University
Philadelphia, Pennsylvania

David Roland–Holst
OECD Development Center
Paris, France

Thomas F. Rutherford
Department of Economics
University of Colorado
Boulder, Colorado

Carl-Erik Schulz
Department of Economics
University of Tromsø
Tromsø, Norway

V. Kerry Smith
Department of Agricultural
  and Resource Economics
North Carolina State University
Raleigh, North Carolina

M. Scott Taylor
Department of Economics
University of Wisconsin
Madison, Wisconsin

James A. Tobey
United States Department of Agriculture
Economic Research Service
Research and Technology Division

Alistair Ulph
Department of Economics
University of Southampton
Southampton, U.K.

Randall M. Wigle
Department of Economics
Wilfred Laurier University
Waterloo, Ontario, Canada

Table of Contents

1 Introduction and Overview of the Economics of International Trade and the Environment[1]

Amitrajeet A. Batabyal and Hamid Beladi

We describe the theoretical and empirical contributions that rigorous economic analysis can make to improve our understanding of the salient issues relating to environmental protection in the presence of international trade. We do this by analyzing and summarizing the intellectual contributions of nineteen theoretical and empirical papers about the nexuses between environmental and trade policy.

Key words: economic theory, environmental policy, game theory, trade policy

JEL classifications: F10, F13, Q20

A. PRELIMINARIES

There is no gainsaying the fact that the subject of trade, particularly international trade, has been central to economic thinking for well over two centuries. Beginning with the seminal work of Adam Smith (1776) and continuing with the well-known work of David Ricardo (1817), economists have generally considered unfettered international trade to be a source of many gains. For instance, with regard to trade[2] with more efficient countries, economists have used the notion of *comparative advantage* to demonstrate that two nations can trade to their mutual advantage even when one of these two nations is more efficient than the other at producing everything. In addition to this, economists have shown that international trade is salutary because it allows nations to export goods whose production makes relatively heavy use of resources that are plentiful nationally while importing goods whose production makes heavy use of resources that are scarce nationally. Finally, economists have pointed out that international trade permits nations to specialize in producing a narrower range of goods, thereby permitting them to enjoy the greater efficiencies of large-scale production.[3]

Despite this demonstration of the many gains from international trade, in recent times the desirability of free trade has been questioned by several groups of people. Environmentalists in particular, disheartened by the General Agreement on Tariffs and Trade (GATT) ruling in favor of Mexico and free trade and against the United States and the apparent welfare of the dolphin, have been aggressive in pointing out what they believe to be the many problems with free trade.[4] Some, such as D. Morris (1990), have even referred to free trade as the great destroyer.[5]

---

[1] We thank Keith Criddle for generously providing funds at appropriate times and Qing Xu for competent research assistance. Batabyal acknowledges financial support from the Utah Agricultural Experiment Station, Utah State University, Logan, UT 84322–4810, by way of grant UTA 024. The usual disclaimer applies.

[2] In the rest of this chapter, we shall use the terms "international trade" and "trade" interchangeably.

[3] For more on the gains from trade, see Krugman and Obstfeld (1994), Ethier (1995), and Rauscher (1997).

[4] See Whalley (1991) for additional details on this issue.

[5] For more on the environmentalist perspective on free trade, see Ekins (1989), Arden–Clarke (1992), the debate between Bhagwati (1993) and Daly (1993) in *Scientific American*, and Ropke (1994).

Why do environmentalists and other like-minded people object to free trade? To comprehend this, consider three issues that have been raised by the opponents of free trade.[6] First, there is the *specialization* issue. It has been pointed out that with free trade, some nations may end up specializing in the production of pollution-intensive goods. Not only will this lead to greater environmental degradation in these nations but it is also likely to lead to substantial welfare losses in the same nations.

Second, it has been claimed that unfettered international trade will encourage trade in *hazardous substances*, with the developed nations of the world typically exporting such substances to the developing nations of the world. The recipients of such substances are generally ill-prepared to handle them; moreover, it has been said that this kind of trade will only encourage cost-conscious Northern corporations to export environmental problems to the nations of the South.

Finally, there is the *interjurisdictional competition* issue. Because all governments are interested in attracting mobile factors of production to their own nations and because it is costly to comply with stringent environmental regulations, governments may deliberately lower their environmental regulations in order to attract these mobile factors of production. This is likely to lead to suboptimal levels of environmental regulation throughout the world.

Recognizing the salience of these issues pertaining to international trade and the environment, the chapters in this book explore, from an economic standpoint, many of the questions that are germane to increasing our knowledge of environmental policy in the presence of international trade, and trade policy in the presence of environmental externalities. What can economic *theory* tell us about the connections between environmental and trade policy? This is the general question that is addressed in Chapters 2 through 11 that comprise Part B of this book. The tools of game and microeconomic theory are used efficaciously by the authors to analyze diverse issues such as the effects of international trade in waste products in the presence of illegal disposal, the nature of environmental policy when market structure and plant locations are endogenous, and ecological dumping.

The authors of Chapters 12 through 20 *apply* economic theory to practical settings to ascertain, *inter alia*, the extent to which this theory can inform actual policy decisions about problems at the interface of international trade and the environment. This section of the book focuses on topics such as the impact of industrial pollution abatement on a nation's balance of trade, the German tax initiative in which carbon taxes with exemptions were used to combat carbon dioxide (CO_2) emissions, and the effects of the North American Free Trade Agreement (NAFTA) on Mexico's environmental policies.

B. THEORY

Given recent discussions about the desirability of instituting environmental policies to deal with transboundary pollution, it is salient to ascertain how transboundary pollution flows, production, factor prices, and the terms of trade are affected by alternate pollution control policies. This question is addressed comprehensively by John Merrifield in Chapter 2. Merrifield analyzes the effects of production taxes and best available control technology standards in a two-country, static, general equilibrium model. Unsurprisingly, this analysis shows that in an international setting, neither country is able to use policies *unilaterally* to deal with transboundary pollution effectively. More interestingly, comparing the pros and cons of two pollution control instruments, Merrifield shows that a production tax (a domestic policy instrument) can actually have a perverse effect on pollution. In particular, the use of a production tax to control pollution can actually *increase* pollution.

If domestic policies cannot always be relied upon to control external diseconomies, then can one rely on trade policies to control externalities? This question is the subject of the interesting Chapter 3 by Brian Copeland. Copeland analyzes this question in the context of international trade

---

[6] For a more detailed description of these and other related issues, see Chapter 1 in Rauscher (1997).

in waste products. He shows that there are two circumstances in which the use of trade policies to restrict trade in waste can be welfare improving. First, when the waste disposal sector is *not* taxed optimally, a policy that restricts foreign waste disposal is optimal in a second-best sense. Second, in the presence of illegal waste disposal, a trade tax, when used to supplement a production tax, can improve welfare. This is because the trade tax reduces both the flow of waste *and* the fraction of waste that is illegally disposed.

Because Copeland works with a single-country model, his analysis does not account for the *strategic* aspects of the use of a trade tax to control waste disposal. The strategic aspects of environmental policy are nicely studied by James Markusen et al. in Chapter 4. These authors use a two-country, two-firm, three-good model with increasing returns and pollution to examine the links between environmental policy, plant location, and market structure. Their model permits polluting firms to alter the number and locations of their plants in response to specific environmental policies, and general equilibrium is found as the solution to a two-stage game. In this setting, two key results are obtained. First, it is shown that when firm-specific fixed costs are high (low) and plant-specific fixed costs are low (high), a multi-plant (single-plant) market structure is likely to emerge. Second, the authors convincingly argue that when setting pollution taxes, regulators need to account for the endogeneity of the market structure to environmental policies.[7]

Like Chapter 4, Chapter 5 also focuses on the strategic aspects of environmental policy in an international setting. In this important chapter, Michael Rauscher tells us that two interpretations can be given to the notion of *ecological dumping*. With these two interpretations in place, Rauscher identifies the economic motives for engaging in ecological dumping. His analysis tells us that ecological dumping can be rationalized by appealing either to strategic trade policy arguments or to lobbying arguments. Rauscher favors the latter argument. As he explains, even though it is not always true, in actual policy settings most exporting producers believe that less stringent environmental regulations will *help* them. This provides a rationale for employing lobbyists who will press for *relaxed* environmental regulations. In turn, if these export lobbies are more powerful than other lobbies, then this provides a possible explanation for ecological dumping.

Rauscher's less favored strategic trade policy arguments are elaborated upon by Scott Barrett in Chapter 6. In particular, Barrett poses and answers the salient question, "When does it make sense for governments to set weak environmental standards?" Using a model that is a stage game involving two governments and their industries that sell their output in a third market, Barrett shows that the domestic government will want to set weak environmental standards when the domestic industry is a monopoly, the foreign industry is imperfectly competitive, and firms engage in Cournot competition. More significantly, Barrett points out that this finding is *not* robust. Specifically, if the domestic industry is oligopolistic or if firms engage in Bertrand competition, the incentive to weaken environmental standards is itself weakened and may even disappear completely.

From Barrett's analysis in Chapter 6, it is clear that in order to develop optimal environmental policies, one needs to comprehend the *nexuses* between markets and the environment. But, what about institutions such as *property rights*? In particular, what role do property rights play over environmental resources in encouraging or hindering trade between nations? This question is ably addressed by Graciela Chichilnisky in Chapter 7.[8] To conduct her analysis, Chichilnisky uses a two-factor, two-good, two-country model in which the environment — which is one of the factors of production — is owned as unregulated common property in one country (the South) and is owned as private property in the second country (the North). In this setting, Chichilnisky establishes two results. First, she shows that *differences* in the property rights regime in otherwise identical countries is sufficient to create North–South trade. Second, it is shown that this trade will result in

---

[7] An important issue in this setting concerns the incentives that firms have to agglomerate in a single location. For more on this issue, see Ulph and Valentini (1997).

[8] Similar issues have been addressed in an interesting recent paper by Brander and Taylor (1997). However, the Chichilnisky and the Brander and Taylor models are quite different. As such, it is not surprising that some of Chichilnisky's results are not corroborated by the analysis of Brander and Taylor.

excessive use of the environment in the South. To correct this excessive use, Chichilnisky recommends the use of property rights policies rather than taxes in the South.

A North–South world is also the setting of the Chapter 8 analysis of trade and transboundary pollution by Brian Copeland and Scott Taylor. However, here the difference between the Northern and the Southern countries is that the Northern countries are human capital-abundant relative to the Southern countries. In this setting, three significant results are shown to hold. First, in an equilibrium with factor price equalization, Northern countries lose from trade, Southern countries gain from trade, and trade does *not* affect world pollution. Second, in an equilibrium without factor price equalization, pollution in the North declines with trade, pollution in the South rises with trade, and world pollution is *higher* in the presence of free trade. These two results are valid when there is a *large* number of countries. As one would expect, when there is a small number of countries, the possibility that countries will want to use environmental policy *strategically*, i.e., to improve their terms of trade, must be considered. In this small-numbers case, Copeland and Taylor show that whereas the Southern countries would prefer that environmental policy *not* be used as an instrument of trade policy, the Northern countries would like to have a regime that *permits* environmental policy to be used as an instrument of trade policy.

Additional issues relating to this small-numbers case are analyzed by Alistair Ulph in Chapter 9. Specifically, Ulph revisits the subject of Chapter 5, ecological dumping. However, unlike Michael Rauscher in Chapter 5, Ulph favors a strategic trade policy interpretation of ecological dumping. He uses a partial equilibrium model in which there are two producers of a homogeneous good, and each firm is located in a different country. Because Ulph's focus is on *symmetric* equilibria, both producers and the relevant countries are identical. This construct is used by Ulph to obtain two interesting results about the nature of strategic policy formulation. First, it is shown that permitting producers to act strategically *diminishes*, but does not eliminate, the incentives for governments to loosen environmental policy. Second, allowing governments to act strategically only *increases* the incentives that producers have to act strategically.

Because Ulph's analysis is based on a number of specific assumptions, it is possible to question the generality of his results. For instance, one can ask what happens to the results when producers and governments interact with each other over time. More generally, what insights do *dynamic* models provide about the connections between environmental and trade policy? This question is competently studied by Edward Barbier and Carl–Erik Schulz in Chapter 10. Barbier and Schulz use an augmented bioeconomic model of species exploitation and habitat conversion and ask whether it makes sense for developed country importers of wildlife products to use trade interventions — such as tariffs and import bans — to influence the exploitation of wildlife and the conversion of natural habitats in developing countries. The comparative statics results presented in this chapter show that, in general, trade interventions *cannot* be relied upon to increase the long run total species stock. In contrast, an international transfer of funds will generally lead to *greater* long run conservation of species and natural habitat.

The analysis of Barbier and Schulz in Chapter 10 tells us that in a *non-strategic* dynamic setting, trade interventions are unlikely to attain desired objectives. Is the same true of trade interventions in a strategic setting? In other words, in an imperfectly competitive game setting, can one make a case for using trade policies to achieve environmental goals? This and related issues are taken up by Amitrajeet Batabyal in his Chapter 11 analysis of the links between environmental and trade policy. Specifically, Batabyal poses and answers two questions. First, can environmental policy, pursued strategically by a country in a Cournot game, immiserize that country when the incidence of pollution is domestic? Second, what are the effects of regulating international pollution with a tariff in a Cournot game in which national governments are affected by international pollution but polluting firms within nations are not?[9] The analysis in this chapter shows that the pursuit of strategic environmental policy *can* immiserize a country. Moreover, it is possible for a country to

---

[9] For additional discussion of these two questions, see Batabyal (1996a) and Xu and Batabyal (2000a, 2000b).

use a tariff to make its own consumers and producers better off. However, as Batabyal points out, this latter result is very dependent on the values of specific parameters of his model.

Chapters 2 through 11 of this book provide us with diverse theoretical perspectives on the economics of international trade and the environment. Collectively, these chapters illustrate the many useful theoretical insights that can be gained by engaging in rigorous microeconomic and game-theoretic analyses of environmental protection in the presence of international trade. A logical question now is this: How can this theoretical knowledge be used to increase our understanding of the practical aspects of environmental and trade policy? It is to this application issue that we now turn.

C. APPLICATIONS

In Chapter 12, David Robison uses a 78-sector statistical model to shed light on two related questions. First, what effects do marginal changes in industrial pollution abatement have on the U.S. balance of trade? Second, is it true that undertaking pollution abatement will reduce a nation's comparative advantage in the production of high abatement cost goods and improve it in the production of low abatement cost goods? With regard to the first question, Robison's statistical analysis tells us that marginal changes in industrial pollution abatement have *reduced* the U.S. balance of trade for virtually every industry analyzed. As far as the second question is concerned, this chapter finds empirical support for the hypothesis that industrial pollution abatement is altering U.S. comparative advantage so that the abatement content of imported goods is rising relative to that of goods exported by the U.S.

An implication of the analysis in Chapter 12 is that high abatement cost industries are likely to move to countries that do not adopt stringent environmental regulations. Is this "industrial flight" hypothesis valid?[10] This question is capably addressed by James Tobey in Chapter 13. Tobey uses the cross-section Heckscher–Ohlin–Vanek (HOV) model to determine the impact of environmental regulations on the pattern of trade. Two different approaches are used. In the first approach, a qualitative variable is employed to represent the strength of environmental regulations in the estimated equation. In the second approach, the variable representing the strength of a country's environmental regulations is omitted from the estimated equation and the signs of the estimated error terms are studied. Tobey's econometric analysis shows that environmental regulations per se have *not* caused the pattern of trade to deviate from the predictions of the HOV model. This result is interesting because it contradicts an implication of one of David Robison's central findings in Chapter 12. In addition to this, Tobey's analysis finds no evidence to support the "industrial flight" hypothesis.

Generally speaking, international *cooperation* is necessary to resolve disputes involving trade and the environment. However, as is well known, this cooperation will often not be forthcoming.[11] Consequently, in such situations, individual countries may want to pursue environmental and/or trade policies unilaterally. The econometric analysis of Chapter 13 tells us that the outcome of *unilateral* environmental regulations need *not* be deleterious.[12] The subject of unilateral policy-making is clearly an important one and it needs to be studied in detail, which it is by Stefan Felder and Thomas Rutherford in Chapter 14. In this chapter, Felder and Rutherford use a six-region dynamic general equilibrium model to conduct a detailed empirical analysis of the economic consequences of a unilateral cutback of carbon dioxide emissions by the OECD countries. The authors show that unilateral cuts create incentives for free riding by non-participating regions. Second, unilateral cuts lead to carbon leakage. Although this suggests that unilateral carbon abatement policies are *damaging*, it is important to note that this analysis does *not* take the benefits of reduced greenhouse gases into account. Consequently, it is still possible that when all the relevant

---

[10] This hypothesis has aroused a great deal of interest in the literature on international trade and the environment. For additional details, see Leonard (1984) and the papers cited in Batabyal (1991).

[11] For more on this, see Haas et al. (1993), Batabyal (1996b, 2000a), and the collection of papers in Batabyal (2000b).

[12] Also see Footnote 9 and Batabyal (1991, 1993).

effects have been considered, unilateral policies will have a salubrious effect on environmental quality and on national welfare.

Unilateral environmental policies receive some attention from Carlo Perroni and Randall Wigle in Chapter 15 as well, but the primary focus of these authors is on the following salient question: Does trade liberalization necessarily have a negative impact on environmental quality, or is it possible to treat environmental protection and trade liberalization as separate objectives? To answer this question, Perroni and Wigle analyze a general equilibrium model that is calibrated to a 1986 world data set. This analysis leads to two important conclusions. First, it is shown that although trade liberalization does have a noticeable effect on the environment, the cause-effect link between trade liberalization and environmental degradation is *weak*.[13] Moreover, this weak link is likely to disappear when countries institute apposite environmental policies. Second, the authors point out that environmental and trade policies are not necessarily interdependent. Contrary to conventional wisdom, this means that optimally set environmental policies are *unlikely* to be immiserizing in an open economy.

The analysis in the previous chapter suggests that meaningful conclusions about the nexuses between environmental and trade policies can only be drawn by comprehending the *specific* circumstances of each country's economy in relation to world markets. This suggestion is also made in Chapter 16. Here, Kerry Smith and Andres Espinosa first observe that a key weakness of extant empirical models of trade and the environment is that these models have assumed *separable* preferences. They then extend a standard computable general equilibrium model and assess the extent to which the *assumptions* made in four different models have influenced their conclusions about the welfare effects of trade liberalization. *Inter alia*, this assessment is persuasive in documenting the far-reaching implications of separability. As Smith and Espinosa note, if one assumes separable preferences, one will be *unable* to recognize that externalities influence *and* are influenced by final good choices.

The suggestion that it is salient to comprehend the specific circumstances of a country's economy in relation to world markets is followed by the authors of Chapters 17 through 19 as well. Specifically, the subject of Chapter 17 is West Germany. In this chapter, Christoph Böhringer and Thomas Rutherford use a 58-sector general equilibrium model of the West German economy that is calibrated to 1990 data, and they pose and answer three questions. First, do exemptions magnify the costs of unilaterally imposed carbon taxes? Second, if the objective is to protect jobs, then is it better to use direct wage subsidies or tax exemptions? Third, how do tax exemptions affect export performance? Böhringer and Rutherford's analysis of a static model yields interesting answers to these three questions. Exemptions *do* increase the costs of carbon taxation, and it is significantly *cheaper* to protect jobs with a wage subsidy. Finally, it is shown that exports from the tax exempted sectors *decline*. These findings lead the authors to conclude that increases in emission reduction targets will pose serious adjustment problems for energy and export intensive sectors of the West German economy.

In Chapter 18, Hiro Lee and David Roland–Holst point out that when it comes to studying issues pertaining to trade and the environment, Indonesia is worthy of analysis, not only because Indonesia has a comparative advantage in polluting industries, but also because Indonesia's trade has historically conferred asymmetric environmental effects on its trading partners. As such, Lee and Roland–Holst use a two-country computable general equilibrium model of Indonesia and Japan, and analyze the nature and the effects of Indonesian trade. This analysis leads to two striking conclusions. First, the authors point out that unilateral trade liberalization by Indonesia will *increase* emissions of virtually all industrial pollutants. This notwithstanding, it is noted that when uniform pollution taxation is combined with trade liberalization, it is possible to reduce industrial pollution *and* maintain or even increase welfare. These two conclusions show that welfare enhancement and environmental quality improvement need not be contradictory goals.

---

[13] This conclusion has not received universal support. In a recent theoretical paper, Copeland and Taylor (1997) have shown that the cause-effect link between trade and environmental degradation can be "strong." In other words, trade can lead to a cycle of increased pollution, lower environmental quality, and lower real incomes.

Mexico and the quality of its environment are the subject of Chapter 19. In this chapter, Bryan Husted and Jeanne Logsdon ask a simple but important question: what effect has the North American Free Trade Agreement (NAFTA) had on the formulation and the implementation of environmental policy in Mexico? To answer this question, the authors provide empirical evidence and discuss the nature of environmental policy in Mexico before and after 1990, the year in which the drive to make NAFTA a reality began in earnest. The authors point out that Mexico's environment was in bad shape in the pre-1990 era. However, in the early 1990s, the time period in which NAFTA debates were vigorous, both environmental policy-making and enforcement improved. In particular, environmental programs were *not* subjected to budget cuts, even when the nation was going through the financial crisis of 1995. This and measures taken to make Mexico's environmental performance transparent to outsiders lead the authors of this chapter to the following conclusion: although it is still too early to make definitive statements about the total effect of NAFTA on Mexico's environment, the available evidence does suggest that the NAFTA experience has left an inexpungible *positive* mark on environmental policy in Mexico.

NAFTA and the more general question of the impact of trade on the environment are looked at from a different angle by Lewis Gale and Jose Mendez in the concluding Chapter 20. In a well known paper, Gene Grossman and Allan Krueger (1993) demonstrated the existence of an inverted-U relationship between pollution and per capita income. In addition to this, Grossman and Krueger also suggested that patterns of specialization have more to do with traditional sources of comparative advantage and less to do with cross-country differences in environmental standards. In this chapter, Gale and Mendez reexamine the causes of these two results. Unlike Grossman and Krueger, who used a single proxy to capture the effects of both scale and technique on environmental quality, Gale and Mendez use two proxies and find support for the Grossman and Krueger contention that first scale and then technique effects account for the inverted-U relationship. With regard to the Grossman and Krueger point about the pattern of specialization, the econometric analysis of Gale and Mendez shows that cross-country differences in endowments *do* have an impact on the environment. Specifically, pollution increases with the capital abundance of a country and falls with increases in land and labor abundance.

D. CONCLUSIONS

The chapters in this book effectively describe the theoretical and empirical contributions that rigorous economic analysis can make in improving our understanding of the causes of and solutions to a variety of problems concerning the conduct of environmental policy in the presence of international trade. These chapters also provide us with a state of the art perspective on what is currently known about the theoretical and empirical nexuses between environmental and trade policy. The task for researchers now is to use the findings contained in this book to design and implement efficient environmental policies that will attain *environmental* policy goals. At the very least, this will assuage the increasingly acrimonious nature of international discussions about issues at the interface of international trade and the environment.

REFERENCES

Arden–Clarke, C., South–North terms of trade: environmental protection and sustainable development, *Int. Environ. Aff.*, 4, 122, 1992.

Batabyal, A. A., Environmental policy and unilateral control in an open economy, *Nat. Resource Modeling*, 5, 445, 1991.

Batabyal, A. A., Should large developing countries pursue environmental policy unilaterally? *Indian Econ. Rev.*, 28, 191, 1993.

Batabyal, A. A., Game models of environmental policy in an open economy, *Ann. Reg. Science*, 30, 185, 1996a.

Batabyal, A. A., An agenda for the design and study of international environmental agreements, *Ecol. Econ.*, 19, 3, 1996b.

Batabyal, A. A., Introduction, in *The Economics of International Environmental Agreements*, Batabyal, A. A., Ed., Ashgate, Aldershot, U.K., 2000a.

Batabyal, A. A., Ed., *The Economics of International Environmental Agreements*, Ashgate, Aldershot, U.K., 2000b.

Bhagwati, J., The case for free trade, *Sci. Am.*, 275, 42, 1993.

Brander, J. A. and Taylor, M. S., International trade between consumer and conservationist countries, *Resour. Energy Econ.*, 19, 267, 1997.

Copeland, B. R. and Taylor, M. S., The trade-induced degradation hypothesis, *Resour. Energy Econ.*, 19, 321, 1997.

Daly, H. E., The perils of free trade, *Sci. Am.*, 275, 50, 1993.

Ekins, P., Trade and self-reliance, *Ecologist*, 19, 186, 1989.

Ethier, W. J., *Modern International Economics*, 3rd ed., W. W. Norton, New York, 1995.

Grossman, G. M. and Krueger, A. B., Environmental impacts of a North American free trade agreement, in *The U.S.–Mexico Free Trade Agreement*, Garber, P., Ed., MIT Press, Cambridge, 1993.

Haas, P. M., Keohane, R. O., and Levy, M. A., Eds., *Institutions for the Earth: Sources of Effective International Environmental Protection*, MIT Press, Cambridge, 1993.

Krugman, P. and Obstfeld, M., *International Economics: Theory and Policy*, 3rd ed., Harper Collins College Publishers, New York, 1994.

Leonard, H. J., *Are Environmental Regulations Driving U.S. Industries Overseas?* The Conservation Foundation, Washington, D.C., 1984.

Morris, D., Free trade: the great destroyer, *Ecologist*, 20, 190, 1990.

Rauscher, M., *International Trade, Factor Movements, and the Environment*, Clarendon Press, Oxford, 1997.

Ricardo, D., *The Principles of Political Economy and Taxation*, John Murray, Albermarle Street, London, 1817.

Ropke, I., Trade, development, and sustainability: a critical assessment of the "free trade dogma," *Ecol. Econ.*, 9, 13, 1994.

Smith, A., *An Inquiry into the Nature and the Causes of the Wealth of Nations*, Strahan and Cadell, London, 1776.

Ulph, A. and Valentini, L., Plant location and strategic environmental policy with inter-sectoral linkages, *Resour. and Energy Econ.*, 19, 363, 1997.

Whalley, J., The interface between environmental and trade policies, *Econ. J.*, 101, 180, 1991.

Xu, Q. and Batabyal, A. A., Price competition, pollution, and environmental policy in an open economy, forthcoming, *Ann. Reg. Science*, 2000a.

Xu, Q. and Batabyal, A. A., A Bertrand Model of Trade and Environmental Policy in an Open Economy, unpublished manuscript, Utah State University, Logan, 2000b.

2 The Impact of Selected Abatement Strategies on Transnational Pollution, the Terms of Trade, and Factor Rewards: A General Equilibrium Approach

John D. Merrifield[1]

Widespread publicity of international conflict over transfrontier pollutants, such as acid deposition and the November 1986 chemical spill in the Rhine River, has increased the public's awareness of the need to devise abatement strategies appropriate for transnational pollution. This chapter develops a general equilibrium approach for analyzing the economic impacts of selected abatement strategies. The model includes internationally mobile goods, capital and pollution flows, and flexible output and factor prices. The general equilibrium approach provides policymakers with important information that is usually not revealed by a partial equilibrium approach. That is demonstrated by applying the model to the North American acid deposition issue. © 1988 Academic Press, Inc.

A. INTRODUCTION

Governments throughout the world have grappled with the issue of transnational pollution for some time. A long-time problem has been Europe's Rhine River; a recipient of wastes from several countries, including a major chemical spill in November 1986 (*Time*, 11/24/86). In 1909, the United States (U.S.) and Canada negotiated the Boundary Waters Treaty to control pollution in the lower Great Lakes and the St. Lawrence Seaway (Walter, 1975). The *Trail Smelter* case (Walter, 1976) put transfrontier pollution in the headlines in the 1920s. Pollution of the Great Lakes is still a subject of U.S.–Canadian negotiation and cooperation, resulting in an agreement in 1983 to limit phosphorous pollution (*USA Today*, 10/1/83).

Acid deposition, more often referred to as acid rain (wet acid deposition), has received widespread publicity in North America.[2] Pictures of decaying structures (such as the Statue of Liberty), reports of fish kills, possible soil productivity losses, and the high cost of abatement have aroused[3]

---

[1] Based on Dr. Merrifield's dissertation research, conducted at the University of Wyoming in Laramie.

[2] Acid deposition is also a major issue in Europe (Ohlendorf, 1984). Other current North American transnational pollution problems include air pollution in Ciudad Juarez–El Paso (U.S.–Mexico), Gulf of Mexico oil spills (mostly U.S.–Mexico), Lower Colorado River Basin salinity (U.S.–Mexico), and water pollution in the Great Lakes (U.S.–Canada).

[3] According to a Louis Harris poll, 90% of the American people think acid rain is a serious problem. Awareness of the problem has doubled since 1980 (*Coal Outlook*, 1984c).

---

the general public's awareness. The high cost of action and inaction,[4] and the income and geographic distributional effects of proposed abatement strategies, have made acid deposition a hot political issue. Controversy within the Reagan administration and the Congress, demands for action by affected domestic groups[5] and by the Canadian government have, until recently, prompted only numerous studies of the problem. In March 1987, President Reagan endorsed a $2.5 billion, five-year effort to find cleaner methods of burning coal.

The primary purpose of this chapter is to develop and illustrate the properties of a model that shows how transnational pollution flows, production, the terms of trade, and factor prices are affected by two abatement strategies. A welfare index summarizes the impact of both strategies on national welfare. After discussing the general properties of the model, it is applied to the North American acid deposition issue. The acid deposition issue is employed primarily to put the discussion of the model on a more concrete basis. A secondary objective of the chapter is to demonstrate how a general equilibrium (GE) model can reveal some important aspects of a transnational pollution issue, which a partial equilibrium analysis often fails to discover. No attempt is made to provide a thorough economic analysis of acid deposition or all the proposed abatement strategies.

Production taxes and regulations requiring a certain amount of pollution abatement equipment at each source (Best Available Control Technology standards) are analyzed because existing practices in the U.S. suggest that such strategies will be considered for any pollution problem, and are likely to be implemented again in the future (USGAO, 1983).

The model is an advance over previous work in this area. Pollutants, goods, and capital are internationally mobile, and prices of factors and goods are determined endogenously. Both countries are explicitly included in the model, and each has some price-setting power. In contrast, previous work in this area generally has assumed that some prices are constant, with frequently only one country explicitly included in the model, and usually only one model component (goods, factors, or pollutants) is internationally mobile. Section B provides a brief review of previous related work, including trade models with externalities and public goods, and GE models with pollution.

Following the literature review, a GE trade model with transnational pollution is introduced (Section C) and discussed in some detail. The supply and demand sides of the economy are included, but the supply side is emphasized. An increase in the economic activity of either country raises the pollution flow of both countries (pollution is a pure public bad). The scarcity of capital services and the pollution flow are directly related. Capital (defined to include natural resources) is damaged by the pollutant. The model's endogenous variables are the factor rewards, the terms of trade, each country's total output, each country's share of the available stock of capital, and the pollution flow.

In Section D, general properties of the model are discussed. Then the model is used to analyze the abatement strategies in the context of the acid deposition issue. For a more realistic analysis, pollution flows are allowed to vary between the two countries; pollution is no longer a pure public bad. Comparative static results of changes in production taxes and abatement equipment standards are presented. The welfare implications of each strategy for each country are discussed in Section E.

Together, the results indicate that Canada has more unilateral policy options likely to increase its welfare. That result was unexpected, because Canada is a much smaller country and receives a much larger share of its acid deposition from the U.S. than the U.S. receives from Canada. It occurs because Canadian emissions account for a larger share of either country's acid deposition than its share of the total internationally mobile capital stock. Also, the imposition of a production tax or a tightening of the abatement equipment standard by either country or jointly by both countries is more likely to increase Canadian than U.S. welfare. Each country's net creditor — debtor status is the major determinant of that result. A summary and conclusions are presented in Section F.

---

[4] The only published estimate of abatement benefits for the U.S. is a 1978 figure of $5 billion annually (Crocker, 1983).

[5] Six northeastern states (Maine, New York, Rhode Island, Vermont, Massachusetts, and Connecticut) and four environmental groups (no specifics given) have sued the Environmental Protection Agency to use existing authority to curtail sulfur emissions (*Coal Outlook*, 1984b). In addition, the National Governors Association voted 24–10 to back immediate action to control acid rain (*Coal Outlook*, 1984a).

B. LITERATURE REVIEW

The effort to advance economic theory and improve policy-making is continuing on at least two fronts. This section will show how existing work with GE models with pollution but not international trade[6] (Comolli, 1977; Forster, 1977; Forster, 1981; Yohe, 1979), trade models with pollution (Asako, 1979; Pethig, 1976), and the work of McGuire (1982) and Siebert et al. (1980), including trade and pollution in a GE setting, have increased our understanding of transnational pollution.

The following pages will be devoted to a somewhat detailed discussion of the articles mentioned above. This section concludes with a discussion of how those articles have contributed to and facilitated further advances by this research effort.

1. GENERAL EQUILIBRIUM MODELS WITH POLLUTION

GE models are now being applied to the problems of pollution damage and pollution control because of growing interest in the incidence (who bears the burden) and distributional effects of pollution abatement policies. The comparative advantage of GE models is their ability to reflect the distributional effects of policy changes. Furthermore, it is often possible to summarize the comparative static results with a welfare index (Section E).

Of the four papers to be discussed here, only the Yohe (1979) and Forster (1981) articles address the income distribution-incidence question directly. Distributional-incidence implications are extracted from the models of Comolli (1977) and Forster (1977) by Forster (1981).

Comolli developed a static model and a dynamic model to consider the comparative statics of more stringent abatement requirements and the adjustment process set in motion by requiring increased abatement effort. The distinctive features of the Comolli model arise from the assumptions about pollution damage and the mechanism for pollution control. Pollution is a byproduct of producing the first good (X_1). It can accumulate over time, and the stock (not flow) of pollution enters the production function of the second good as an input with a negative marginal product. The pollutant either does not directly enter the utility functions of individuals, or if it does, it is assumed that goods consumption and the pollution stock are additively separable arguments within the utility function. In other words, the pollution level is taken as given by each consumer, and so does not affect the consumption mix decision. Abatement occurs when a larger share (β) of X_1 output goes to pollution control. The value of β is set by the government.

The mechanism that reduces pollution is the key to the distributional-incidence implications of the Comolli model. As the amount of X_1 available to consumers is reduced, its relative price is bid up, which then increases the return of the factor used intensively in its production. That result is obtained from both the static and the dynamic versions of Comolli's model. If, instead, X_2 is diverted for the pollution abatement effort, the return of X_1's intensive factor falls, while the return of X_2's intensive factor rises.

In Forster (1977), pollution is a result of producing and consuming both goods. The stock of pollutants is an argument in the social welfare function, but it is not in the production function of either good. As in Comolli's model, individual consumers take the stock of pollution as given. Without government regulation, the aggregate purchases of consumers will not be the socially optimal amount.

For the "more polluting commodity," the producer's price is too high and the consumer's price is too low. The authorities correct the market failure with a tax on the more polluting commodity. The reduction in the producer's price reduces the price of the input used intensively to produce the more polluting commodity. With a different policy tool for increasing pollution abatement, a result opposite Comolli's is derived. The reversal occurs because Forster's per-unit tax reduces the producer's price, while Comolli's hike in β raises the producer's price.

---

[6] Trade does occur between sectors within the single country.

Yohe (1979) adapts the model of Batra and Casas (1976) to look at how selected abatement strategies affect factor rewards. Pollution occurs because X_1 producers require the environment as a waste disposal sink. The ability to pollute is a specific production factor of X_1 producers, which makes pollution their intensive factor. In contrast to the Comolli and Forster models, the Yohe model is static, and why abatement is desirable is not indicated. Pollution does not explicitly damage producers or consumers in the model, but the authorities specify the maximum rate of pollution.

Early in this paper, Yohe assumes that output prices are fixed. That assumption, along with incomplete specialization, would normally[7] break the link between factor endowments (including the permissible rate of pollution in Yohe's model) and factor prices. However, with more factors than goods,[8] factor prices are affected by a tightening of the pollution standard. The shadow price of waste emissions rises, the non-intensive factor (of both goods) reward falls, and the return to X_2's intensive factor rises. Tightening the pollution standard reduces the supply of an important X_1 input, and thereby reduces the supply of X_1. That releases capital and labor to X_2 producers. If labor (capital) is X_2's intensive factor, the shift of some capital and labor from X_1 to X_2 will raise (lower) K_2/L_2 and increase (decrease) the return to labor while reducing (raising) the price of capital.

Yohe's result is similar to Forster's (1977) but occurs for different reasons. In Forster's model, price changes alter factor rewards. In Yohe's model, with prices fixed, a change in the availability of the environmental factor, plus the difference in the number of goods and factors, alters factor prices. When output prices are flexible, the effect of a tighter pollution standard on factor rewards is qualitatively identical to the result with fixed output prices but the size of the effect will vary with the price elasticity of demand for each good.

Forster (1981) also introduces extensions of a model similar to Yohe's. Forster (1981) first considers a situation in which waste emissions are not homogeneous. Labor, initially the only other productive input, is mobile between industries. With a mobile factor and two specific factors, the model is similar to Jones' (1971) three-factor, two-good model. The two factors employed in each industry are continuously substitutable. With flexible factor prices and the other restrictions, labor is fully employed. The authorities regulate the flow of pollution (static model, no stock accumulation) by selling transferable pollution permits to each industry.

With fixed-output prices, fewer pollution permits for either industry means a rise in the market price of permits in both industries. It also reduces the price of labor by increasing the labor-pollution ratio in each sector. The sector facing the permit cutback reduces its output and its demand for labor. The full employment assumption requires that the wage rate fall so that the sector not directly affected by the permit reduction can expand its labor force.

Additional results are obtained from the model by varying the pollution control mechanism. When the authorities levy an emissions' tax, factor rewards respond to an increase in one of the taxes like they did to a reduction in the number of one sector's pollution permits. Again, the price of labor falls, but since the government now controls the price of pollution in each sector, the return to emissions does not change. With an emissions' tax, the government no longer directly controls the rate of pollution. An attempt to reduce the waste emissions of one sector can increase the emissions of the other sector. For instance, if the government raises the tax on the X_1 industry's emissions, X_1 falls, some workers move to X_2 industry jobs, and the marginal product of X_2 emissions rises. With no change in X_2's emissions' tax or the price of X_2, the X_2 industry releases more pollutants so as to reestablish equality between the cost of polluting and the value of its marginal product.

Intersectoral trade has an important role in the papers just discussed, but the effect of introducing international trade was not considered. Selected models combining international trade and pollution are discussed next.

---

[7] See Jones (1965) for an explanation of that result.

[8] Jones (1971) has demonstrated some of the ramifications for factor rewards of having more factors than goods.

2. International Trade Models with Pollution

Papers such as Pethig (1976) and Asako (1979) are of greatest interest. The Pethig model was set up to show that the expected welfare gains from trade may not materialize, or they may be accentuated when the pollution intensiveness of the two goods varies. A fixed consumption ratio assures that the international ratio of X_1 to X_2 production cannot change. Only the location of production can change. Pollution is not physically mobile between countries, but it can be indirectly imported or exported. For instance, if the pollution-intensive commodity (X_1) is exported by country A, increased trade could lower A's welfare. Additional pollution damages could offset the usual material gains from trade. If A exports X_2, an expansion of trade would add the benefits of reduced pollution to the material gains from trade. The country that exports the pollution-intensive good may not gain from trade, while the other country always gains. Pollution concerns provide an incentive for one country to expand trade, and for the other country to restrict trade. However, if the same good is pollution intensive in both countries, the fixed worldwide output ratio prevents both countries from simultaneously tightening their environmental policies.

With binding pollution control constraints, the country with the least restrictive controls will export the pollution-intensive good. In effect, the Ricardian basis for trade is reinforced or replaced by a factor-endowments basis for trade. The country with the lowest shadow (with a standard) or real (with an emissions' tax) price for pollution becomes the one relatively well-endowed with the pollution factor. It follows that the two countries' welfare is interdependent, because a change in one country's environmental policy will affect the location of production, and thereby affect the pollution level and welfare of both countries.

Asako's results are similar to Pethig's, despite some differences in their models. In both models, only commodity trade links the two countries. However, in Asako's model, the foreign country's existence is implicit rather than explicit. Other significant differences are Asako's policy tools and his treatment of pollution. Pollution is a joint product, rather than an input. Labor is the only explicit productive factor in the model. In Asako's model, the authorities pursue their environmental quality objectives by altering the pattern of trade with a tax-subsidy approach. By directly linking trade policy and environmental quality concerns, the Pethig result that it may be unwise to expand trade if the export good is the more polluting commodity is also obtained by Asako.

3. Static General Equilibrium Trade Models

Most relevant to this research is the work of McGuire (1982). The setup is similar to Yohe's (1979). Permission to pollute is an input only for the X_1 industry of both countries. Unlike Yohe, damages are included in the model. The marginal social cost of pollution is assumed to be a known constant. With no regulation, pollution is "employed" such that its marginal product is zero. Trade occurs because of differences in factor endowments. Prior to the introduction of environmental regulation,[9] the model is identical to the Heckshcer–Ohlin (H–O) model. Consequently, with no regulation, prices are the same in both countries. Demand factors are discussed occasionally but are not explicitly included in the model.

To analyze the effects of a reduction in the permissible rate of pollution, McGuire further assumes that capital and emissions and labor and emissions are equally substitutable. In other words, the elasticities of substitution, σ_{LT} and σ_{KT}, are equal (in McGuire's notation T denotes the pollution input). Thus, when regulation of T occurs, the production function's shape is not altered. The only change is a renumbering of the isoquants. Before introducing international trade into the analysis, the impact of regulation on a closed economy is considered. The results are identical to Yohe's analysis and contrary to Forster's (1981) homogeneous pollution model. Because only one sector "employs" pollution in Yohe and McGuire, regulation prompts the factor rewards of K and L

---

[9] McGuire (1982) mentions that for some issues it may matter whether pollution is internationally mobile, but the model does not formally incorporate that feature.

to move in opposite directions, with the factor used intensively in the non-polluting, and hence unregulated, industry gaining absolutely in terms of both goods.

When international trade is added to the analysis, it is important to distinguish between coordinated and uncoordinated regulation. Coordinated regulation means that the shadow price of permission to pollute is uniform in the two countries. The factor price equalization result will not be disturbed by coordinated regulation. The autarchy result that the intensive factor of the unregulated industry benefits and the other factor suffers is also retained. With uncoordinated regulation, factor price equalization is lost. For a small country facing fixed world commodity prices, pollution restrictions depress the price of the factor used intensively to produce the polluting good, so that when output is reduced, the released factors can be absorbed by the unregulated industry. The other factor reward rises. If the country is large enough to influence world prices, the change in factor rewards is definitely reversed from the small country case for all but the country enacting the regulations. Since McGuire's pollution control mechanism does not separate producer and consumer prices, producer prices will be bid up when regulation curtails production of the polluting good. This latter result is identical to the factor incidence outcome derived from Comolli's model by Forster (1981). The importance of the choice of pollution control mechanism to the factor incidence result was illustrated earlier by the differences in the Comolli and Forster (1977) results.

For the bulk of the paper, McGuire assumes that productive factors are internally immobile. Just prior to the concluding remarks, McGuire provides a discussion of the intuitive results of uncoordinated regulation when factors move freely between the two countries. With H–O assumptions, in particular the identical technology assumption, the international equality of factor prices cannot be restored until all of the polluting industry has departed the country imposing the strictest environmental regulations. With uncoordinated regulation of pollution, factor prices cannot be equalized so long as both countries are incompletely specialized. Thus, if even one factor is internationally mobile, uncoordinated regulation leads to complete specialization in at least one country, and perhaps both. If the pollutant is an international public bad, uncoordinated regulation is futile, because their only effect is to relocate the source of emissions.

An extension of Pethig's work (Siebert et al., 1980) incorporates mobile capital and international trade in a general equilibrium model. Pollution is not internationally mobile, and as in Pethig, the expenditure ratio for the two goods is fixed. Permission to pollute is an input and is thus desired by producers. However, pollution-caused damages are not explicitly incorporated. The authorities pursue the implicit public good, environmental quality, with either an emissions tax or an emissions standard. H–O type conclusions are drawn from the model. The country with the least restrictive environmental laws exports the pollution-intensive good. The impact of tighter pollution control laws on capital flows and income distribution depends on relative factor intensities of the goods and the abatement technology.

4. ANOTHER STEP FORWARD

The model presented in detail in Section C contains elements of many of the models discussed in this section. Because of its static[10] nature and its treatment of pollution as an input, the model most closely resembles the work of Yohe (1979) and McGuire (1982).

The model allows commodity and input prices to be determined endogenously, and goods, pollution, and capital are internationally mobile, yet the assumptions are generally less restrictive (except for its static nature). In the models just discussed, only McGuire considers the impact of factor mobility. Another significant improvement over previous models is the ability to consider the economic impact of tighter abatement equipment standards. Because of perceived low monitoring costs, equipment standards appear to be much more popular abatement strategies than emissions

---

[10] Contrary to models such as Forster (1977) and Comolli's (1977) dynamic model, damages are related to the rate of pollution, rather than to the size of an accumulated stock of pollution. The model provides a steady state, zero discount rate perspective.

taxes or salable pollution permits, the strategies most often considered by economists. This section has shown that the economic consequences of an abatement effort are very sensitive to changes in the abatement strategy.

C. THE MODEL[11]

The essence of an analysis of transnational pollution is the impact of abatement strategies on prices, and hence on the movement of goods and productive resources between the countries. Price changes affect the profit functions of producers and the welfare of households. Pollution damages are also likely to alter prices. Therefore, the model's features include endogenous input (capital[12] and labor), and goods' prices and capital mobility between countries. As Siebert et al. (1980) point out: "Pollutants ambient to the environment may affect the quality of inputs in production processes, the production procedures, produced commodities, capital goods (wealth damages), etc.". Because damages to capital (broadly defined) by transnational pollutants are well known, the model incorporates a damage function in which the scarcity of capital services varies directly with the pollution flow.[13] The model is static, so there is no mechanism that permits pollutants to accumulate with time. The total final demand output of each country is treated as a single composite good in the model. Country A produces X_1, consumes some of it, and exports the balance to country B. X_1 serves as a numeraire good with a constant price of one. B produces X_2 consumes some of it, and exports the remainder of its output to A. The price of X_2 is P, where P is a relative price ($P_2/P_1 = P_2/1 = P$) determined endogenously. The two economy-wide production functions are:

$$X_i = X_i\big(g(R)K_i, L_i, E_i, Z_i\big) \tag{2.1}$$

where K_i = capital used to produce X_i, $i = 1, 2$; L_i = labor used to produce X_i; E_i = emissions released to the environment in the course of producing X_i; Z_i = pollution abatement equipment employed by producers of X_i, where Z is a pure intermediate good; R = transnational pollutant flow; and $g(R)K$ = capital scarcity varies directly with the transnational pollutant flows ($g' < 0$).

For individual firms (Equation 2.1) is homogeneous degree one. Total output is exhausted by factor payments at the firm level (zero economic profits),

$$C_{L1}W_A + C_{K1}r + C_{Z1}P_{Z1} = 1 \tag{2.2}$$

$$C_{L2}W_B + C_{K2}r + C_{Z2}P_{Z2} = P \tag{2.3}$$

$$C_{LZ1}W_A + C_{KZ1}r = P_{Z1} \tag{2.4}$$

$$C_{LZ2}W_B + C_{KZ2}r = P_{Z2}, \tag{2.5}$$

where C_{ij} = input i per output j; $i = L, K, Z$; $j = X_1, X_2, Z_1, Z_2$; W_A = price of labor in country A; r = price of capital; and P_{Zi} = price of Z_i employed as an intermediate good by X_i producers; $i = 1, 2$. The linear homogeneity assumption eliminates any complications that could result from variability

---

[11] For notational simplicity and to emphasize the general applicability of the model, the countries are referred to as A and B rather than U.S. and Canada. Many equations and complete mathematical derivations of the model are relegated to the Appendix.

[12] In a two-factor model, capital must include natural resources, which can certainly be thought of as part of a country's "capital" stock (i.e., non-labor resource base).

[13] A damage function for labor could be included without sacrificing mathematical certainty in the general case as long as the elasticity of productivity with respect to pollution flows is smaller for labor than for capital.

in firm size or numbers in the initial equilibrium. With an equilibrium as a starting point, it also follows that each firm is using the best available technology.

Several restrictions apply to (Equation 2.1). Capital can be substituted for labor, and Z can be substituted for emissions, but capital cannot be substituted for Z. It follows that the input substitution elasticities σ_{KE}, σ_{KZ}, σ_{LE}, and σ_{LZ} are equal to zero. For the acid deposition issue, recent research[14] suggests that opportunities to marginally reduce emissions through production process modifications are very limited. The alternative processes in use or available represented major capital investments that would have only permitted the firms to install less expensive new abatement equipment. For the small changes in policy variables the model is able to consider, it is not overly heroic to assume that individual firms cannot substitute productive capital or labor for emissions' or abatement equipment. With $\sigma_{KE} = \sigma_{LE} = \sigma_{KZ} = \sigma_{LZ} = 0$, output cannot be increased by employing additional E or Z without also acquiring additional K or L.

Economy wide, doubling of inputs will not result in doubling of output. That conclusion follows from the feedback effect of the externality on the scarcity of capital. When many firms seek to increase output, emissions increase significantly and some of the service flow they expect from their capital stock fails to materialize. Because private marginal products can be shown to equal social average products (Markusen, 1983), the externality will not block the exhaustion of total output at the firm level. Consequently, the simplifying features of linear homogeneity indicated by the preceding paragraphs are not ruled out by the externality's impact on capital scarcity.

In this model, both governments enforce a minimum ratio of Z to output for each firm. Thus, although additional Z is by itself a nonproductive factor, government regulation ensures that an increase in X_i ($i = 1, 2$) will require an equal percentage increase in Z_i. With homogeneous degree one production functions, the expansion path is linear. With constant input prices, an effort to increase output by N percent means an $N\%$ increase in the use of all inputs, including Z. It will be assumed that an unregulated firm would use the environment for waste disposal to the point where the marginal product of emissions is zero.

While individual firms are unable to substitute capital or labor for Z, for the economy as a whole, they are indirect substitutes for Z.[15] Indirect substitution occurs because increasing the demand for Z will divert capital and labor away from the production of final demand goods X_1 and X_2. A change in the government-controlled ratio of Z_i to X_i is analogous to Comolli's shift in β. Both divert resources from final demand to pollution abatement:

$$Z_i = Z_i\big(g(R)K_{Zi}, L_{Zi}\big), \qquad i = 1, 2 . \tag{2.6}$$

With flexible factor prices and the ability to substitute capital for labor, productive inputs will be fully employed in both countries:

$$C_{L1}X_1 + C_{LZ1}Z_1 = L_A \tag{2.7}$$

$$C_{L2}X_2 + C_{LZ2}Z_2 = L_B \tag{2.8}$$

$$C_{K1}X_1 + C_{KZ1}Z_1 = K_A \tag{2.9}$$

---

[14] Summarized in the final report of Work Group 3B formed under the U.S.–Canada Memorandum of Intent on Transboundary Air Pollution (Memorandum, 1981).

[15] Z is assumed to be a non-traded good. While it is certainly conceivable that Z could be exported by one of the countries, that feature was omitted from the model. Adding trade in Z to the model would have added much more in the way of analytic complexity to the model than increased realness would warrant. That conclusion is based on the fact that most industrialized nations devote but 1 to 2% of their resources to pollution abatement (USDC, 1975).

$$C_{K2}X_2 + C_{KZ2}Z_2 = K_B \tag{2.10}$$

$$\overline{C}_{Z1}X_1 = Z_1 \tag{2.11}$$

$$\overline{C}_{Z2}X_2 = Z_2 \tag{2.12}$$

$$K_A + K_B = \overline{K} \tag{2.13}$$

$$\kappa_A + \kappa_B = \overline{K}, \tag{2.14}$$

where κ_i = capital owned in country i; i = A, B,

$$C_{L1} = C_{L1}(W_A, r) \tag{2.15}$$

$$C_{LZ1} = C_{LZ1}(W_A, r) \tag{2.16}$$

$$C_{K1} = C_{K1}(W_A, r, R). \tag{2.17}$$

For C_{L2}, C_{KZ2}, and C_{K2} replace W_A with W_B in (Equation 2.15)–(Equation 2.17). Labor is assumed to be immobile between countries and fixed in quantity.[16] Input-output rations are a function of input prices. The capital-output ratios also vary inversely with pollution flows.[17] The total capital endowment available to the economic system which A and B comprise is assumed fixed, but at the margin, capital is assumed to be freely mobile between A and B. Capital ownership is immobile internationally.

The demand for X_1 and X_2 is assumed to depend on their relative price (P) and total income ($Y = Y_A + Y_B$). Only one demand equation is included explicitly in the model, because, according to Walras' law, when all but one market clears, the remaining market must also clear. The X_2 market was arbitrarily selected as the one to explicitly include in the model:

$$X_2 = X_2(P, Y) \tag{2.18}$$

$$Y = Y_A + Y_B \tag{2.19}$$

$$Y_A = X_1 + r(\kappa_A - K_A) \tag{2.20}$$

$$Y_B = PX_2 + r(\kappa_B - K_B). \tag{2.21}$$

---

[16] That assumption is consistent with the immigration restrictions existing between most nations. Where intranational labor mobility is restricted by cultural ties, incomplete information, or high transport costs, that assumption may still be reasonable for analyses of regions within a nation.

[17] For many transnational pollutants, including acid deposition, no direct health effects have been identified. Indirect effects, so far, are only speculative and seem limited to certain subpopulations. In addition, the health impacts would probably not be observed until long after the initial exposure to the pollutant.

Tastes are assumed to be identical. The identical tastes assumption allows demand to be expressed as a function of the combined income of A and B. With no net savings, identical tastes (equal income elasticities) ensure a balance of payments between the two countries. Homotheticity is also assumed, but is not a critical assumption. Homotheticity simplifies the mathematics and rules out cases where income elasticities are negative (inferior goods) or large (luxury goods).

The model is completed with the addition of the transnational pollution relationships. The pollution flow depends on emissions emanating from both countries. As suggested by (Equation 2.1) and (Equation 2.6), the emissions released by each country depend on output and on the pollution abatement equipment standard:

$$R = R(E_A, E_B) \tag{2.22}$$

$$E_A = E_A(X_1, Z_1) \tag{2.23}$$

$$E_B = E_B(X_2, Z_2). \tag{2.24}$$

Many of the preceding assumptions are necessary only for mathematical certainty (unambiguously signed matrix determinant) in the general case. For simulation purposes, the static model could be constructed without all but the competitive markets, constant returns to scale, and factor supply assumptions in most cases. In addition, each country's single composite good could be disaggregated into at least two composite goods; the ones responsible for the pollution of interest and all other goods. Merrifield (1984) showed that the model could be extended to consider abatement strategies other than production taxes and abatement equipment standards (e.g., emission taxes, fuel switching, and tariffs), and that more than two countries could be explicitly included.

Total differentiation of each equation (except Equation 2.1 and Equation 2.6), and some substitutions, produces an eight-equation, operational model.[18] The mathematical operations that transform the equations above into the eight equations below are explained in the Appendix:

$$X_1^* - \gamma_{L1} W_A^* + \gamma_{L1} r^* = L_A^* - \lambda_{LZ1} C_{Z1}^* \tag{2.25}$$

$$X_2^* - \gamma_{L2} W_B^* + \gamma_{L2} r^* = L_B^* - \lambda_{LZ2} C_{Z2}^* \tag{2.26}$$

$$X_1^* - \gamma_{K1} W_A^* - \gamma_{K1} r^* + \varepsilon R^* - K_A^* = -\lambda_{KZ1} C_{Z1}^* \tag{2.27}$$

$$X_2^* - \gamma_{K2} W_B^* - \gamma_{K2} r^* + \varepsilon R^* + \psi K_A^* = -\lambda_{KZ2} C_{Z2}^* \tag{2.28}$$

$$\theta_{L1} W_A^* + \theta_{K1} r^* = -\theta_{V1} V_A^* \tag{2.29}$$

$$\theta_{L2} W_B^* + \theta_{K2} r^* - P^* = -\theta_{V2} V_B^* \tag{2.30}$$

$$\pi_1 X_2^* + \pi_1 X_1^* - \alpha_P' P^* = 0 \tag{2.31}$$

---

[18] The determinant of the equation system is unambiguously positive. No additional assumptions are required to obtain that result.

$$\delta_{X1}X_1^* + \delta_{X2}X_2^* - R^* = -\delta_{Z1}C_{Z1}^* + \delta_{Z2}C_{Z2}^* \tag{2.32}$$

where $i^* = di/i$ – the percentage change in i; $\psi = K_A/K_B$; V_i = production tax on output of country i; $i =$ A, B; $\gamma_{L1} = \lambda_{L1}\rho_{K1}\sigma_{LK}^1 + \lambda_{LZ1}\rho_{KZ1}\sigma_{LK}^1$; $\gamma_{L2} = \lambda_{L2}\rho_{K2}\sigma_{LK}^2 + \lambda_{LZ2}\rho_{KZ2}\sigma_{LK}^2$; $\gamma_{K1} = \lambda_{K1}\rho_{L1}\sigma_{LK}^1 + \lambda_{KZ1}\rho_{LZ1}\sigma_{LK}^1$ $+ \lambda_{K2}\rho_{L2}\sigma_{L2}^2 + \lambda_{KZ2}\rho_{LZ2}\sigma_{LK}^2$; σ_{LK}^i = elasticity of substitution of K for L in the production of i; $i = X_1$, X_2, Z_1, and Z_2; ρ_{ij} = *net* or direct share of input i of the revenues generated from the production of j; $i = L$, K; $j = X_1$, X_2, Z_1, and Z_2; λ_{ij} = *net* or direct share of a country's endowment of i employed in the production of j; $i = L$, K; $j = X_1$, X_2, Z_1, and Z_2; θ_{ij} = *gross* or direct plus indirect share of input i of the revenues generated from the production of j; π_i = share of total income accounted for by X_i production; α = price elasticity of demand for X_2, $\alpha < 0$; α'_P = income compensated price elasticity of demand for X_2, $\alpha'_P < 0$; δ_{Xi} = partial elasticity of the pollution flow with respect to X_i production, $i = 1, 2$; ϕ_{Xi} = partial elasticity of country B's pollution flow with respect to X_i production, $i = 1, 2$; δ_{Zi} = partial elasticity of the pollution flow (country A's flow for acid deposition example) with respect to C_{Zi}, the abatement equipment standard for X_i, $i = 1, 2$; ϕ_{Zi} = like δ_{Zi} but for country B's pollution flow; and ε = elasticity of capital productivity with respect to the pollution flow.

The eight unknowns or endogenous variables are capital stock of country A (K_A), X_1, X_2, wages (W_A, W_B), price of capital (r), P, R. The policy variables to be examined with the model are changes in abatement equipment standards and production taxes. In the next section, comparative static results will be presented. Welfare implications are discussed after the comparative statics section.

D. COMPARATIVE STATIC RESULTS

The two control strategies will be discussed first from a general perspective, and then in the context of the acid deposition issue.

1. THE GENERAL CASE

A general property of the model is that with internationally mobile goods, capital, and pollutants, only the equipment standard strategy has an unambiguous impact on pollution flows. That conclusion alone is remarkable. An attempt to reduce pollution by way of a new tax on the output of polluting industries could actually increase pollution! The ambiguity exists because the change in pollution flows due to a reduction in one country's emissions may be more than offset if capital movements increase the other country's output and emissions. In fact, a number of the comparative static results are ambiguously signed. The sign depends on international differences in the pollution and capital intensiveness of production functions, international differences in input substitution elasticities, and the sensitivity of the capital stock to pollution damage.

The change in pollution flows which results from a new production tax imposed by either country depends on which country produces the more pollution-intensive good. Assume country A employs $n\%$ of the total, internationally mobile capital stock, and its emissions account for $m\%$ of pollution flows. If m is greater (less) than n, A's (B's) output is pollution intensive. A new tax on X_2 (B's output) will result in a net increase in pollution flows. A new tax on X_2 (A's output) reduces pollution flows.

The changes in the price of capital, the price of labor, the terms of trade, and the outputs of each country that normally follow imposition of a new production tax are reinforced when pollution flows and the scarcity of capital are directly related. For example, if a tax on X_2 is introduced, the price consumers face rises and the price received by producers falls. Factor rewards fall in country B, prompting some capital to migrate to A. X_1 producers demand additional capital because its price has fallen relative to W_A, and because the higher consumer price for X_2 increases the demand for X_1. If the tax has a net negative (positive) effect on pollution flows, the bulk of additional (diminished) capital services are employed in (taken from) A (B).

The comparative statics of a change in the abatement equipment standard are slightly more complicated. A tightening of the standard in either country reduces the pollution flow[19] and is likely to increase the relative scarcity of the good produced in the country tightening its standard. A country could reduce the relative scarcity of its output by tightening its standard if it is best able to absorb the additional capital services released when pollution is reduced. Unless the production of abatement equipment is tremendously capital intensive, the availability of additional capital services depresses the price of capital and raises the price of labor in both countries. Labor prices may fall if the passive country's share of total income is large and the acting country can more easily substitute capital for labor. Then the drop in the price of the acting country's output necessitates a reduction in its payments to labor. Changes in the output of either country are uncertain because the diversion of capital and labor into pollution abatement contradicts the reduced scarcity of capital services.

The discussion of comparative static results becomes more concrete and is thus easier to follow if the model is applied to a particular transnational pollution issue. Information relevant to the North American acid deposition controversy is used to help interpret the comparative static results discussed in subsection D2.

2. THE NORTH AMERICAN ACID DEPOSITION ISSUE

Preliminary research suggests that long range transport of oxides of sulfur and nitrogen generated in both countries is responsible for acid deposition in each country. The sulfur oxides are more important than the oxides of nitrogen. Canada receives at least three units of airborne sulfur from the U.S. for each unit the U.S. receives from Canada (Galloway and Whelpdale, 1980). U.S. sulfur emissions are responsible for about 50% of Canada's sulfur deposition (Galloway and Whelpdale, 1980), while Canadian emissions are to blame for only 10 to 20% (Robinson, 1982; Forster, personal communication, 1984) of U.S. sulfur deposition.

Distributive shares data (CEA, 1983; Ministry of Supply and Services, 1981) were also useful in reducing the ambiguity of comparative static results. For the period 1976–79, labor's share of total output was 78% in the U.S. and 76% in Canada, with the remainder going to capital. The U.S.'s capital stock is 10 times larger than Canada's (Forster, 1983). Because it is unreasonable to assume that acid deposition flows are identical in the two countries, an additional equation and endogenous variable are added to the model. R is replaced by R_A and R_B, where R_A and R_B are still functions of the emissions of both countries. The determinant of the nine equation system remains unambiguously positive. The mathematical expression for each comparative static result is in the Appendix.

If Canada levies a production tax on its output, X_2, the relative selling price increases, which lowers X_2 and raises X_1. As in Forster (1977), the pollution abatement effort reduces factor rewards. The price of capital falls, so that the capital leaving Canada can be absorbed in the U.S. The loss (addition) of capital in Canada (the U.S.) reduces (raises) the productivity of labor and thereby lowers (increases) the wage rate. Because of the difference in the size of the U.S. and Canada, the effect on Canadian acid deposition of the reduction in X_2 is much more significant than the much smaller increase in X_1. Canadian acid deposition is reduced by the tax on X_2. The Canadian tax has a small but negative effect on acid deposition in the U.S. The effect of the reduction in Canadian emissions is not quite offset by the increase in U.S. emissions.

If the U.S. imposes a tax on X_1, most of the comparative static results are the reverse of those that follow a tax on X_2. However, there are some similarities. The price of capital falls, so that capital leaving one country can be absorbed by the other. The price of capital will fall by much more as a result of a tax on X_1, than a tax on X_2. The smaller Canadian economy will be required to absorb a relatively large amount of capital. Again, there is little or no change in U.S. acid

---

[19] The unambiguity of this result is lost if pollution flows are not identical (not a public bad) in the two countries, clearly the case for acid deposition in the U.S. and Canada, which is considered in subsection D2.

deposition. The effect of a reduction in one country's emissions is offset by an increase in the other country's emissions. In percentage terms, X_2 rises much more than X_1 falls. As a result, the tax on X_1 increases acid deposition in Canada, while leaving it virtually unchanged in the U.S.

If the two countries act jointly, and simultaneously levy a production tax, X_1 and X_2 production will not be altered.[20] Consequently, the terms of trade will not change. However, the price received by producers is reduced by the taxes, which means that the price of capital and wages in both countries fall. With no change in X_1 or X_2, acid deposition is unchanged in both countries.

In Canada, a stricter abatement equipment standard means an increase in C_{Z2}, the ratio of Z_2 to X_2. The initial direct effect acts to reduce acid deposition in two ways. The additional Z_2 directly reduces the release of Canadian emissions, and the production of additional Z_2 means the diversion of some capital and labor from producers of X_2, thereby further reducing emissions. Furthermore, the initial tightening of Canada's capital market causes some capital to leave the U.S., thereby curtailing U.S. emissions. That impact is more significant the easier the U.S. firms can substitute labor for capital, and the greater the difficulty in making input substitutions in Canada. The reduction in acid deposition in both countries reduces the scarcity of capital. If the reduction in K scarcity is significant, X_1 and X_2 go up, instead of down.

The key to understanding the impact of an increase in C_{Z2} is the realization that comparative static results are a combination of first- and second-round effects. In the first-round, production and use of additional Z_2 reduces emissions and acid deposition in both countries. X_1 and X_2 are reduced, and the price of capital increases to shift some capital from the U.S. to Canada. That, in turn, reduces U.S. wages. The change in the terms of trade, and the fate of Canadian wages, depends on whether the relative scarcity of X_2 increases or decreases. Since X_2 producers lose workers and capital to Z_2 producers, while X_1 producers lost only capital, it seems likely that the relative scarcity of X_2 will increase. However, that cannot be stated with certainty. If labor is much more readily substituted for capital in the U.S. than in Canada, the relative scarcity of X_1 increases.

The first-round may be offset, overwhelmed, or complemented by the second-round effects. The curtailment of acid deposition reduces the scarcity of capital services, thereby lowering the price of capital, increasing the production of both goods, and increasing emissions in both countries.

Therefore, there is some degree of uncertainty about each of the comparative static results of an increase in C_{Z2}. In general, the sign of the results depends on differences in σ_{LK} between the two nations, the capital intensiveness of Z_2, and the sensitivity of capital productivity to changes in acid deposition (ε_A and ε_B). As suggested earlier, when A and B are the U.S. and Canada, respectively, the composite goods X_1 and X_2 include many of the same goods and services. With similar technologies available in both countries, σ_{LK} is not likely to differ significantly between them. Furthermore, since most industrialized nations only devote 1 to 2% of their resources to pollution control (USDC, 1975), the degree of capital intensiveness of X_2 is likely to be a trivial factor in determining the sign of a comparative static result. From that additional information, it follows that an increase in C_{Z2} will shift the terms of trade in Canada's favor (the price effect of Comolli's (1977) increase in β was the same), lower the price of capital, and raise the wages of workers in both countries. Acid deposition falls in both countries.

The remaining comparative static results (X_1 and X_2) depend on values of ε_A and ε_B, in the initial equilibrium. For small values of ε_A and ε_B output decreases, and for larger values production of X_1 and X_2 increases. To gain some insight on what constitutes a small or large value of the ε's, some numerical illustration is useful. If the price elasticity of demand and input substitution elasticities are unity, the critical value of the ε's is 0.17. If ε_A and ε_B are less than 0.17, a 1% increase in C_{Z2} reduces X_2 production. If a 1% reduction in acid deposition in each country reduces the scarcity of each country's capital stock by 0.17%, the 1% increase in C_{Z2} leaves X_2 production

---

[20] Households receive the tax proceeds in lump-sum payments. With unchanged relative prices, consumers purchase the same amount of X_1 and X_2 because the income effect of the tax proceeds offsets the effect of the increase in the absolute price of both goods.

unchanged. Since only a small part of the capital services of either nation is affected by acid deposition, it would seem, *a priori*, that the true value[21] of the ε's is well below one, close to the critical value of 0.17 suggested by this illustration.

Recall that for the U.S., a production tax does not reduce acid deposition. Fortunately, an increase in C_{Z1} cannot fail to reduce U.S. acid deposition. The first-round effects of the increase in C_{Z1} are analogous to the Canadian case, with second-round effects again either complementing, offsetting, or overwhelming the first-round effects. Using the same information with regard to the capital intensiveness of pollution abatement equipment and similarity of σ_{LK} in the U.S. and Canada to analyze the comparative static results of the increase in C_{Z1}, the following results emerge. The terms of trade shift in favor of the U.S., the price of capital falls, wages of U.S. workers rise, and there is little or no change in Canadian wages. The relative scarcity of X_1 increases, because its producers lose L and K to Z_1 producers, while X_2 producers lose only K. Pollution cleanup reduces damage to capital, thereby increasing the effective supply of capital. In the U.S., additional capital and an improvement in the terms of trade raises wages. For Canadian wages, the terms of trade effect and additional capital work in opposite directions.

The R_B, X_1, and X_2 comparative statics depend on the ε's. Unless ε_A is greater than 7, an increase in C_{Z1} reduces R_B. It seems unlikely that a 1% decrease in U.S. acid deposition would reduce capital scarcity by over 7%. The critical value of the ε's for the X_1 and X_2 comparative static results is less than 0.1, a number probably closer to the true values of the ε's.

If both countries agree to raise their abatement equipment standards, there will be no change in the terms of trade if the σ_{LK} are similar in the two countries. The reduced scarcity of capital services lowers the price of capital, raises wages in both countries, and reduces acid deposition in both countries.

E. WELFARE ANALYSIS

A government's decision to adopt a particular abatement strategy depends less on the expected reduction in pollution than on the expected increase in welfare. The strategy that reduces pollution the most may not be the welfare-maximizing strategy. Pollution affects only production directly, so welfare is defined as

$$U_i = U_i\left(D_{1i}, D_{2i}\right), \qquad i = A, B \tag{2.33}$$

where U_i = welfare of country i; D_{1i} = quantity of X_1 consumed in country i, i = A, B; and D_{2i} = quantity of X_2 consumed in country i, i = A, B. Changes in A's welfare ($d\omega_A$) and B's welfare ($d\omega_B$) can be written:

$$d\omega_A = r^*\left(\kappa_A - K_A\right)\left(r/Y_A\right) - P^*\left(PD_{2A}/Y_A\right) \tag{2.34}$$

$$d\omega_B = r^*\left(\kappa_B - K_B\right)\left(r/Y_B\right) + P^*\left(X_2 - D_{2B}\right)\left(P/Y_B\right) \tag{2.35}$$

For the complete derivation of (Equation 2.34) and (Equation 2.35) see the Appendix. Country A will seek a strategy that reduces P and increases r if it is a net creditor, or reduces r otherwise. The opposite is true for country B. The data (CEA, 1983; Ministry of Supply and Services, 1981) suggest that Canada was a net debtor and that the U.S. was a net creditor.[22]

---

[21] There is no reason why ε_A and ε_B should be equal. This discussion is only intended to suggest what general range ε_A and ε_B should be in for X_1 and X_2 to be uncertain.

A tax on X_2 will raise Canadian welfare. The resulting increase in P and reduction in r raise the Canadian welfare index. Furthermore, acid deposition in Canada is reduced. A reduction in the wages of Canadian workers is also an outgrowth of the tax. A potentially undesirable distributional consequence of the tax could be eliminated by redistributing the tax receipts to workers, rather than in the lump-sum manner assumed so far.

In the U.S., the welfare implications of a production tax are less clear-cut, because for the U.S., the drop in r has a detrimental effect on welfare. The wages of American workers are also reduced. The welfare implications are further complicated, because acid deposition in the U.S. is not greatly affected by the tax.

If the two countries jointly impose a production tax; that is, a new, small production tax is introduced simultaneously for X_1 and X_2, U.S. welfare is reduced and Canadian welfare increases. The terms of trade are not affected, so the welfare implications are the result of the reduction in the price of capital. The production taxes drive a wedge between the producer and consumer price of the goods, thereby reducing the return to labor and capital in both countries. Those results could be altered by changing the formula for redistributing the tax revenues.

A Canadian decision to increase C_{Z2} is also likely to raise the Canadian welfare index, but with less certainty than the new tax on X_2. Recall that the increase in P, the reduction in r, and the raise for Canadian workers are certain to follow the hike in C_{Z2} only if U.S. and Canadian σ_{LK}'s are similar, and if Z_2 production is not a nationally significant capital-intensive process. The changes in P and r which follow an increase in C_{Z1}, again, appear to have contradictory welfare implications for the U.S., although with the same reasons for uncertainty as in Canada. Joint action again favors Canada. With similar σ_{LK}'s, the simultaneous increase in C_{Z1} and C_{Z2} has little or no impact on the terms of trade, but it reduces the price of capital.

Which of the two strategies discussed so far will be favored in each country, either acting independently or in preparing a negotiating position for joint action? In Canada, the production tax might be favored. The new tax on X_2 unambiguously increases welfare. While the welfare implications of an increase in C_{Z2} are less certain (welfare could actually be reduced), the potential for Canada to significantly improve its welfare is greater. Recall expression (Equation 2.35). Introducing a production tax will reduce X_2 and D_{2B} proportionately, resulting in a smaller absolute difference between them. The larger the price elasticity of demand for X_2, the larger the drop in X_2 and D_{2B} following the introduction of the tax. If the marginal damage to capital (the ε's) is large in the neighborhood of the initial equilibrium, the increase in C_{Z2} could increase P and X_2 simultaneously and reduce r while some foreign-owned capital leaves, the best possible outcome. However, if production technologies in the two countries are such that σ_{LK}^2 is significantly greater than σ_{LK}^1, the terms of trade will not be shifted in Canada's favor by the increase in C_{Z2}. Whether Canada will favor the production tax or the tighter abatement equipment standard depends on risk aversion and prior knowledge of the value of the ε's, the price elasticity of demand for X_2, and input substitution elasticities. If the price elasticity of demand for X_2 were known to be large, and if σ_{LK}^2 were significantly greater than σ_{LK}^1, the best course of action for Canada might be to refrain from taking any unilateral abatement efforts and instead pressure the U.S., through diplomatic channels, to reduce its emissions. As the net importer of acid deposition precursers, Canada does have the advantage of the so-called diplomatic high ground in demanding action from the U.S.

The U.S. decision is more difficult and, in another sense, easier to make. It is more difficult because both the tax and the hike in C_{Z1} will lower the price of capital. Although the terms of trade are definitely improved by the production tax, and very probably improved by the increase in C_{Z1},

---

[22] While a precise measure of the importance of capital payments in the total welfare equation depends on the year of the data and whether stock or flow data are used, it seems safe to say that net income from foreign investment is at least 10% of the U.S. capital owner's total income. Likewise, at least 10% of Canada's net capital income goes to foreigners. Consequently, changes in the price of capital are important in each country's welfare outlook. As a side note, it appears as if ballooning government budget deficits caused the U.S. to join the ranks of net debtor nations in 1984. The older data suffice for the illustrative purposes to which they are being applied.

the drop in r prevents either strategy from producing an unambiguous welfare improvement. The decision is made easier because with an increase in C_{Z1}, acid deposition is reduced, and there is at least some chance that P and r will both move in the desired direction. If the marginal damage to capital is high, X_1 could increase, even while its relative price increases. If the U.S.'s apparent recent switch to net debtor status is permanent, the decision becomes even less difficult.

F. SUMMARY AND CONCLUSIONS

The model discussed has shown the importance of capital flows and trade in determining the best strategy to reduce transfrontier pollution. An open economy does not make it impossible for either country to take effective unilateral action against a transnational pollutant, but each country must recognize that some policies could increase, rather than decrease, pollution flows. Foreign emissions could be increased by the policy enough to offset the effect of reduced domestic emissions. The model has also shown that the distributional and terms of trade effects vary with the abatement strategy. Along with the effect on pollution flows, those factors will weigh heavily in the decision on which abatement strategy is best.

Assuming the σ_{LK} do not differ greatly between the U.S. and Canada, and that ε_A and ε_B are much less than one, the following general conclusions emerge. Canada is more likely to benefit (or less likely to suffer) from the implementation of unilateral or joint pollution abatement efforts. Even though Canada is the smaller country and it receives a larger share of acid deposition from the U.S. than the U.S. receives from Canada, both abatement options will reduce Canada's acid deposition, while only the hike in C_{Z1} reduces U.S. acid deposition. While the first result appears to be consistent with the observed behavior of the Canadian and U.S. federal governments, the conclusion stated in the preceding sentence is not. The conclusion drawn from this model that Canada appears to have more unilateral abatement options is due primarily to two factors: the international mobility of capital, and the fact that Canada's capital stock is only a tenth as large as the U.S. capital stock, even though Canadian emissions account for more than a tenth of the acid deposition occurring in both countries.

The welfare index adopted in Section E suggests that changes in the terms of trade and the price of capital by themselves capture welfare changes. However, politicians will certainly continue to weigh the distributional effects of policy. In addition, if factors such as the existence value of historic structures, visibility, and biodiversity (which are not explicitly included in the model) are significant in the decision-making process, the impact of an abatement strategy on pollution flows is important independent of the welfare changes measured by $d\omega_A$ and $d\omega_B$.

Certainly, this research is not the final word on transnational pollution abatement strategies, either generally or on the more specific North American acid deposition issue. Rather, it offers a beginning, by expanding the use of GE trade models to new analytical territory, and for policymakers it suggests several questions that should be given additional, more in-depth attention.

MATHEMATICAL APPENDIX

1. Transformation of a zero profit equation into percentage change form
    a. Begin with (Equation 2.3) and (Equation 2.5)
    b. Totally differentiate each:

$$dC_{L2}W_B + C_{L2}dW_B + dC_{K2}r + C_{K2}dr + C_{Z2}dP_{Z2} = dP \qquad (2.36)$$

$$dC_{LZ2}W_B + C_{LZ2}dW_B + dC_{KZ2}r + C_{KZ2}dr = dP_{Z2} \qquad (2.37)$$

c. Divide through by P in (Equation 2.36) and P_{Z2} in (Equation 2.37) and multiply each term by i/i = one, where i = the change variable of that term:

$$C_{L2}^* \rho_{L2} + \rho_{L2} W_B^* + C_{K2}^* \rho_{K2}^* + \rho_{K2} r^* + \rho_{Z2} P_{Z2}^* = P^*$$ (2.38)

$$C_{LZ2}^* \rho_{LZ2} + \rho_{LZ2} W_B^* + C_{LZ2}^* \rho_{KZ2} + \rho_{KZ2} r^* = P_{Z2}$$ (2.39)

d. Note that for homogeneous production functions

$$C_{L2} \rho_{L2}^* + C_{K2}^* \rho_{K2} = C_{LZ2}^* \rho_{LZ2} + C_{KZ2}^* \rho_{KZ2} = 0$$

then substitute (Equation 2.39) for P_{Z2} in (Equation 2.38),

$$\theta_{L2} W_B^* + \theta_{K2} r^* - P^* = 0$$ (2.40)

where $\theta_{K2} = \rho_{L2} + \rho_{Z2}\,\rho_{LZ2}$ and $\theta_{k2} = \rho_{K2} + \rho_{Z2}\,\rho_{KZ2}$.
2. *Transformation of a full employment equation into percentage change form*
   a. Begin with (Equation 2.10): $C_{K2}\,X_2 + C_{KZ2}\,Z_2 = K_B$.
   b. Substitute (Equation 2.12) in (Equation 2.10) and obtain

$$C_{K2}X_2 + C_{KZ2}C_{Z2}X_2 = K_B$$ (2.41)

c. Totally differentiate:

$$dC_{K2}X_2 + C_{KZ2}X_2 + dC_{KZ}2C_{Z2}X_2 + C_{KZ2}dC_{Z2}X_2 + C_{KZ2}C_{Z2}dX_2 = dK_B$$ (2.42)

d. Divide every term by K_B and multiply each term by one i/i, where i for each term is the change variable:

$$\lambda_{L2}C_{K2}^* + X_2^* + \lambda_{KZ2}C_{Z2}^* + \lambda_{KZ2}C_{KZ2}^* = K_B^*$$ (2.43)

e. Input-output ratios are a function of input prices, and in the case of capital, they are also a function of the pollution flow:

$$C_{K2}^* = \rho_{L2}\sigma_{lk}W_B^* + \rho_{K2}\sigma_K r^* + \beta_2 R^*$$ (2.44)

$$C_{KZ2} = \rho_{LZ2}\sigma_{LK}W_B^* + \rho_{KZ2}\sigma_{KK} r^* + \beta_{Z2}R^*$$ (2.45)

If all factor prices change in the same proportion, and the pollution flow does not change ($R^* = 0$), $C_{K2} = C_{KZ2} = 0$. That implies the following:

$$\rho_{K2}\sigma_{KK} = -\rho_{L2}\sigma_{LK}$$ (2.46)

$$\rho_{KZ2}\sigma_{KK} = -\rho_{LZ2}\sigma_{LK}$$ (2.47)

f. After substituting (Equation 2.46) and (Equation 2.47) into (Equation 2.44) and (Equation 2.45), (Equation 2.44) and (Equation 2.45) replace C_{K2} and C_{KZ2} in (Equation 2.43). Also, ψK_A replaces K_B,

$$\psi K_A^* + X_2^* + \gamma_{K2} r^* + \varepsilon R^* = -\lambda_{KZ2} C_{Z2}^* \tag{2.48}$$

where $\varepsilon = \lambda_{K2} \beta_2 + \lambda_{KZ} 2 \beta_{Z2}$.

3. *Transformation of the demand equations into a single percentage change equation*
   a. Begin with (Equation 2.18)–(Equation 2.21).
   b. Totally differentiate and rewrite in percentage change form.
   c. Transformed (Equation 2.20) and (Equation 2.21) are then substituted for Y_A^* and Y_B^* in transformed (Equation 2.19), which is then substituted for Y^* in transformed (Equation 2.18):

$$\pi_1 X_2^* - \pi_1 X_1^* - \alpha_p P^* = 0 \tag{2.49}$$

4. *Transformation of the pollution flow equations into a single percentage change equation*
   a. Begin with (Equation 2.22), (Equation 2.23), and (Equation 2.24).
   b. Totally differentiate each, divide through by R, E_A, and E_B, respectively, and divide each term by i/i, where i is each term's change variable.
   c. Substitute E_A and E_B from the transformed versions of (Equation 2.23) and (Equation 2.24), respectively, into transformed (22):

$$\delta_{X1} X_1^* + \delta_{X2} X_2^* - R^* = -\delta_{Z1} C_{Z1}^* - \delta_{Z2} C_{Z2}^* \tag{2.50}$$

5. *Derivation of the welfare index for country A*
   a. Begin with (Equation 2.51) and (Equation 2.52)

$$U_A = U_A \left(D_{1A}, D_{2A} \right) \tag{2.51}$$

$$Y_A = D_{1A} + P D_{2A} = r \kappa_A + W_A L_A \tag{2.52}$$

   b. Totally differentiate (Equation 2.51) and divide through by

$$dU_A / dD_{1A} = U_{1A}$$

$$dU_A / U_{1A} = d\omega_A = dD_{1A} + U_{2A} / U_{1A} \, dD_{2A} \tag{2.53}$$

In equilibrium, the ratio of marginal utilities (U_{2A}/U_{1A}) is equal to $P(P_2/P_1)$, the price ratio.

   c. Totally differentiate (Equation 2.52):

$$dY_A = dD_{1A} + P dD_{2A} + D_{2A} dP = dr \kappa_A + dW_A L_A \tag{2.54}$$

   d. Note the $d\omega_A$ in (Equation 2.53) equals the left two terms in the middle of (Equation 2.54). Substitute $d\omega_A$ into (Equation 2.54) and divide through by Y_A:

$$Y_A = d\omega_A/Y_A + dPD_{2A}/Y_A = dr\kappa_A/Y_A + dW_A L_A/Y_A \tag{2.55}$$

e. Redefine $d\omega_A = d\omega_A/Y_A$, reduce (Equation 2.55) to the two sides of the right equal sign, and convert to percentage change form:

$$d\omega_A = r^*\left(r\kappa_A/Y_A\right) + W_A^*\left(W_A L_A/Y_A\right) - P^*\left(PD_{2A}/Y_A\right) \tag{2.56}$$

f. Use (Equation 2.29) to define W_A^* in terms of r^*:

$$W_A^* = -\theta_{K1}/\theta_{L1} r^* \tag{2.57}$$

g. Substitute (Equation 2.57) for W_A in (Equation 2.56) and then simplify:

$$d\omega_A = r^*\left(r\kappa_A/Y_A\right) - \left(rK_A/W_A L_A\right)\left(W_A L_A/Y_A\right)r^* - P^*\left(PD_{2A}/Y_A\right) \tag{2.58}$$

$$d\omega_A = \left(r/Y_A\right)\left(\kappa_A - K_A\right)r^* - P^*\left(PD_{2A}/Y_A\right) \tag{2.59}$$

COMPARATIVE STATIC RESULTS

6. *A new production tax is levied on X_2. Assume $\theta_{V2} = 0.1$*

a. $R_B^* = \alpha_P \cdot 0.01 \cdot \left[\psi\gamma_{L2}\phi_{XB}\left(\gamma_{K1} + \gamma_{L1}\right) - \phi_{XA}\gamma_{L1}\left(\gamma_{K2} + \gamma_{L2}\right)\right]V_B^*$ inverse relationship

b. $R_A^* = (-0.01)\cdot\alpha_P\left[-\varepsilon_B\gamma_{L1}\gamma_{L2}\left(\phi_{XA}\delta_{XB} - \phi_{XB}\delta_{XA}\right) + \delta_{XA}\gamma_{L1}\left(\gamma_{L2} + \gamma_{K2}\right)\right.$

$\left.-\psi\gamma_{L2}\delta_{XB}\left(\gamma_{K1} + \gamma_{L1}\right)\right]V_B^*$

c. $X_2^* = \alpha_P\gamma_{L2}\cdot 0.01\cdot\left[\phi_{XA}\varepsilon_B\gamma_{L1} + \psi\left(\delta_{XA}\varepsilon_A\gamma_{L1} + \gamma_{K1} + \gamma_{L1}\right)\right]V_B^*$ inverse relationship

d. $P^* = 0.01\cdot\pi_1\left[\gamma_{L1}\left(\gamma_{K2} + \gamma_{L2} + \phi_{XB}\gamma_{L2}\varepsilon_B + \psi\gamma_{L2}\delta_{XB}\varepsilon_A + \gamma_{L2}\phi_{XA}\varepsilon_B\right)\right.$

$\left.+\psi\gamma_{L2}\left(\delta_{XA}\varepsilon_A\gamma_{L1} + \gamma_{K1} - \gamma_{K1}\right)\right]V_B^*$ positive relationship

e. $W_B^* = \alpha_P \cdot 0.01 \cdot \left[\theta_{L1}\left(\phi_{XB}\gamma_{L2}\varepsilon_B + \gamma_{L2} + \gamma_{K2}\right) + \phi_{XA}\varepsilon_B\gamma_{L1}\right.$

$\left.+\psi\left(\delta_{XA}\varepsilon_A\gamma_{L1} + \gamma_{L1} + \gamma_{K1} + \delta_{XB}\gamma_{L2}\varepsilon_A\gamma_{L1}\right)\right]V_B^*$ inverse relationship

f. $X_1^* = \alpha_P\theta_{L1}\cdot(-0.01)\cdot\left[\gamma_{L2} + \gamma_{K2} + \phi_{XB}\gamma_{L2}\varepsilon_B + \psi\gamma_{L2}\varepsilon_A\delta_{XB}\right]V_B^*$ positive relationship

g. $W_A^* = \alpha_P\theta_{K1}\cdot(-0.01)\cdot\left[\gamma_{L2} + \gamma_{K2} + \phi_{XB}\gamma_{L2}\varepsilon_B + \psi\gamma_{L2}\varepsilon_A\delta_{XB}\right]V_B^*$ positive relationship

h. $r^* = \alpha_P\theta_{L1}\cdot(-0.01)\cdot\left[\gamma_{L2} + \gamma_{K2} + \phi_{XB}\gamma_{L2}\varepsilon_B + \psi\gamma_{L2}\varepsilon_A\delta_{XB}\right.$ inverse relationship

i. $K_A^* = \alpha_P\cdot 0.01\cdot\left[\gamma_{L2}\left\{\varepsilon_B\left(-\delta_{XB}\varepsilon_A\gamma_{L1} + \phi_{XB}\left(\delta_{XA}\varepsilon_A\gamma_{L1} + \gamma_{K1} + \gamma_{L1}\right)\right)\right.\right.$

$\left.\left.-\delta_{XA}\varepsilon_A\gamma_{L1} - \gamma_{K1} - \gamma_{L1}\right\} - \gamma_{K1}\left(\varepsilon_A\delta_{XA}\gamma_{L1} + \gamma_{K1} + \gamma_{L1}\right)\right]V_B^*$ positive relationship

7. An increase in C_{Z2}^*

a. $R_B^* = C_{Z2}^* \gamma_{LZ2} \left[\alpha_P \{ \phi_{XB} \theta_{L1} \gamma_{K2} + \theta_{L2} \gamma_{L1} (\psi \phi_{XB} - \phi_{XA}) \right.$

$\left. + \psi \theta_{L2} (\varepsilon_A \gamma_{L1} (\phi_{XB} \delta_{XA} - \phi_{XA} \delta_{XB}) + \phi_{XB} \gamma_{K1}) \} - \gamma_{L1} \gamma_{K2} \pi_1 (\phi_{XB} + \phi_{XA}) \right]$

$- C_{Z2}^* (\lambda_{KZ2} + \delta_{Z2} \varepsilon_A \psi) \left[-\alpha_P (\phi_{XB} \theta_{L1} \gamma_{L2} + \phi_{XA} \theta_{L2} \gamma_{L1}) \right.$

$\left. + \gamma_{L2} \gamma_{L1} \pi_1 (\phi_{XB} + \phi_{XA}) \right] - \phi_{Z2} C_{Z2}^* \left[\alpha_P \{ \theta_{L1} (\gamma_{K2} + \gamma_{L2}) + \psi (\theta_{L2} (\gamma_{K1} + \gamma_{L1}) \right.$

$\left. + \varepsilon_A (\delta_{XA} \theta_{L2} \gamma_{L1} + \delta_{XB} \theta_{L1} \gamma_{L2})) \} - \pi_1 \{ \gamma_{L1} (\varepsilon_A \psi \gamma_{L2} \delta_{XB} + \gamma_{K2} + \gamma_{L2}) \} \right]$

b. $X_2^* = \lambda_{LZ2} C_{Z2}^* \left[\alpha_P \{ \phi_{L1} \gamma_{K2} + \phi_{XA} \varepsilon_B \theta_{L2} \gamma_{L1} + \psi \theta_{L2} (\gamma_{L1} + \gamma_{K1} + \delta_{XA} \varepsilon_A \gamma_{L1}) \} - \pi_1 \gamma_{K2} \gamma_{L1} \right]$

$- C_{Z2}^* \gamma_{L2} (\pi_1 \gamma_{L1} - \alpha_P \theta_{L1}) (\lambda_{K2} + \delta_{Z2} \varepsilon_A \psi + \phi_{Z2} \varepsilon_B)$

c. $X_1^* = -C_{Z2}^* \lambda_{LZ2} \left[\alpha_P \{ \gamma_{L1} \theta_{L2} (1 + \varepsilon_B \phi_{XB} + \psi \delta_{XB} \varepsilon_A \gamma_{L1}) \} + \gamma_{K2} \pi_1 \gamma_{L1} \right]$

$- \gamma_{L1} C_{Z2}^* (\lambda_{KZ2} + \delta_{Z2} \varepsilon_A \psi + \phi_{Z2} \varepsilon_B) (\pi_1 \gamma_{L2} - \alpha_P \theta_{L2})$

d. $W_A^* = -\lambda_{LZ2} \theta_{K1} C_{Z2}^* \left[\alpha_P \theta_{L2} \{ 1 + \varepsilon_B \phi_B + \psi \varepsilon_A \delta_{XB} \} + \pi_1 \gamma_{K2} \right]$

$- \theta_{K1} C_{Z2}^* (\lambda_{KZ2} + \delta_{Z2} \varepsilon_A \psi + \phi_{Z2} \varepsilon_B) (\pi_1 \gamma_{L2} - \alpha_P \theta_{L2})$

e. $r^* = -W_A^* \theta_{L1} / \theta_{K1}$

f. $W_B^* = \lambda_{LZ2} C_{Z2}^* \left[-\alpha_P \theta_{L1} \theta_{K2} (1 + \varepsilon_B \phi_{XB} + \psi \delta_{XB} \varepsilon_A) \right.$

$\left. + \pi_1 \{ \gamma_{L1} (1 + \varepsilon_B \phi_{XB} + \psi \delta_{XB} \varepsilon_A + \phi_{XA} \varepsilon_B + \psi \delta_{XA} + \psi) + \theta_{L1} \gamma_{K2} + \psi \gamma_{K1} \} \right]$

$- C_{Z2}^* (\lambda_{KZ2} + \delta_{Z2} \varepsilon_A \psi + \phi_{Z2} \varepsilon_B) (-\alpha_P \theta_{L1} \theta_{K2} + \pi_1 (\gamma_{L1} - \gamma_{L2} \theta_{L1}))$

g. $P^* = \lambda_{LZ2} \pi_1 C_{Z2}^* \left[\gamma_{L1} \theta_{L2} (1 + \varepsilon_B) + \theta_{L1} \gamma_{K2} + \psi \theta_{L2} \gamma_{L1} (1 + \varepsilon_A) + \psi \theta_{L2} \gamma_{K1} \right]$

$- \pi_1 C_{Z2}^* (\lambda_{KZ2} + \delta_{Z2} \varepsilon_A \psi + \phi_{Z2} \varepsilon_B) (\theta_{L2} \gamma_{L1} - \theta_{L1} \gamma_{L2})$

h. $K_A^* = -\lambda_{LZ2} C_{Z2}^* \left[\alpha_P \{ \varepsilon_B \theta_{L2} \gamma_{L1} (\phi_{XA} \delta_{XB} \varepsilon_A + \phi_{XB} + \phi_{XB} \delta_{XA} \varepsilon_A) + \gamma_{K1} \theta_{L2} (1 + \phi_{XB} \varepsilon_B) \right.$

$\left. + \gamma_{L1} \theta_{L2} (1 + \delta_{XA} \varepsilon_A) - \delta_{XB} \varepsilon_A \theta_{L1} \gamma_{K2} \} + \gamma_{L1} \pi_1 \gamma_{K2} (\varepsilon_A + 1) + \gamma_{K2} \pi_1 \gamma_{K1} \right]$

$- C_{Z2}^* (\lambda_{KZ2} + \phi_{Z2} \varepsilon_B) \left[-\alpha_P \{ \theta_{L2} \gamma_{L1} (1 + \delta_{XA} \varepsilon_A) + \theta_{L2} \gamma_{K1} + \delta_{XB} \gamma_{L2} \theta_{L1} \varepsilon \} \right.$

$\left. + \gamma_{L2} \gamma_{L1} \pi (1 + \varepsilon_A) + \gamma_{L2} \gamma_{K1} \pi_1 \right] + \delta_{Z2} \varepsilon_A C_{Z2}^*$

$\left[-\alpha_P \{ \theta_{L2} (\delta_{XA} \varepsilon_A \gamma_{L1} \gamma_{L1} + \gamma_{K1}) + \delta_{XB} \varepsilon_A \theta_{L1} \gamma_{L2} \} + \pi_1 \gamma_{L1} \gamma_{L2} (1 + \varepsilon_B) + \pi_1 \gamma_{L2} \gamma_{K2} \right]$

i. $R_A^* = \lambda_{LZ2} C_{Z2}^* \big[\alpha_P \{ -\delta_{XA} \gamma_{L1} \theta_{L2} (1 + \phi_{XB} \varepsilon_B)$

$\qquad + \delta_{XB} (\theta_{L1} \gamma_{K2} + \phi_{XA} \varepsilon_B \theta_{L2} \gamma_{L1} + \psi \theta_{L2} (\gamma_{K1} + \gamma_{L1})) \}$

$\qquad - \gamma_{K2} \pi_1 \gamma_{L1} \big] + C_{Z2}^* (\lambda_{KZ2} + \phi_{Z2} \varepsilon_B) \big[\alpha_P (\delta_{XB} \theta_{L1} \gamma_{L2} + \delta_{XA} \theta_{L2} \gamma_{L1}) - \gamma_{L2} \gamma_{L1} \pi_1 \big]$

$\qquad - \delta_{Z2} C_{Z2}^* \big[\alpha_P \{ \theta_{L1} \gamma_{L2} (1 + \phi_{XB} \varepsilon_B) + \theta_{L1} \gamma_{K2} + \psi \theta_{L2} \gamma_{K1} + \theta_{L2} \gamma_{L1} (\phi_{XA} \varepsilon_B + \psi) \}$

$\qquad - \pi_1 \{ \gamma_{L1} \gamma_{L2} (1 + \varepsilon_B + \psi) + \gamma_{L1} \gamma_{K2} + \gamma_{L2} \psi \gamma_{K1} \} \big]$.

8. *A production tax is levied on* X_1. *Assume* $\theta_{V1} = 0.01$

a. $R_A^* = \alpha_P \cdot 0.01 \cdot \big[\delta_{XA} (\phi_{XB} \varepsilon_B \gamma_{L2} + \gamma_{L2} + \gamma_{K2}) - \delta_{XB} \phi_{XA} \varepsilon_B \gamma_{L2} - \psi \delta_{XB} \gamma_{L2} (\gamma_{K1} + \gamma_{L1}) \big] V_A^*$

b. $R_B^* = \alpha_P \cdot 0.01 \cdot \big[\gamma_{L1} \phi_{XA} (\gamma_{K2} + \gamma_{L2}) - \psi \phi_{XB} \gamma_{L2} (\gamma_{K1} + \gamma_{L1})$

$\qquad - \psi \gamma_{L1} \varepsilon_A \gamma_{L2} (\phi_{XA} \delta_{XB} - \phi_{XB} \delta_{XA}) \big] V_A^*$

c. $K_A^* = \alpha_P \cdot 0.01 \cdot \big[\gamma_{L1} \gamma_{L2} \{ \varepsilon_A \varepsilon_B (\phi_{XB} \delta_{XA} - \phi_{XA} \delta_{XB}) + \varepsilon_B \phi_{XB} + \gamma_{K2} + \gamma_{L2}$

$\qquad + \varepsilon_A \delta_{XA} (\gamma_{K2} + \gamma_{L2}) \} + \gamma_{K1} (\gamma_{L2} + \gamma_{K2} + \phi_{XB} \varepsilon_B \gamma_{L2}) \big] V_A^*$ inverse relationship

d. $X_1^* = \alpha_P \gamma_{L1} \cdot 0.01 \cdot \big[\gamma_{L2} (\phi_{XB} \varepsilon_B + 1 + \psi \delta_{XB} \varepsilon_A) + \gamma_{K2} \big] V_A^*$ inverse relationship

e. $X_2^* = \alpha_P \gamma_{L2} \cdot (-0.01) \cdot \big[\gamma_{L1} (\varepsilon_B \phi_{XA} + \psi + \psi \varepsilon_A \delta_{XA}) + \psi \gamma_{K1} \big] V_A^*$ positive relationship

f. $W_A^* = \alpha_P \cdot (0.01) \cdot \big[\gamma_{L2} (\phi_{XB} \varepsilon_B + 1) + \phi_{XA} \gamma_{L1} \varepsilon_B \theta_{L2} + \gamma_{K2}$

$\qquad + \psi \{ \theta_{L2} (\delta_{XA} \gamma_{L1} \varepsilon_Z + \gamma_{L1} + \gamma_{K1}) + \delta_{XB} \varepsilon_A \gamma_{L2} \} \big] V_A^*$

$\qquad - \pi_1 \big[\gamma_{L1} \gamma_{L2} (1 + \varepsilon_B + \psi \varepsilon_A + \psi) + \gamma_{L1} \gamma_{K2} + \gamma_{L2} \gamma_{K1} \big] V_A^*$

g. $W_B^* = \alpha_P \theta_{K2} \cdot (-0.01) \cdot \big[\gamma_{L1} (\varepsilon_B \phi_{XA} - \psi \phi_{XA} \varepsilon_A + \psi) + \psi \gamma_{K1} \big] V_A^*$

$\qquad - \pi_1 \big[\gamma_{L1} \gamma_{L2} (1 + \varepsilon_B + \psi + \psi \varepsilon_A) + \gamma_{L1} \gamma_{K1} \psi \big] V_A^*$

h. $r^* = \alpha_P \theta_{L2} \cdot 0.01 \cdot \big[(\varepsilon_B \phi_{XA} + \psi \delta_{XA} \varepsilon_A + \psi) \gamma_{L1} + \psi \gamma_{K1} \big] V_A^*$

$\qquad - \pi_1 \cdot 0.01 \cdot \big[\gamma_{L1} \gamma_{L2} (1 + \varepsilon_B + \psi + \psi \varepsilon_A) + \gamma_{L1} \gamma_{K2} + \gamma_{L2} \gamma_{K1} \psi \big] V_A^*$ inverse relationship

i. $P^* = -\pi_1 \cdot 0.01 \cdot \big[\gamma_{L1} \gamma_{L2} (\varepsilon_B + 1 + \psi + \psi \varepsilon_A)$

$\qquad + \gamma_{L1} \gamma_{K2} + \gamma_{L2} \gamma_{K1} \psi \big] V_A^*$ inverse relationship.

9. An increase in C_{Z1}

a. $R_A^* = -\lambda_{LZ1} C_{Z1}^* \left[-\alpha_P \left\{ \theta_{L1} \varepsilon_B \gamma_{L2} \left(\delta_{XA} \phi_{XB} - \delta_{XB} \phi_{XA} \right) + \theta_{L1} \delta_{XA} \left(\gamma_{K2} + \gamma_{L2} \right) \right. \right.$

$\left. \left. + \psi \left(\delta_{XA} \theta_{L2} \gamma_{K1} - \delta_{XB} \theta_{L1} \gamma_{L2} \right) \right\} - \psi \gamma_{L2} \gamma_{K1} \pi_1 \right]$

$- \lambda_{KZ1} \psi C_{Z1}^* \left[-\alpha_P \left(\delta_{XA} \theta_{L2} \gamma_{L1} + \delta_{XB} \theta_{L1} \gamma_{L2} \right) + \gamma_{L2} \gamma_{L1} \pi_1 \right]$

$+ \delta_{Z1} C_{Z1}^* \left[-\alpha_P \left\{ \varepsilon_B \left(\phi_{XA} \theta_{L2} \gamma_{L1} + \phi_{XB} \theta_{L1} \gamma_{L2} \right) + \theta_{L1} \left(\gamma_{K2} + \gamma_{L2} \right) + \psi \vartheta_{L2} \left(\gamma_{K1} + \gamma_{L1} \right) \right\} \right.$

$\left. + \pi_1 \left\{ \gamma_{L1} \left(\gamma_{K2} + \phi_{XB} \gamma_{L2} \varepsilon_B + \gamma_{L2} \phi_{XA} \varepsilon_B \right) + \psi \gamma_{L2} \left(\gamma_{K1} + \gamma_{L1} \right) \right\} \right]$

$- \left(\phi_{Z1} \varepsilon_B C_{Z1}^* \left[\gamma_{L1} \gamma_{L2} \pi_1 - \alpha_P \left(\delta_{XA} \theta_{L2} \gamma_{L1} + \delta_{XB} \theta_{L1} \gamma_{L2} \right) \right] \right)$ inverse relationship

b. $R_B^* = -\lambda_{LZ1} C_{Z1}^* \left[-\alpha_P \left\{ -\phi_{XA} \theta_{L1} \left(\gamma_{L2} + \gamma_{K2} \right) + \psi \left(-\phi_{XA} \theta_{L2} \gamma_{K1} + \delta_{XB} \varepsilon_A \theta_{L1} \gamma_{L2} \right. \right. \right.$

$\left. \left. \left. + \phi_{XB} \gamma_{L1} \theta_{L2} \left(1 + \delta_{XA} \varepsilon_A \right) \right) \right\} + \psi \pi_1 \gamma_{L2} \gamma_{K1} \right] - \psi C_{Z1}^* \left(\lambda_{KZ1} + \delta_{Z1} \varepsilon_A \right)$

$\left[-\alpha_P \left(\phi_{XB} \theta_{L1} \gamma_{L2} + \phi_{XA} \theta_{L2} \gamma_{L1} \right) + \gamma_{L2} \pi_1 \gamma_{L1} \right]$

$- \phi_{Z1} \left[\alpha_P \left\{ \theta_{L1} \left(\gamma_{K2} + \gamma_{L2} \right) + \psi \left(\theta_{L2} \delta_{XA} \varepsilon_A \gamma_{L1} + \gamma_{L1} + \gamma_{K1} \right) + \delta_{XB} \varepsilon_A \theta_{L1} \gamma_{L2} \right) \right\}$

$- \pi_1 \left\{ \gamma_{L1} \left(\gamma_{K2} + \gamma_{L2} + \psi \gamma_{L2} \delta_X X_B \varepsilon_A \right) + \gamma_{L2} \psi \left(\delta_{XA} \varepsilon_A \gamma_{L1} + \gamma_{K1} + \gamma_{L1} \right) \right\} \right]$

c. $K_A^* = \lambda_{LZ1} C_{Z1}^* \left[\alpha_P \left\{ \varepsilon_A \theta_{L1} \left(\delta_{XA} \left(\phi_{XB} \varepsilon_B \gamma_{L2} + \gamma_{L2} + \gamma_{K2} \right) - \delta_{XB} \phi_{XA} \varepsilon_B \gamma_{L2} \right) \right. \right.$

$\left. + \theta_{L1} \left(\phi_{XB} \varepsilon_B \gamma_{L2} + \gamma_{L2} + \gamma_{K2} \right) - \phi_{XA} \varepsilon_B \theta_{L2} \gamma_{K1} \right\}$

$+ \pi_1 \left\{ \gamma_{K1} \theta_{K1} \left(\gamma_{K2} + \gamma_{L2} + \phi_{XB} \gamma_{L2} \varepsilon_B \right) + \theta_{L1} \left(\phi_{XB} \varepsilon_B \gamma_{L2} \gamma_{K1} + \gamma_{K1} \left(\gamma_{K2} + \gamma_{L2} \right) \right) \right.$

$\left. + \gamma_{L2} \gamma_{K1} \phi_{XA} \varepsilon_B \right\} \right] - C_{Z1}^* \left(\lambda_{KZ1} + \delta_{Z1} \varepsilon_A \right) \left[\alpha_P \left\{ \theta_{L1} \left(\theta_{XB} \varepsilon_B \gamma_{L2} + \gamma_{K2} \right) + \phi_{XA} \gamma_{L1} \theta_{L2} \varepsilon_B \right\} \right.$

$\left. - \pi_1 \gamma_{L1} \left(\left(\phi_{XB} \gamma_{L2} \varepsilon_B + \gamma_{L2} + \gamma_{K2} \right) - \gamma_{L2} \phi_{XA} \varepsilon_B \right) \right] C_{Z1}^* \phi_{Z1} \varepsilon_B$

$\left[-\alpha_P \left\{ \theta_{L2} \left(\gamma_{K1} + \gamma_{L1} + \delta_{XA} \varepsilon_A \gamma_{L1} \right) + \delta_{XB} \varepsilon_A \theta_{L1} \gamma_{L2} \right\} + \gamma_{L2} \gamma_{L1} \pi_1 \left(\varepsilon_A + 1 \right) + \gamma_{L2} \pi_1 \gamma_{K1} \right]$

d. $X_1^* = -\lambda_{LZ1} C_{Z1}^* \left[-\alpha_P \left\{ \theta_{L1} \left(\gamma_{K2} + \gamma_{L2} + \phi_{XB} \varepsilon_B \gamma_{L2} \right) + \psi \left(\phi_{L2} \gamma_{K1} + \delta_{XB} \varepsilon_A \theta_{L1} \gamma_{L2} \right) \right\} \right.$

$\left. + \pi_1 \gamma_{L2} \psi \gamma_{K1} \right] - \gamma_{L1} C_{Z1}^* \left(\lambda_{KZ1} \psi + \delta_{Z1} \varepsilon_A \psi + \phi_{Z1} \varepsilon_B \gamma_{L1} \right) \left(\pi_1 \gamma_{L2} - \alpha_P \theta_{L2} \right)$

e. $X_2^* = -\lambda_{LZ1} C_{Z1} \left[\alpha_P \gamma_{L2} \theta_{L1} \left(\varepsilon_A \phi_{XA} + \psi + \psi \varepsilon_A \delta_{XA} \right) + \pi_1 \gamma_{L2} \psi \gamma_{K1} \right]$

$- C_{Z1}^* \gamma_{L2} \left(\lambda_{KZ1} \psi + \delta_{Z1} \varepsilon_A \psi + \phi_{Z1} \varepsilon_B \right) \left(\pi_1 \gamma_{L1} - \alpha_P \theta_{L1} \right)$

f. $P^* = -\lambda_{LZ1} \pi_1 C_{Z1}^* \left[\theta_{L1} \left(\phi_{XB} \varepsilon_B \gamma_{L2} + \gamma_{L2} + \gamma_{K2} \right) + \psi \left(\theta_{L2} \gamma_{K1} + \delta_{XB} \varepsilon_A \theta_{L1} \gamma_{L2} \right) \right.$

$\left. - \gamma_{L2} \theta_{L1} \left(\varepsilon_B \phi_{XA} + \psi + \psi \varepsilon_A \delta_{XA} \right) \right] + \pi_1 C_{Z1}^* \left(\lambda_{KZ1} \psi + \delta_{Z1} \psi \varepsilon_A + \phi_{Z1} \varepsilon_B \right) \left(\theta_{L1} \gamma_{L2} - \theta_{L2} \gamma_{L1} \right)$

g. $r^* = -\lambda_{LZ1}\theta_{L1}C_{Z1}^*\left[-\alpha_P\theta_{L2}\left(\varepsilon_B\phi_{XA} + \psi\left(1 + \delta_{XA}\varepsilon_A\right)\right)\right.$

$\left. + \pi_1\gamma_{L2}\left(1 + \phi_{XB}\varepsilon_B + \psi\delta_{XB}\varepsilon_A + \varepsilon_B\phi_{XA} + \psi + \psi\varepsilon_A\delta_{XA}\right) + \pi_1\gamma_{K2}\right]$

$\quad + \theta_{L1}C_{Z1}^*\left(\lambda_{KZ1}\psi + \delta_{Z1}\psi\varepsilon_A + \phi_{Z1}\varepsilon_B\right)\left(\pi_1\gamma_{L2} - \alpha_P\theta_{L2}\right)$

h. $W_A^* = -\theta_{K1}/\theta_{L1}r^*$

i. $W_B^* = -\lambda_{LZ1}C_{Z1}^*\left[\alpha_P\theta_{K2}\theta_{L1}\left(\varepsilon_B\phi_{XA} + \psi\left(1 + \delta_{XA}\varepsilon_A\right)\right) + \pi_1\gamma_{L2}\left\{\theta_{L1}\left(1 + \phi_{XB}\varepsilon_B\right)\right.\right.$

$\left.\left. + \psi\delta_{XB}\varepsilon_A\gamma_{L2}\theta_{L1} + \varepsilon_B\theta_{L1}\phi_{XA} + \psi\theta_{L1}\left(1 + \varepsilon_A\delta_{XA}\right)\right\} + \pi_1\left\{\theta_{L1}\gamma_{K2} + \psi\gamma_{K1}\right\}\right]$

$\quad - C_{Z1}^*\left(\lambda_{KZ1}\psi + \delta_{Z1}\varepsilon_A\psi + \phi_{Z1}\varepsilon_B\right)\left[-\alpha_P\theta_{L1}\theta_{K2} + \pi_1\left(\gamma_{L1} - \gamma_{L2}\theta_{L1}\right)\right].$

ACKNOWLEDGMENTS

This chapter would not have been possible without the generous assistance of several individuals. Special thanks are due Shelby Gerking, Jack Mutti, and Bruce Forster for their numerous helpful suggestions and moral support. David Brookshire, Ralph d'Arge, and Larry Ostresh also provided helpful comments and encouragement. Susan Kalima, Lori Tobias, Joe Krismer, and Frank Goodale typed and printed the manuscript, a formidable task with the numerous equations and symbols. Remaining shortcomings and errors are the sole responsibility of the author.

REFERENCES

Asako, K., Environmental pollution in an open economy, *Econ. Rec.*, 55, 359, 1979.

Batra, R. and Casas, F., A synthesis of the Heckscher–Ohlin and the neoclassical models of international trade, *J. Int. Econ.*, 6, 21, 1976.

Comolli, P., Pollution control in a simplified general equilibrium model with production externalities, *J. Environ. Econ. Manage.*, 4, 289, 1977.

Council of Economic Advisors (CEA), Economic report of the President, U.S. Government Printing Office, Washington, D.C., 1983.

Crocker, T. D., What economics can currently say about the benefits of acid deposition control, in Proceedings Michigan State University Institute of Public Utilities, Fourteenth Annual Conference, 724, 1983.

Forster, B., Pollution control in a two-sector dynamic general equilibrium model, *J. Environ. Econ. Manage.*, 4, 305, 1977.

Forster, B., Environmental regulation and the distribution of income in simple general equilibrium models, in *Economic Perspectives*, Ballabon, M. B., Ed., Harwood Academic Publishers, New York, 1981.

Forster, B., Acid rain in North America: an international externality, Research Report or Occasional Paper No. 84, Department of Economics, University of Newcastle, 1983.

Galloway, J. and Whelpdale, D., an atmospheric sulfur budget for eastern North America, *Atmos. Environ.*, 14, 409, 1980.

Jones, R., The structure of simple general equilibrium models, *J. Polit. Econ.*, 73, 557, 1965.

Jones, R., A three-factor model in theory, trade, and history, in *Trade, the Balance of Payments and Growth*, Bhagwati, J., Ed., North Holland, Amsterdam, 1971.

Markusen, J., Factor movements and commodity trade as complements, *J. Int. Econ.*, 14, 341, 1983.

McGuire, M., Regulation, factor rewards, and international trade, *J. Public Econ.*, 17, 335, 1982.

Memorandum of intent on transboundary air pollutants, Interim Report of Work Groups, 1981.

Merrifield, J. D., The Impact of Selected Abatement Strategies on Transnational Pollution, the Terms of Trade, and Factor Rewards: A General Equilibrium Approach, Ph.D. thesis, University of Wyoming, Laramie, 1984.

Ministry of Supply and Services, *Canada Yearbook, 1980–1981*, Bryant Press Limited, Ottawa, 1981.

NGA backs acid rain controls, *Coal Outlook*, 8, 3, 1984a.

Northeast states blast EPA SO2 stance, *Coal Outlook*, 8, 1, 1984b.

Ohlendorf, P., Unexplained deaths in the forest, *MacLeans*, 97, 50, 1984.

Panel probes Reagan's acid rain view, *Coal Outlook*, 8, 4, 1984c.

Pethig, R., Pollution, welfare, and environmental policy in the theory of comparative advantage, *J. Environ. Econ. Manage.*, 2, 160, 1976.

Robinson, R., Recognizing the true cost of acid rain, in *Acid Rain*, Canadian–American Center, SUNY Buffalo, 1982.

Siebert, H., Eichberger, J., Gronych, R., and Pethig, R., *Trade and Environment: A Theoretical Enquiry*, Elsevier, Amsterdam, 1980.

U.S. Department of Commerce (USDC), *The Effects of Pollution Abatement on International Trade–III*, U.S. Government Printing Office, Washington, D.C., 1975.

U.S. General Accounting Office (USGAO), *A Market Approach to Air Pollution Control Could Reduce Compliance Costs Without Jeopardizing Clean Air Goals*, U.S. Government Printing Office, Washington, D.C., 1983.

Walter, I., Ed., *International Economics of Pollution*, Macmillan, London, 1975.

Walter, I., Ed., *Studies in International Environmental Economics*, John Wiley & Sons, New York, 1976.

Yohe, G.W., The backward incidence of pollution control — Some comparative statics in general equilibrium, *J. Environ. Econ. Manage.*, 6, 187, 1979.

3 International Trade in Waste Products in the Presence of Illegal Disposal[1]

Brian R. Copeland

This chapter develops a simple model of international trade in waste disposal services and investigates the welfare effects of restricting such trade. While the first-best policy is to allow free trade in all goods and services and to use internal taxes and regulation to control the externalities associated with waste disposal, such a policy increases the incentives for firms to evade taxes and regulations and dispose of their wastes illegally. In this case, the optimal tax can be lower than it would be in the absence of illegal activity, and trade restrictions on waste disposal can be welfare-improving.

A. INTRODUCTION

The disposal of unwanted and, in many cases, toxic waste material is one of the serious and growing problems of our time. As traditional waste disposal sites fill up or are displaced by urban growth, and as people become concerned about the undesirable effects of living near waste dumps, the option of sending waste outside of its region of origin for disposal is increasingly considered. However, exports cannot take place without someone being an importer, and those seeking to find disposal sites outside their own jurisdiction often meet with much resistance.[2] Recent publicized examples include concern over the disposal of European wastes in Africa, attempts by New York City to find a recipient for shipments of its garbage, the controversy over the location for nuclear waste disposal sites in the United States (an issue of regional trade in waste), and claims that toxic waste from the United States have been shipped to Canada in fuel oils. Since waste disposal generates negative externalities, the movement of waste products between countries results in an international transfer of externalities, suggesting that concerns over such trade may be well founded.

There is a large body of literature on international externalities, focusing primarily on two mechanisms for their transmission.[3] In one strand of the literature, pollution generated in one country spills over into another (e.g., Scott, 1976, Markusen, 1975, Conrad and Scott, 1988, and Copeland, 1990). In another strand, the external effects of pollution are confined to the country of origin, but trade, pollution control policies, and tariffs can affect the location of the pollution-generating industry, thereby resulting in indirect trade in pollution (e.g., Baumol and Oates, 1975; D'Arge and Kneese, 1972; Pethig, 1976; and Siebert et al., 1980). Some papers allow for both international spillovers and changes in the location of pollution-intensive industries (McGuire, 1982; Merrifield, 1988).

This chapter considers another mechanism for the international transfer of externalities: trade in the waste material itself. Unlike the first strand of the literature, the externalities from waste disposal are assumed to be confined to the country where the waste is disposed. Unlike the second strand, it is assumed that the waste can be exported without having to move the waste-producing industry.

---

[1] Part of this chapter was written while the author was visiting the economics department at the University of Alberta. The author thanks members of that department for their hospitality and also two anonymous referees for helpful comments and suggestions.
[2] See, e.g., Mitchell and Carson (1986).
[3] See Siebert (1985) and Merrifield (1988) for surveys.

Nevertheless, the approach of this chapter has much in common with the literature on trade in pollution-intensive goods (i.e., the second strand), since the traded waste products generate externalities.

The main issue that we investigate is whether trade restrictions should play a role in the control of the externalities associated with waste disposal. In Section B, we develop a simple model of trade in waste and find that as long as the government can expect perfect compliance with its policies, the first-best policy is to allow free trade in waste disposal and correct the externality with internal regulations and/or taxes. On the other hand, if the government is unable or unwilling to implement the first-best policy, then restricting the entry of foreign waste can improve welfare.

However, even if the government does have the political will to optimally regulate waste disposal, such a policy can increase the incentives for firms to evade the regulations and taxes and dispose of their wastes illegally. Moreover, the consequences of illegal disposal often take time to emerge, thus making enforcement difficult. A similar problem occurs in the regulation of pollution and has been studied by various authors (e.g., Downing and Watson, 1974; Harford, 1978, 1987; Martin, 1984; Linder and McBride, 1984; and Lee, 1984). Sullivan (1987) has recently investigated illegal toxic waste disposal in a closed-economy partial equilibrium model.[4]

Hence, in Section C we extend the model to consider the case where firms evade regulations and taxes. Our model differs from the papers cited above in that it considers the consequences of evasion in a general equilibrium framework and investigates the role of international trade policy in regulating waste disposal in the presence of evasion. Our principle conclusions are, first, that in the presence of evasion, the optimal tax on waste disposal may be lower than in the absence of evasion, and second, that restrictions on trade in waste products, when used in conjunction with internal regulation, can be welfare improving.

B. LEGAL WASTE DISPOSAL

1. THE MODEL

We consider a small, open economy. Two primary factors are supplied inelastically: land (T) and labor (L). To focus on the role of waste disposal, we consider only two production activities: a final consumption good, X, with fixed world price p_x, and waste or garbage disposal services, G, with fixed world price p_g. Waste is a byproduct of the production process for good X.[5] A firm that produces good X must therefore pay for waste disposal services. The technology for X is given by

$$\tilde{H}\left(x, g_x, l_x, t_x\right) = 0,$$

where l_x is the labor input, t_x is the land input, and g_x is the level of waste generated. $\tilde{H}$ is assumed to exhibit constant returns to scale. We assume that this can be inverted to yield

$$x = H\left(l_x, t_x, g_x\right),$$

and that H is quasi-concave, increasing in l_x and t_x, and nondecreasing in g_x.

---

[4] There is a large body of literature on the economics of illegal activity, in addition to that cited above. Much of it is based on the assumption that individuals equate the marginal benefits of increased illegal activity to their marginal cost, with Becker (1968) being the seminal paper. Particularly relevant to our problem is the economics of tax evasion, surveyed recently by Cowell (1985). Much of the work on tax evasion is carried out in a partial equilibrium model; hence an important recent paper is that of Kesselman (1989), who analyzes the general equilibrium consequences of tax evasion. In the international trade literature, attention has focused on smuggling (see Bhagwati, 1981 for a survey and Martin and Panagariya, 1984 for a recent contribution) and illegal immigration (Ethier, 1986; Bond and Chen, 1987). In the fisheries literature, Sutinen and Anderson (1985) and Milliman (1986) consider optimal management in the presence of illegal evasion of fisheries regulations and taxes.

[5] Alternatively, we could model the disposal of wastes associated with consumption. For simplicity, we concentrate on the production side.

Profit maximization by X producers requires maximizing net revenue from the joint products less the cost of primary inputs. We may, however, treat this as a two-stage maximization problem, first choosing l_x, t_x, and g_x to maximize profits for a given level of x, and then choosing x. The first stage is equivalent to treating waste disposal as an intermediate input into the production process for X and minimizing the cost of the primary and intermediate inputs. The unit cost function corresponding to H is

$$c^x\left(w, r, p_g^d\right) = \min_{\{l_x, t_x, g_x\}}\left[wl_x + rt_x + p_g^d g_x \quad \text{s.t.} H\left(l_x, t_x, g_x\right) = 1\right],$$

where w is the wage, r is the rent on land, and p_g^d is the domestic price of disposing one unit of waste. In the absence of any taxes, $p_g^d = p_g$, since we ignore transportation costs.

Treating waste disposal as an input gives us three inputs and allows at most one pair of inputs to be complements. We do not rule out this possibility, but we assume that land and labor are substitutes in the production of X.

The output of the waste disposal industry is measured in units of waste disposed. The production function for waste disposal is $G(l_g, t_g)$, which we assume exhibits constant returns to scale. The corresponding unit cost function is denoted $c^g(w, r)$.

One important aspect of waste disposal is that it may generate externalities. Toxic chemicals disposed of on one plot of land may leach into rivers and adversely affect agents on other plots of land. Waste disposal may also generate visual or air pollution. To model these aspects of waste disposal, we assume that the government regulates waste disposal and that the cost function c^g reflects the costs of legally disposing of waste.[6] However, regulation cannot contain all externalities, hence the level of waste disposal activity within an economy will enter the utility function of the representative agent. Thus utility is given by $U(x, g)$, where x is the consumption of good X, and g is the level of waste disposal service provided in the domestic economy (which reflects waste generated in the production of X plus net foreign waste disposed of in the domestic economy). Note that the consumer faces a trivial decision problem: all income is spent on X, as the external effects of G are passively consumed.

Assuming that the economy does not specialize in either X or G, equilibrium requires zero profits in each sector,

$$p^x = c^x\left(w, r, p_g^d\right), \tag{3.1}$$

$$p_g^d = c^g(w, r), \tag{3.2}$$

and full employment,

$$c_w^x x + c_w^g g = L, \tag{3.3}$$

$$c_r^x x + c_r^g g = T, \tag{3.4}$$

where L and T are the endowments of labor and land, respectively, and where c_j^i is the partial derivative of c^i with respect to input price j, which by Shephard's Lemma yields the unit demand for the factor with return j in sector i.

---

[6] The regulation of *techniques* of disposal is assumed to be exogenous and is perhaps set by scientific standards. We consider the use of taxes and tariffs to regulate the *level* of waste disposal activity. In Section C, we allow for both legal and illegal waste disposal and assume that illegal disposal generates larger externalities.

The net export vector for waste disposal services can be determined from the above by noting that the level of domestic waste disposal is the sum of waste produced by the domestic X sector, g_x, and net foreign waste imported for disposal, g_e. From Shephard's Lemma, the amount of waste produced by the domestic X industry is[7]

$$g_x = c_p^x x .\tag{3.5}$$

Thus,

$$g_e = g - g_x = g - c_p^x x .\tag{3.6}$$

If $g_e > 0$, some foreign waste is disposed of domestically. This is an exported service, but the waste is physically imported into the domestic country. If $g_e < 0$, then the full demand for factors for domestic X production (i.e., the direct demand for factors by the X industry plus the indirect derived demand for factors to dispose waste generated by X) exceeds domestic supply, and hence, some domestic waste is disposed of abroad.

Not surprisingly, the pattern of trade depends on relative factor intensities and factor abundance. For concreteness, we assume throughout that waste disposal is land intensive relative to X. Then it is straightforward to show that the domestic economy exports waste disposal services if the full (direct and indirect) land/labor ratio in X is less than the economy's land/labor ratio. In other words, if the economy is sufficiently land abundant, it processes both foreign and domestic waste material. Note that the pattern of trade in this model is independent of demand since there is only one final consumption good, and since the demand for waste disposal services is a derived demand.

2. Tariffs, Taxes, and Welfare

Because waste disposal yields a negative externality, a free market results in overproduction of waste disposal services. From the literature on optimal intervention in the presence of distortions (e.g., Bhagwati, 1971), we know that the first-best policy requires a tax on the output of the waste disposal sector.[8] That is, producers in the X industry should pay the world price for waste disposal, while firms in the waste disposal sector will receive the world price less the tax.[9] The optimal tax can be easily calculated and is

$$\tau_g = -\frac{\partial U/\partial g}{\partial U/\partial x} \equiv -\text{MRS}_{x,g}\tag{3.7}$$

(evaluated at the optimal point), which reduces the return to the waste disposal industry by the amount of the marginal damage imposed on consumers.

While a disposal tax is optimal, governments may be unable or unwilling to implement it, perhaps because of political constraints. An alternative to taxing waste disposal is to intercept foreign waste at the border and restrict its entry with either trade taxes or quantitative restrictions.

---

[7] Note that here we use a subscript "p" to denote partial differentiation with respect to the third argument of the cost function, p_g^d.

[8] Note that as our model is specified, the waste itself causes the externality. Toxic waste disposal is probably the most important example of this. In other types of disposal, however, such as the burning of waste paper, a by-product of the waste (e.g., smoke) causes the externality. In that case, the optimal policy requires a tax on the smoke, or on the particles in the smoke that causes the damage. We have adopted the toxic waste model in the paper for purposes of exposition.

[9] Alternatively, one can think of both domestic and foreign producers of X as paying a tax on their wastes processed in the domestic economy; in this case, the domestic producer price of waste disposal must again be equal to the world price less the tax.

Let us therefore examine the effects of a tax on foreign waste processed in the domestic economy. Although such a policy taxes the foreign waste as it enters the domestic economy, it is nevertheless an export tax, since it taxes a sector providing services to foreigners.

The utility of the representative agent is $U(I/p^x, g)$, where I is national income. Normalize by treating X as the numeraire and write income as

$$I = x + \left[p_g(1-t) - \tau_g \right] g - p_g(1-t) g_x + t p_g g_e + \tau_g g,$$

where t is the ad-valorem tax rate on foreign waste and τ_g is a specific production tax on the waste disposal industry. Income is the value of final output at domestic producer prices less the cost of waste disposal to the X industry (to avoid double counting) plus the two types of tax revenue. Note that in the presence of an export tax, arbitrage forces the domestic price of waste disposal services down to $p_g(1-t)$.

Totally differentiating U and simplifying yields the first-order condition for t (assuming no terms-of-trade effects),

$$\frac{dU/dt}{\partial U/\partial x} = t p_g \frac{dg_e}{dt} + \left[\tau_g + \text{MRS}_{x,g} \right] \frac{dg}{dt} = 0, \tag{3.8}$$

where $\text{MRS}_{x,g} \equiv U_g / U_x < 0$. It can easily be shown that $dg/dt < 0$ and $dg^e/dt < 0$; that is, an export tax on waste disposal causes both the waste disposal sector and exports of waste disposal services to contract, since the export tax lowers the domestic producer price in that sector.

First, note that if the production tax is chosen optimally, then the second term in (Equation 3.8) vanishes, yielding the result that the optimal export tax is zero. Next suppose that the government does not impose a production tax on the waste industry, so that $\tau_g = 0$. Then (Equation 3.8) may be solved for the optimal trade tax:

$$t = -\frac{\text{MRS}_{x,g} \, dg/dt}{p_g \, dg_e/dt} > 0.$$

Thus, a tax on domestic processing of foreign waste may be used as a second-best instrument, since, by reducing the inflow of foreign waste, it reduces the total amount of waste processed domestically. However, unlike the production tax, it creates an input distortion by lowering the domestic relative price of waste disposal. This renders the trade tax inferior to a production tax. One particularly unattractive feature of a trade tax is that it may have the undesirable side effect of increasing the waste intensity of domestic X-producing firms, depending on the net effect of the lower domestic price of waste disposal caused by the export tax and the induced factor price changes.

C. ILLEGAL WASTE DISPOSAL

1. THE MODEL

The above analysis assumes that governments can achieve the first-best outcome by using their powers of regulation and taxation. However, in practice, it can be difficult to achieve compliance, since firms or individuals may face strong incentives to dispose of their wastes surreptitiously. We consider two important such incentives. First, firms may attempt to avoid paying a tax imposed to correct the externality, and second, illegal disposal may be cheaper than legal disposal because government regulations designed to minimize damage to the environment may be ignored. Both of these motives for evasion complicate the regulation problem, since increases in taxes or tightening of regulations may lower the tax base and send more of the disposal activity into the gray economy, where it cannot be regulated.

For simplicity, we assume that all firms are identical and maximize expected profits.[10] We continue to assume constant returns to scale. The unit cost of illegal disposal is ci(w,r), and the unit cost of legal disposal is c1(w,r). It is assumed that illegal disposal is less costly than legal disposal since it is unregulated,[11] and in particular that ci = bc1, with b < 1.[12] However, if caught disposing waste illegally, the firm pays a fine of F per unit dumped illegally. Resources devoted to enforcement and the level of enforcement effort are assumed to be fixed,[13] and the fine is treated as exogenous.[14]

The disposing firm faces a probability $\rho(\Lambda)$ of its illegal activity being detected, where Λ is the fraction of its waste disposed of illegally. For most of the analysis, we assume that $\rho'(\Lambda) > 0$ and $\rho''(\Lambda) \geq 0$, but we also consider some examples where ρ is constant. Note that the assumption that $\rho'(\Lambda) > 0$ implies that the probability of being caught increases with either (1) an increase in the amount of illegal activity, with legal disposal held constant; or (2) a reduction in the amount of legal activity, with illegal disposal held constant. That is, we implicitly assume that legal disposal can be used to (partially) mask illegal disposal. One enforcement strategy that yields a detection probability of the form $\rho(\Lambda)$ is a system where firms face a given probability of being checked for illegal activity during a given unit of time and where, because illegal disposal activity is distributed in either time or space, the probability of illegal activity being detected during a check depends on the fraction of disposal that is illegal.[15]

---

[10] An alternative formulation would have managers maximizing the expected utility of profits, where the psychic costs of illegal activity enter the utility function.

[11] If firms have to take evasive activity to avoid prosecution, it is possible that $c^i > c^1$ in some cases. The model can be extended to cover this case, but it is ignored for ease of exposition. In the case of pollution, Lee (1984) finds that the presence of avoidance costs tends to lower the tax on pollution. The case examined in the text seems to be the relevant one for waste disposal.

[12] This assumption ensures that legal and illegal disposal have the same factor intensity. The model can be extended to allow for different factor intensities, at the cost of added complexity.

[13] The model can easily be extended to allow for an endogenous enforcement decision, if one assumes that an increase in enforcement effort requires an increased cost in terms of consumption goods (but not an increased commitment of factor inputs). For example, existing enforcement officers can be offered a higher "bounty" for capturing illegal disposers. In this case, the government sets the marginal cost of enforcement equal to the marginal benefit. Although this gives the government an additional instrument to control illegal waste disposal, the principal conclusions of the model will not be significantly affected. For example, both a tax on foreign waste and an increase in enforcement can be used to reduce illegal waste disposal but the option of increasing the level of enforcement will not eliminate the usefulness of a trade tax, since the first-order cost of a small trade tax is zero (since, in the absence of distortions, the optimal trade tax is zero), while the first-order cost of increased enforcement is positive. Hence, trade restrictions can usefully supplement domestic enforcement.

Extending the model to allow for variable levels of *inputs* allocated to enforcement would add more complexity to the model, since an increase in enforcement will affect factor prices and thus have complicated general equilibrium effects on other sectors. If, however, resources devoted to enforcement are a small fraction of total resources in the economy, then the results should not significantly differ from those outlined above.

[14] Endogenizing the fine is more difficult, a problem which is pervasive in the literature on the economics of crime (e.g., see Cowell, 1985). In principle, it is relatively straightforward to ensure compliance with a regulation in these models by setting the fine at a prohibitively high level, or by using the death penalty. In practice, however, there are many reasons why the severity of the punishment tends to be commensurate with the seriousness of the crime: for example, using the death penalty to control tax evasion would encourage evaders to kill tax collectors. Hence, the optimal punishment is properly analyzed in the context of the entire legal system. Our assumption of an exogenous fine is thus an approximation, but one which seems to be more reasonable than allowing it to be an unconstrained choice variable.

[15] The assumption that the probability of detection depends on the *fraction* of illegal activity has also been used by Martin (1984) to study the enforcement of pollution laws (although his is a closed economy, partial equilibrium model) and Martin and Panagariya (1984) to study smuggling in a general equilibrium model. This is a special case of the more general specification $\rho(g^1, g^i)$. We have adopted the specification used in the text because it allows us to capture the dependence of the detection probability on both legal and illegal activity and yet also allows us to use a standard general equilibrium trade model incorporating constant returns to scale. In the more general formulation, the size of the firm will affect the detection probability, and hence we would have to depart from constant returns to scale to obtain a non-infinitesimal optimal firm size. This would add considerable complexity to the model, but would not affect the major conclusion of the paper, which is that trade policy can have a second-best role to play in regulating waste disposal in the presence of imperfect enforcement of internal regulations.

While illegal waste disposal is cheaper than legal waste disposal (and thus uses fewer factors), it is nonetheless socially costly. To capture this, we specify the representative utility function as $U(x,g^l,g^i)$, where g^l is the total amount of legal waste disposal in the domestic economy, and g^i is the amount of illegal disposal. It is assumed that

$$\left|U_{g^l}\right| \le \left|U_{g^i}\right|, \tag{3.9}$$

and in cases where illegal waste disposal is strictly cheaper than legal disposal, it is assumed that transferring one unit of waste from legal to illegal disposal will result in a utility loss which is greater than the benefits of increased consumption of X resulting from the cost saving in waste disposal. This assumption is formalized below in Subsection 3.

It can be shown that profit maximization by waste disposers requires minimization of the expected cost of waste disposal. Hence, to dispose of g units of waste, firms must choose the optimal mix of legal and illegal disposal and solve

$$\min_{\{g^l,g^i\}}\left\{g^l\left[c^l(w,r)+\tau_g\right]+g^i\left[c^i(w,r)+\rho(\Lambda)F\right]\ \ \text{s.t.}\, g^l+g^i=g\right\}.$$

For waste disposed legally, firms pay the unit cost c^l plus the tax, while on illegal waste disposal, firms avoid the tax and pay a possibly lower cost c^i, but face some risk of being fined. It should be noted that this function is linearly homogeneous in g, so that a unit cost function $C^G(w,r;\tau_g,F)$ exists and with a change of variables can be written as

$$C^G\left(w,r;\tau_g,F\right) \equiv \min_{\{\Lambda\}}\left\{(1-\Lambda)\left[c^l(w,r)+\tau_g\right]+\Lambda\left[c^i(w,r)+\rho(\Lambda)F\right]\ \ \text{s.t.}\ \le\Lambda\le 1\right\}.$$

The first-order condition for the privately optimal fraction of illegal disposal satisfies (in the case of an interior solution)

$$c^l(w,r)+\tau_g-\left[c^i(w,r)+\rho(\Lambda)F\right]=\Lambda F\rho'(\Lambda), \tag{3.10}$$

which sets the cost saving from an increase in illegal disposal equal to the expected increase in fines.[16]

Free entry results in zero profits,

$$p^x = c^x\left(w,r,p_g^d\right), \tag{3.11}$$

$$p_g^d = C^G\left(w,r;\tau_g,F\right), \tag{3.12}$$

where we recall that $p_g^d = p_g(1-t)$ is the domestic price of waste disposal services. We also require full employment,

---

[16] A corner solution with no illegal activity results if $c^l(w,r) + \tau_g < c^i(w,r) + \rho(0)F$, that is, when the fine is very high or when the tax and cost differential is low. Also, legal disposal will vanish if $c^l(w,r) + \tau_g -[c^i(w,r) + \rho(1)F] > F\rho'(1)$, which would be the case if fines or detection probabilities were low relative to the cost saving from illegal disposal.

$$c_w^x x + \left[(1-\Lambda)c_w^1 + \Lambda c_w^i\right]g = L , \qquad (3.13)$$

$$c_r^x x + \left[(1-\Lambda)c_r^1 + \Lambda c_r^i\right]g = T , \qquad (3.14)$$

where L and T are to be interpreted as factor endowments, less the fixed level of resources devoted to enforcement.

2. COMPARATIVE STATICS OF PRODUCTION

Our primary interest is to determine the welfare effects of tax and trade policies in the presence of illegal disposal. Since the welfare effects of various policies depend critically on output responses, we begin our analysis by investigating the response of the fraction of illegal activity to changes in policy. Differentiating (Equation 3.10) yields

$$d\Lambda = \frac{(1-\beta)\left(c_w^1 dw + c_r^1 dr\right) + d\tau_g}{(2\rho' + \Lambda p'')F} , \qquad (3.15)$$

where we have used the assumption that $c^i(w,r) = \beta c^1(w,r)$. Under the same assumption, differentiation of (Equation 3.12) yields

$$dp_g^d = \left(1 - \Lambda + \Lambda\beta\right)\left(c_w^1 dw + c_r^1 dr\right) + (1-\Lambda)d\tau_g .$$

By combining, we obtain

$$d\Lambda = \frac{(1-\beta)dp_g^d + \left[1 - (1-\beta)(1-\Lambda)\right]d\tau_g}{(1 - \Lambda + \Lambda\beta)(2\rho' + \Lambda p'')F} . \qquad (3.16)$$

Since the denominator is positive, (Equation 3.16) implies that an increase in the tax on legal disposal *increases* the fraction of disposal which is illegal (i.e., $d\Lambda/d\tau_g > 0$). This follows directly from cost minimization. For given factor prices, the tax increases the cost of legal disposal, while leaving the cost of illegal disposal unaffected. Hence, it is optimal for a firm to dispose a higher fraction of its waste illegally. Finally, because the factor intensity of the two methods of waste disposal is the same, the indirect effects of any induced factor price changes are not strong enough to offset the direct effect.

On the other hand, an increase in the export tax on waste disposal services *reduces* the fraction of illegal disposal (i.e., from (Equation 3.16), $d\Lambda/dt < 0$, since, in this case, $dp_g^d = -p_g dt$). This is because the reduction in the price of waste disposal induced by the export tax reduces the absolute difference between the return to illegal versus legal disposal, and with the fine rate held constant, this reduces the relative incentive for illegal activity.

We now turn to the effects of tax changes on the levels of legal and illegal waste disposal. In the Appendix, it is shown that (with $\hat{L} = \hat{T} = \hat{p}_x = 0$)

$$\hat{g} = e_p^g \hat{p}_g^d - e_\tau^g (1-\Lambda) d\tau_g \Big/ p_g^d + e_\Lambda^g d\Lambda , \qquad (3.17)$$

where $e_p^g > 0$, $e_\tau^g > 0$, and

$$e_\Lambda^g = \frac{(1-\beta)}{(1-\Lambda+\Lambda\beta)} > 0 .$$

Although the coefficient of $\hat{p}_g^d$ is positive, and that of $d\tau_g$ is negative, the total effects depend also on the response of Λ, given by (Equation 3.16).

Let us first consider the effects of export tax changes on waste disposal services. By using (Equation 3.16) in (Equation 3.17) and noting again that for an export tax change, $dp_g^d = -p_g dt$, we see that an increase in the export tax on waste disposal services reduces both aggregate waste disposal (i.e., $dg/dt < 0$) and illegal waste disposal (i.e., $dg^i/dt < 0$). Total domestic waste disposal falls because the export tax lowers the return to that sector. Illegal waste disposal falls since an export tax directly reduces the return to both legal and illegal disposers. Moreover, as noted above, the export tax reduces Λ, the fraction of illegal disposal, and, hence, when indirect effects are taken into account, the export tax actually penalizes illegal disposal relatively more than legal disposal. The effect of an export tax on g^1 is ambiguous, however. While the total amount of waste disposal activity declines, the fraction of that activity, which is legal, increases. Note that if both legal and illegal disposal have the same cost function, so that tax evasion is the only motive for illegal activity, then an export tax does not affect the ratio of legal to illegal disposal (i.e., $d\Lambda/dt = 0$), and g^1 unambiguously declines in response to an export tax.

Next consider the effects of a disposal tax. First, such a tax will unambiguously reduce the level of legal disposal, because it is a direct tax on this activity. To see this, note that $\hat{g}^1 = \hat{g} - d\Lambda/(1 - \Lambda)$, and use (Equation 3.17) to obtain

$$\frac{dg^1}{d\tau_g} = -\frac{g(1-\Lambda)e_\Lambda^g}{p_g^d} - \frac{\beta dg\Lambda/d\tau_g}{(1-\Lambda+\Lambda\beta)(1-\Lambda)} < 0 . \tag{3.18}$$

However, the effect on both illegal and total waste disposal is ambiguous. In the case of illegal disposal,

$$\frac{dg^i}{d\tau_g} = -\frac{g(1-\Lambda)e_\tau^g}{p_g^d} + \frac{gd\Lambda/d\tau_g}{(1-\Lambda+\Lambda\beta)\Lambda} . \tag{3.19}$$

On one hand, the tax on legal disposal causes a shift to illegal disposal (the second term above) and this tends to raise g^i. But, on the other hand, for constant Λ, the tax tends to reduce g (the first term above), and this also tends to reduce g^i, since the probability of illegal activity being detected depends on the ratio of illegal disposal to total disposal. Note that if the supply elasticity e_τ^g is small, then the second effect will dominate.

The ambiguous response of the total level of waste disposal to an increase in the disposal tax is perhaps more surprising. When Λ is held constant, g falls, but the tax results in a shift to illegal disposal, an effect which mitigates the decline of g. Since illegal waste disposal has a smaller factor input requirement than legal disposal, the replacement of legal activity with illegal activity frees up factors, some of which may be absorbed by the waste disposal industry. Hence, we cannot rule out a perverse expansion of this sector in response to the tax.

As an example, let us consider the case where the probability of detection, ρ, is exogenous and independent of the fraction of illegal disposal. Suppose that initially, with $\tau_g = \tau_g^0$, we have

$$c^1 + \tau_g^0 = c^i + \rho F - \varepsilon ,$$

for $\varepsilon > 0$ small. Then all disposal is done legally, since the expected cost of illegal disposal (including fines) is greater than the cost of legal disposal. Now suppose the tax increases to τ_g^1, such that

$$c^1 + \tau_g^1 = c^i + \rho F + \varepsilon .$$

Then setting the tax at this level will cause firms to shift *all* of their disposal to the illegal mode. The change in g is approximately

$$\hat{g} = -e_\tau^g (1 - \Lambda) d\tau_g / p_g^d + e_\Lambda^g d\Lambda,$$

where (with slight abuse of notation) differentials represent discrete changes.

Let us consider the effect on g in two steps. First hold Λ constant (at zero) and consider the effect of the small increase in τ_g. This is measured by the first term, and for small ε, this first effect approaches zero. Next allow Λ to change. This will free up factors since all legal disposal ceases. Note also that since the factor input ratios for legal and illegal disposal are the same, virtually all of these factors will be absorbed by illegal disposal. That is, the full employment conditions are satisfied for the same factor combination, regardless of the level of Λ. Hence in this example, an increase in the tax on legal waste disposal increases the total amount of waste disposal by a factor of approximately $1/\beta$.

On the other hand, if both legal and illegal disposal have the same cost function, so that tax evasion is the only motive for illegal activity, then $e_\Lambda^g = 0$, and g unambiguously declines in response to the tax.

Finally, we should note the asymmetry between the effects of an export tax and a disposal tax. The export tax reduces the level and fraction of illegal disposal, while having an ambiguous effect on the level of legal disposal. A disposal tax reduces the level and fraction of legal disposal, while having an ambiguous effect on illegal disposal. A disposal tax specifically targets legal disposal and hence has a direct negative impact on that sector. An export tax, on the other hand, is a blunter instrument, directly affecting both legal and illegal disposal. However, since it reduces the absolute differential between the return to legal versus illegal disposal, the illegal sector is hit relatively harder because the penalty in the event of detection is held constant. In the literature on optimal intervention in the presence of distortions (e.g., Harford, 1978), trade policy is typically found to be an inappropriate intervention instrument precisely because it is a blunt instrument, often solving one problem by creating another. In our second-best problem, however, we can exploit the ability of the export tax to simultaneously affect both the legal and illegal sectors, and hence use it as a means of compensating for imperfect enforcement.[17]

3. WELFARE EFFECTS OF TAX CHANGES

We now turn to the welfare analysis of policy changes. With p_x as the numeraire, the utility of the representative consumer is $U(I, g^1, g^i)$, where I is national income. Then

$$dU/U_x = dI + MRS_{x,1} dg^1 + MRS_{x,i} dg^i, \tag{3.20}$$

where $MRS_{x,1} = U_{g^1}/U_x$ and $MRS_{x,i} = U_{g^i}/U_x$ are the marginal rates of substitution between the consumption good and legal and illegal waste disposal, respectively. Income is given by

$$I = x + p_g g_e - (1 - \gamma)\rho(\Lambda)Fg^i, \tag{3.21}$$

---

[17] Although not considered explicitly in the text, it is worth briefly considering the effect of changes in the enforcement variables on waste disposal. An increase in the level of the fine will reduce total waste disposal, the fraction of illegal waste disposal, and the level of illegal waste disposal. This is because the fine acts as a direct tax on illegal disposal, thus reducing the level of illegal activity, and also thereby increases the overall unit cost level in the waste disposal sector, thus reducing total output in that sector. In addition, an increase in the detection probability (i.e., an exogenous upward shift in $\rho(\Lambda)$ that leaves $\rho'(\Lambda)$ unaffected) will also decrease overall waste disposal, and the level and fraction of illegal disposal, for similar reasons.

where γ is the fraction of the firm's fine that is not dissipated through court procedures and other collection costs and is available to be returned to consumers by the government. In general, we would expect that γ would be significantly less than 1, reflecting the fact that a legal sanction imposed on a firm does not result in a simple transfer to the government.

It is more convenient to write income in the equivalent form

$$I = x + \left[p_g(1-t) - \tau_g\right]g^1 + \left[p_g(1-t) - \rho(\Lambda)F\right]g^i - p_g(1-t)g_x + tp_gg_e + \tau_gg^1 + \gamma\rho(\Lambda)Fg^i \quad ,(3.22)$$

which consists of expected producer revenue, less the value of the intermediate input, plus government revenues. Totally differentiating and making use of the profit-maximizing conditions for producers yields

$$dI = tp_g dg_e + \tau_g dg^1 + \gamma\rho(\Lambda)F dg^1 + \gamma\rho'(\Lambda)Fg^i d\Lambda . \tag{3.23}$$

Hence, we have

$$dU/U_x = tp_g dg_e + \left(\tau_g + MRS_{x,1}\right)dg^1 + MRS'_{x,i}dg + \gamma\left[\rho(\Lambda)F dg^i + \rho'(\Lambda)Fg^i d\Lambda\right]. \tag{3.24}$$

The first term measures the change in the trade distortion. The second measures the effect of a change in the level of legal waste disposal. Note that this effect vanishes if the tax on legal waste disposal is just enough to offset the consumption externality. The third effect measures the external damage to consumers from an increase in illegal disposal, and the final term is equivalent to $\gamma d[\rho(\Lambda)Fg^i]$ and thus measures the net change in fines collected by the government. Note that if an increase in illegal disposal generates an increase in fines that just offsets the loss in consumer welfare from the change, then the last two terms vanish.

We can use (Equation 3.24) to analyze policy changes, but first we need to formalize our assumption that illegal waste disposal is more socially damaging than legal disposal. To do this, rewrite (Equation 3.24) as follows, making use of the identities $g^1 = (1 - \Lambda)g$, and $g^i = \Lambda g$:

$$dU/U_x = tp_g dg_e + \left[(1-\Lambda)\left(\tau_g + MRS_{x,1}\right) + \Lambda\left(\gamma\rho(\Lambda)F + MRS_{x,i}\right)\right]dg$$

$$+ \left[\gamma\rho(\Lambda)F + MRS_{x,i} - \left(\tau_g + MRS_{x,1}\right) + \gamma\rho'(\Lambda)F\Lambda\right]gd\Lambda \tag{3.25}$$

The assumption that the social benefits achieved through the lower cost of illegal disposal are not sufficient to offset the increased severity of the externality is equivalent to assuming that for a given level of waste disposal g, an increase in the fraction of waste disposed of illegally ($d\Lambda > 0$) is welfare decreasing. By setting $dg_e = dg = 0$ in (Equation 3.25), we see that this assumption implies

$$\gamma\rho(\Lambda)F + MRS_{x,i} - \left(\tau_g + MRS_{x,1}\right) + \gamma\rho'(\Lambda)F\Lambda < 0 . \tag{3.26}$$

To interpret this condition, use the first-order condition for optimal choice of Λ to write (Equation 3.26) as

$$c^1 - c^i < \frac{MRS_{x,1} - MRS_{x,i}}{\gamma} + \frac{(1-\gamma)\tau_g}{\gamma} \tag{3.27}$$

In the special case where all fines paid by the firm are available to be returned to consumers (i.e., $\gamma = 1$), this simplifies to

$$c^1 - c^i < MRS_{x,1} - MRS_{x,i}, \tag{3.28}$$

which says that the cost saving from using illegal instead of legal dumping is not sufficient to offset the induced loss in utility from the stronger external effect. In (Equation 3.27) with $\gamma < 1$, the social cost of increased illegal dumping is greater, since in addition to the external effect, some of the fines are dissipated.

We now turn to policy analysis. Let us first consider the optimal tax on waste disposal. By using (Equation 3.24) and setting $dU/d\tau_g = 0$, we obtain

$$\tau_g = -MRS_{x,1} - \left[\frac{MRS_{x,i}\left(dg^i/d\tau_g\right) + \gamma\left(d[\rho(\Lambda)Fg^i]/d\tau_g\right)}{dg^1/d\tau_g} \right]. \tag{3.29}$$

Let us compare this with the optimal tax in the absence of illegal disposal. In that case, the tax is chosen to just offset the externality, i.e., $\tau_g = -MRS_{x,1}$. With illegal disposal, the tax must attempt to control both g^i and g^1, and hence there is an additional term, reflecting the influence of the tax on the level of illegal disposal. The tax reduces the level of legal disposal ($dg^1/d\tau_g < 0$) and increases the *fraction* of illegal disposal, but, as we saw earlier, the absolute level of illegal disposal may either rise or fall in response to an increase in the tax. Let us consider both cases.

If the tax results in a reduction in the levels of both legal and illegal disposal, then the second term in (Equation 3.29) is positive, and hence the tax should be higher than that required to just offset the externality from legal dumping.

In the more interesting case, the tax increases the level of illegal disposal, as it will, for example, if $\rho(\Lambda)$ is very flat. Then it is possible that the tax should be less than that required to offset the externality from legal disposal. We have $\tau_g < -MRS_{x,1}$ if

$$MRS_{x,i}\frac{dg^i}{d\tau_g} + \gamma\frac{d[\rho(\Lambda)Fg^i]}{d\tau_g} < 0,$$

that is, if the increased social damage caused by illegal disposal is greater than the increase in the level of fines collected.

As an example, let us again consider the extreme case where the probability of detection, ρ, is independent of Λ. Suppose that in the absence of any tax we have

$$c^1(w,r) < c^i(w,r) + \rho F.$$

Then all disposal is done legally, since the expected cost of illegal disposal is greater than the cost of legal disposal. However, suppose that the optimal tax, $\tau_g = -MRS_{x,1}$ (calculated on the assumption of no illegal disposal) is such that

$$c^1(w,r) + \tau_g > c^i(w,r) + \rho F.$$

Then setting the tax at this level will cause firms to shift *all* of their disposal to the illegal mode. Hence, there is an upper bound on the tax,

$$\tau_g < c^i(w,r) + \rho F - c^i(w,r)$$

and the optimal tax is lower in the presence of illegal disposal than in its absence.

Note that fines act as a tax on illegal disposal. Under our assumption (Equation 3.26), the optimal tax on illegal disposal should be prohibitive. If it is not, then local behavior of the fine revenue will affect the optimal tax on legal disposal. In particular, if either the fine (F) or the fraction of the fines not dissipated through the court system (γ) is small, then the government will face a difficult regulatory problem: it may have to use a lower tax rate and tolerate excessive legal disposal in order to avoid triggering an increase in the more damaging problem of illegal disposal. In fact, if, at $\tau_g = 0$, the total level of waste disposal, g, increases in response to a tax, then a subsidy, rather than a tax, on legal waste disposal may be called for.[18]

4. Welfare Effects of Trade Policy

We now consider the role of trade barriers as a regulatory device. Since the government has two targets in this problem, the levels of legal and illegal disposal, it requires at least two instruments to adequately control them. If penalties are not sufficiently flexible, it may be possible to use trade barriers in addition to taxes to influence the level of illegal disposal.

Let us suppose that the domestic economy processes foreign waste in equilibrium. We also assume that there is no smuggling of foreign waste material into the country.[19] To determine whether a tax on foreign waste can usefully supplement a production tax, we calculate dU/dt, given that the production tax is set optimally, and evaluate it at $t = 0$,

$$\left.\frac{dU/dt}{U_x}\right|_{t=0} = \left[(1-\Lambda)\left(\tau_g + MRS_{x,1}\right) + \Lambda\left(\gamma p(\Lambda)F + MRS_{x,i}\right)\right]dg/dt$$

$$+ \left[\gamma p(\Lambda)F + MRS_{x,i} - \left(\tau_g + MRS_{x,1}\right) + \gamma p'(\Lambda)F\Lambda\right]gd\Lambda/dt, \qquad (3.30)$$

where the tax, τ_g, is given by (Equation 3.29). By simplifying (Equation 3.30) using (Equation 3.29), we obtain

$$\left.\frac{dU/dt}{U_x}\right|_{t=0} = \frac{g\left[\gamma p(\Lambda)F + MRS_{x,i} + \gamma p'(\Lambda)F\Lambda(1-\Lambda)\right]\cdot\left[(d\Lambda/dt)(dg/d\tau_g) - (d\Lambda/d\tau_g)(dg/dt)\right]}{dg^1/d\tau_g}.$$

Using (Equation 3.17), we have

$$\frac{d\Lambda}{dt}\frac{dg}{d\tau_g} - \frac{d\Lambda}{d\tau_g}\frac{dg}{dt} = -\frac{g(1-\Lambda)e_\tau^g}{p_g^d}\frac{d\Lambda}{dt} + ge_p^g\frac{d\Lambda}{d\tau_g} > 0. \qquad (3.31)$$

---

[18] Sullivan (1987) also finds grounds for subsidizing toxic waste disposal, although in a different framework. In his model, there are two waste disposal sectors, one legal and the other illegal, and it is assumed that the legal sector always complies with the law. Hence, subsidizing the legal sector increases the rate at which waste is disposed of legally. In contrast, in analyzing pollution control, Martin (1984) finds that the tax should be higher than marginal damage. However, he considers the case where consumers are indifferent between legal and illegal pollution and has a partial equilibrium model which does not yield the general equilibrium effects due to the freeing up of resources as firms switch to cheaper, illegal disposal methods.

[19] This can be justified by arguing that it is easier to block the flow of material into a country than to monitor the material once it is inside the country, since border patrols are in place for customs and immigration purposes. However, a natural extension of the analysis would be to allow for evasion of both domestic taxes and trade taxes. If smuggling does not consume domestic resources, and as long as an increase in the trade barrier reduces the total flow of foreign waste into the domestic country, we would not expect the results to be significantly changed by the presence of smuggling (see also Footnote 20).

Using (Equation 3.26) and (Equation 3.29), we have

$$\frac{\left[\gamma p(\Lambda)F + MRS_{x,i} + \gamma p'(\Lambda)F\Lambda(1-\Lambda)\right]\left(dg/d\tau_g\right)}{dg^1/d\tau_g} < 0 . \tag{3.32}$$

Hence, in the normal case with $dg/d\tau_g < 0$, (Equation 3.31) and (Equation 3.32) imply that (Equation 3.30) is positive. Finally, note that, given our assumption that the tax is initially set optimally, we cannot have the perverse response, $dg/d\tau_g > 0$, since, in that case, we can use (Equation 3.25) to infer that a reduction in the tax improves welfare, contradicting the initial optimality of the tax.

Thus, supplementing a domestic disposal tax with an export tax is welfare improving. In the presence of illegal waste disposal, firms have an incentive to evade taxes and regulations that the government imposes on waste disposal. In these circumstances, it pays to reduce the flow of waste disposed of domestically by restricting the entry of foreign waste. A tax on foreign waste has the advantage of unambiguously reducing both the total amount of waste disposed of domestically (g) and the fraction of the waste disposed of illegally (Λ).[20] As discussed in subsection B2, however, it may have the undesirable side effect of increasing the amount of waste generated per unit of X produced domestically (since the domestic price of waste disposal falls). However, in the presence of illegal disposal, this effect is more than offset by the reduction in total waste disposed.

Finally, it should be noted that if the country initially sends its waste abroad for disposal, there is an incentive for the government to subsidize the exports of this waste (that is, subsidize the imports of waste disposal services) if firms evade domestic taxes.

D. CONCLUSION

In this chapter we have analyzed the economics of waste disposal in an international context. If one assumes perfect markets and treats waste disposal as a productive activity like any other, then not surprisingly, one finds that trade in waste should be based on comparative advantage, and that trade policy should play no role in efficient regulation of the industry. Needless to say, however, governments and/or environmentalists often tend to object to the importation of foreign waste for disposal.

Two reasons why restricting trade in waste may be welfare improving are considered here. First, if governments do not adequately regulate waste disposal and do not have the ability or political will to optimally tax the waste disposal sector, then restricting disposal of foreign waste will be a second-best policy to control the externality associated with waste disposal. Second, even if the government desires to regulate the waste sector optimally, it may find that firms or individuals evade its regulations. That is, increased taxation and regulation of the waste disposal sector creates incentives for illegal waste disposal. In this case, seemingly optimal taxation of the waste sector

---

[20] Note that we would also expect this result to hold in the presence of smuggling, as long as an increase in the trade restriction lowers the net return to the waste disposal industry in the domestic economy, and as long as, once the waste is inside the country, the detection probability for illegal disposal of smuggled waste is the same as that for legally imported waste. Regarding the first point, one would expect trade restrictions to lower the net return to the waste disposal sector, even in the presence of smuggling, as long as smuggling is costly. An importer of foreign waste will have to either bear the cost of the export tax (if the waste is brought in legally) or bear evasion costs and face a possibility of prosecution (if the waste is smuggled in). In either case, the flow of foreign waste into the domestic country will be reduced, thus reducing the demand for local waste disposal services and hence reducing the domestic price of waste disposal services. Hence we would expect both Λ and g to fall in response to the trade restriction, and thus an export tax will continue to be a useful policy instrument.

On the other hand, the detection probability for illegal disposal may be lower for smuggled waste than for legally imported waste (since there is no record of the existence of the illegal waste). In this case, the model is more complicated, since there is an incentive to smuggle waste even in the absence of a trade restriction. A trade restriction will still lower the return to the waste disposal sector, which will tend to lower Λ. However, the fraction of total waste which enters the country without any record increases, and hence the average detection probability for illegal disposal may fall, thus tending to increase Λ. Hence the net effect on Λ may be ambiguous in this case.

can make things worse, since the unregulated disposal which occurs in response to the tax can be more damaging to the environment than legal disposal. While the first-best policy is to optimally control the rate of illegal disposal, this is not feasible in many cases. In this case, supplementing a production tax with a trade barrier can improve welfare, since the trade tax can both reduce the flow of waste and reduce the fraction of waste which is disposed of illegally.

This analysis has focused on the single-country decision problem. However, once trade barriers are introduced as a means of controlling waste disposal, then one country solves its problems by passing them along to another. Moreover, if all countries attempt to control illegal disposal within their boundaries and either keep foreign waste out or encourage the export of their own waste, then there is an incentive for individuals to attempt to evade the regulations of *all* jurisdictions by dumping their wastes into international waters. Thus, as with other types of international externalities, there is a role for international cooperation in the control of waste disposal.

APPENDIX

In this Appendix, we derive factor price and output responses and, in particular, derive (Equation 3.17). To determine factor price responses, totally differentiate (Equation 3.11) and (Equation 3.12) and use (Equation 3.10) to obtain

$$\theta_{gw}\hat{w} + \theta_{gr}\hat{r} = \hat{p}_g^d - \frac{(1-\Lambda)d\tau_g}{p_g^d},$$

$$\theta_{xw}\hat{w} + \theta_{xr}\hat{r} = \hat{p}_x - \theta_{xp}\hat{p}_g^d,$$

where $\hat{w} = dw/w$, etc., $\theta_{gj} = [(1-\Lambda)jc_j^1 + \Lambda jc_j^i]/C^G$ is the share of input j in the expected cost of waste disposal (including taxes and fines), and θ_{xj} is the share of input j in the cost of x. Solving, we find

$$\hat{w} = -\frac{\hat{p}_g^d\left(\theta_{xr} + \theta_{xp}\theta_{gr}\right) - \hat{p}_x\theta_{gr} - (1-\Lambda)\theta_{xr}\,d\tau_g/p_g^d}{|\theta|} \tag{3.33}$$

and

$$\hat{r} = \frac{\hat{p}_g^d\left(\theta_{xw} + \theta_{xp}\theta_{gw}\right) - \hat{p}_x\theta_{gw} - (1-\Lambda)\theta_{xw}\,d\tau_g/p_g^d}{|\theta|} \tag{3.34}$$

where $|\theta| = \theta_{xw}\theta_{gr} - \theta_{xr}\theta_{gw}$. If both legal and illegal waste disposal are land intensive relative to X, or if based on average factor inputs, waste disposal is land intensive, then $|\theta| > 0$.

Note that the factor responses have the expected signs. An increase in the domestic price of waste disposal services (which is land intensive) increases the return to land and reduces the return to labor. It can also be shown that the real return to land increases. The reverse holds for an increase in the price of X. A tax on waste disposal increases the return to labor and reduces the return to land, since it reduces the return to the waste disposal sector. Note that the impact of the tax is weighted by $1 - \Lambda$, which is the fraction of waste disposed of legally. The smaller the fraction of legal waste disposal, the weaker the effect of the tax.

To find output responses (i.e., to derive Equation 3.17), differentiate (Equation 3.13) and (Equation 3.14) and rearrange to obtain

$$\lambda_{xw}\hat{x} + \lambda_{gw}\hat{g} = \hat{L} + \delta_{wr}^L\left(\hat{w} - \hat{r}\right) + \delta_{wp}^L\left(\hat{w} - \hat{p}_g^d\right) - g\left(c_w^i - c_w^1\right)d\Lambda/L, \tag{3.35}$$

$$\lambda_{xr}\hat{x} + \lambda_{gr}\hat{g} = \hat{T} - \delta_{wr}^T\left(\hat{w} - \hat{r}\right) + \delta_{rp}^T\left(\hat{r} - \hat{p}_g^d\right) - g\left(c_r^i - c_r^1\right)d\Lambda/T, \tag{3.36}$$

where λ_{jk} is the share of factor with return k used in sector j, and where

$$\delta_{wr}^L = \lambda_{xw}\theta_{xr}\sigma_{wr}^x + \lambda_{lw}\theta_{lr}\sigma^1 + \lambda_{iw}\theta_{ir}\sigma^i,$$

$$\delta_{wr}^T = \lambda_{xr}\theta_{xw}\sigma_{wr}^x + \lambda_{lr}\theta_{lw}\sigma^1 + \lambda_{ir}\theta_{iw}\sigma^i,$$

$$\delta_{wp}^L = \lambda_{xw}\theta_{xp}\sigma_{wp}^x, \quad \text{and} \quad \delta_{rp}^T = \lambda_{xr}\theta_{xp}\sigma_{rp}^x,$$

with the σ_{ij}^k being the Allen partial elasticities of substitution between the indicated inputs in sector k (in the case of waste disposal where there are only two inputs, we omit the subscripts). Recall that i and 1 denote the illegal and legal waste disposal industries, respectively.

By solving (Equation 3.35) and (Equation 3.36) and using (Equation 3.33) and (Equation 3.34), we obtain (with $\hat{L} = \hat{T} = \hat{p}_x = 0$)

$$\hat{g} = e_p^g\hat{p}_g^d - e_\tau^g(1 - \Lambda)d\tau_g / p_g^d + e_\Lambda^g d\Lambda, \tag{3.37}$$

where

$$e_p^g = \frac{e_{wr}^g\left(1 - \alpha_t\theta_{xp}\right) + e_{wp}^g\left(\theta_{gr} + \alpha_t\theta_{xr}\right) + e_{rp}^g\left(\theta_{gw} + \alpha_t\theta_{xw}\right)}{|\lambda||\theta|},$$

$$e_\tau^g = \frac{e_{wr}^g\left(1 - \theta_{xp}\right) + e_{wp}^g\theta_{xr} + e_{rp}^g\theta_{xw}}{|\lambda||\theta|},$$

and

$$e_\Lambda^g = -\left[\frac{\lambda_{xw}\left(c_r^i - c_r^1\right)}{|\lambda|T} - \frac{\lambda_{xr}\left(c_w^i - e_w^1\right)}{|\lambda|L}\right]g = \frac{(1 - \beta)}{(1 - \Lambda + \Lambda\beta)} > 0,$$

and where $|\lambda| = \lambda_{xw}\lambda_{gr} - \lambda_{gw}\lambda_{xr} > 0$ if waste disposal is land intensive; $|\lambda||\theta| > 0$; $\alpha_t = [1 - \Lambda)\tau_g + \Lambda\rho F]/p_g^d$ is the share of taxes and expected fines in the expected cost of disposing waste; and $e_{wp}^g = \lambda_{xr}\delta_{wp}^L$, $e_{wr}^g = \lambda_{xw}\delta_{wr}^T + \lambda_{xr}\delta_{wr}^L$, and $e_{rp}^g = \lambda_{xw}\delta_{rp}^T$, are output elasticities with respect to relative input price changes.

To find the sign of e_p^g, let

$$e^{i,1} = \lambda_{xw}\left(\lambda_{1r}\theta_{1w}\sigma^1 + \lambda_{ir}\theta_{iw}\sigma^i\right) + \lambda_{xr}\left(\lambda_{1w}\theta_{1r}\sigma^1 + \lambda_{iw}\theta_{ir}\sigma^i\right) > 0,$$

and use the adding-up constraints on substitution elasticities implied by Euler's theorem to obtain

$$e_g^p |\lambda\| \theta| = \left(1 - \alpha_t \theta_{xp}\right) e^{i,1} + \lambda_{xw} \lambda_{xr} \theta_{xw} \left(\theta_{gr} + \alpha_t \theta_{xr}\right)\left(\sigma_{wr}^x - \sigma_{ww}^x\right)$$

$$+ \lambda_{xw} \lambda_{xr} \theta_{xr} \left(\theta_{gw} + \alpha_t \theta_{xw}\right)\left(\sigma_{wr}^x - \sigma_{rr}^x\right) > 0$$

since $\sigma_{ww}^x < 0$ and $\sigma_{rr}^x < 0$, from the concavity of the cost function (own elasticities are negative), and $\sigma_{wr}^x < 0$, since land and labor are substitutes.

Similarly, we have

$$e_g^p |\lambda\| \theta| = \left(1 - \theta_{xp}\right) e^{i,1} + \lambda_{xw} \lambda_{xr} \theta_{xw} \theta_{xr} \left(2\sigma_{wr}^x - \sigma_{ww}^x - \sigma_{rr}^x\right) > 0 ,$$

where the inequality follows from the concavity of the cost function, which implies that the matrix of elasticities of substitution is negative semi-definite.

REFERENCES

Baumol, W. J. and Oates, W. E., *The Theory of Environmental Policy*, Prentice-Hall, Englewood Cliffs, NJ, 1975.

Becker, G. S., Crime and punishment: an economic approach, *J. Polit. Econ.*, 76, 169, 1968.

Bhagwati, J. N., The generalized theory of distortions and welfare, in *Trade, Balance of Payments, and Growth*, Bhatwati, J. N. et al., Eds., North–Holland, Amsterdam, 1971.

Bhagwati, J. N., Alternative theories of illegal trade: economic consequences and statistical detection, *Welwirt. Arch.*, 117, 409, 1981.

Bond, E. W. and Chen, T., The welfare effects of illegal immigration, *J. Int. Econ.*, 23, 315, 1987.

Conrad, J. M. and Scott, A. D., Transfrontier pollution: cooperative and non-cooperative solutions, discussion paper 88–30, Department of Economics, University of British Columbia, 1988.

Copeland, B. R., Strategic enhancement and destruction of fisheries and the environment in the presence of international externalities, *J. Environ. Econ. Manage.*, 19, 212, 1990.

Cowell, F. A., The economic analysis of tax evasion, *Bull. Econ. Res.*, 37, 163, 1985.

D'Arge, R. C. and Kneese, A. V., Environmental quality and international trade, *Int. Organ.*, 26, 419, 1972.

Downing, P. B. and Watson, W. D., Jr., The economics of enforcing air pollution controls, *J. Environ. Econ. Manage.*, 1, 219, 1974.

Ethier, W. J., Illegal immigration: the host-country problem, *Am. Econ. Rev.*, 76, 56, 1986.

Harford, J. D., Firm behavior under imperfectly enforceable pollution standards and taxes, *J. Environ. Econ. Manage.*, 5, 26, 1978.

Harford, J. D., Self-reporting of pollution and the firm's behavior under imperfectly enforceable regulations, *J. Environ. Econ. Manage.*, 14, 293, 1987.

Kesselman, J. R., Income tax evasion: an intersectoral analysis, *J. Public Econ.*, 39, 137, 1989.

Lee, D. R., The economics of enforcing pollution taxation, *J. Environ. Econ. Manage.*, 11, 147, 1984.

Linder, S. and McBride, M., Enforcement costs and regulatory reform: the agency and firm response, *J. Environ. Econ. Manage.*, 11, 327, 1984.

Markusen, J. R., Cooperative control of international pollution and common property resources, *Q. J. Econ.*, 88, 618, 1975.

Martin, L., The optimal magnitude and enforcement of evadable Pigovian charges, *Public Finance*, 39, 347, 1984.

Martin, L. and Panagariya, A., Smuggling, trade, and price disparity: a crime-theoretic approach, *J. Int. Econ.*, 17, 201, 1984.

McGuire, M., Regulation, factor rewards, and international trade, *J. Public Econ.*, 17, 335, 1982.

Merrifield, J. D., The impact of selected abatement strategies on transnational pollution, the terms of trade, and factor rewards: a general equilibrium approach, *J. Environ. Econ. Manage.*, 15, 259, 1988.

Milliman, S. R., Optimal fishery management in the presence of illegal activity, *J. Environ. Econ. Manage.*, 13, 363, 1986.

Mitchell, R. C. and Carson, R. T., Property rights and the siting of hazardous waste facilities, *Am. Econ. Rev. Pap. Proc.*, 76, 285, 1986.

Pethig, R., Pollution, welfare, and environmental policy in the theory of comparative advantage, *J. Environ. Econ. Manage.*, 2, 160, 1976.

Scott, A. D., Transfrontier pollution: are new institutions necessary?, in *Economics of Transfrontier Pollution*, OECD, Paris, 1976.

Siebert, H., Spatial aspects of environmental economics, in *Handbook of Natural Resource and Energy Economics*, Vol. 1, Kneese, A. V. and Sweeney, J. L., Eds., North–Holland, Amsterdam, 1985.

Siebert, H., Eichberger, J., Gronych, R., and Pethig, R., *Trade and Environment: A Theoretical Enquiry*, Elsevier, Amsterdam, 1980.

Sullivan, A. M., Policy options for toxics disposal: laissez-faire, subsidization, and enforcement, *J. Environ. Econ. Manage.*, 14, 58, 1987.

Sutinen, J. G. and Anderson, P., The economics of fisheries law enforcement, *Land Econ.*, 61, 387, 1985.

4 Environmental Policy When Market Structure and Plant Locations are Endogenous[1]

James R. Markusen, Edward R. Morey, and Nancy D. Olewiler

A two-region, two-firm model is developed in which firms choose the number and regional locations of their plants. Both firms pollute, and market structure is endogenous to environmental policy. There are increasing returns at the plant level, imperfect competition between the "home" and the "foreign" firm, and transport costs between the two markets. At critical levels of environmental policy variables, small policy changes cause large discrete jumps in a region's pollution and welfare as a firm closes or opens a plant, or shifts production to/from a foreign branch plant. The implications for optimal environmental policy differ significantly from those suggested by traditional Pigouvian marginal analysis.

A. INTRODUCTION

Existing analyses of environmental policy tend to follow the Pigouvian tradition of examining the effects of taxes, subsidies, and other policy instruments on marginal price and output decisions of firms.[2] Marginal analysis is perfectly appropriate for a world of constant returns to scale and perfect competition. In such a world, one can deal directly with the reduced form of an industry and its continuous and differentiable supply function. Optimal tax and/or regulation formulas can then be derived and expressed in terms of underlying parameters of demand and supply functions. Typically these formulas equate two marginal effects, such as a pollution tax equating a marginal benefit of pollution reduction, to the marginal cost of that reduction.

In an industry with increasing returns to scale, which in turn are generally associated with imperfect competition, such an analysis is at worst inappropriate, and at best incomplete. Along with the marginal decisions over continuous variables such as prices and outputs, firms in increasing-returns industries make discrete decisions such as whether or not to serve another region or country by exports or by building a branch plant in that region. Environmental regulation in one region

---

[1] We thank James Alm, Mark Cronshaw, Charles Howe, Shannon Ragland, and two anonymous referees for helpful comments. Markusen's portion of the research is funded by NSF Grant SES-9022898.

[2] The early analyses of environmental policy were in a partial equilibrium framework, Pigou (1932) and Meade (1952) being two classic examples. See Baumol and Oates (1975). General equilibrium models with pollution date back to the 1970s. Førsund (1972), Comoli (1977), Forster (1977, 1981), and Yohe (1979) consider environmental policy in a one-region general equilibrium framework. General equilibrium models with trade and pollution have been examined by Markusen (1975a, 1975b), Pethig (1976), Asako (1979), Siebert et al. (1980), McGuire (1982), and Merrifield (1988). Pollution may or may not cross international boundaries in these models. For example, in Pethig (1976), pollution intensiveness of goods varies, but pollution does not cross boundaries. Environmental policy affects the location of production (domestic or foreign), amount of pollution in each country, and welfare. Pethig shows how gains from trade can be offset by losses from domestic pollution damages. Asako (1979), in a slightly different model obtains similar results. Alternatively, Merrifield (1988) examines transfrontier pollution (Canada U.S. regulation of acid deposition). All of these models analyze impacts on the margin and assume pure competition and constant returns to scale.

---

may cause one or more plants in that region to shut down and transfer production to plants in the other region.[3]

Policymakers are quite aware of the possibility that stiff environmental regulations may cause plant closures, yet most formal policy analysis by economists has continued to pursue the marginal approach, even when imperfect competition is added to the model.[4] In such a marginal analysis, market structure, by which we mean the number and locations of plants, is assumed exogenous.[5]

The purpose of this chapter is to readdress environmental issues that have been dealt with before, but in a model that allows firms to enter or exit, and to change the number and location of their plants in response to environmental policies. The model consists of two Regions (A and B) and three goods ($X, Y,$ and Z). A firm incorporated in Region A produces X with increasing returns and a firm in Region B produces Y with the same technology (X and Y may or may not be perfect substitutes). Z is a homogenous good produced in both regions by competitive industries. The production of X and Y generates regional pollution and there are no regional spillovers of pollution. There is no pollution associated with the production of Z.

Each entering firm must incur a firm-specific fixed cost (such as R & D) that is a joint input across their plants, and a plant-specific fixed cost for each plant it opens. Production occurs with constant marginal cost, and there is a transportation cost of shipping output between regions. The decision to serve another region is a discrete choice between the high fixed-cost option of a foreign branch plant or the high variable-cost option of exporting to that market.

General equilibrium is found as the solution to a two-stage game. In stage one, the two firms (X and Y producers) each make a strategy choice over three discrete options: (1) no entry, referred to as the zero-plant strategy; (2) serving both regions from a plant in the home region, referred to as the one-plant strategy; and (3) building plants in both regions, referred to as the two-plant strategy. In stage two, the X and Y firms play a one-shot Cournot output game.[6]

We solve for a subgame perfect equilibrium of this game and show how the equilibrium market structure depends, in part, on environmental policy. In the *positive* analysis of market structure, the policy variable can be interpreted either as (1) a pollution tax when abatement is not possible,[7] or

---

[3] There is, to our knowledge, only a small amount of empirical literature on whether plant locations decisions are influenced by environmental policy (Bartik, 1988; and McConnell and Schwab, 1990). McConnell and Schwab (1990) use a logit model to explain the location decision for 50 new motor vehicle plants in the U.S. They find "some evidence that, at the margin, firms are deterred from locating plants in the most polluted ozone non-attainment areas."

[4] There is some literature on imperfect competition and pollution control, but what exists is generally partial equilibrium and does not treat the market structure as endogenous. Buchanan (1969) examined monopoly and external diseconomies in a partial equilibrium framework. More recent work includes Burrows (1981), Besanko (1987), Misiolek (1988), and Laplante (1989, 1990).

[5] Note that there is literature (Mathur, 1976, Gokturk, 1979, and Forster, 1987) on how a firm's location *within a region* is influenced by a pollution tax on the ambient air quality in the region's urban center. The intent of this literature is to derive sufficient conditions on the technology to imply that an increase in the tax will cause the firm to locate further from the urban center. These papers do not consider the optimal tax, nor consider the possibility that an increase in the tax might cause the firm to go out of business or to relocate to another region with a different government. Tietenburg (1978) considers optimal pollution taxes when there are many firms and where the spatial distribution of these firms is exogenously given. However, he does briefly consider the possibility that firms might migrate as a function of spatially differentiated tax rates.

[6] A large number of papers in international trade theory have focused on the second stage of this type of game: Brander and Spencer (1985), Dixit (1984), Eaton and Grossman (1986), Helpman (1981), Krugman (1979), and Markusen (1981, 1984) are a few examples. To the best of our knowledge, only Horstmann and Markusen (1992) have formally modeled the two-stage game.

[7] Examples of taxes used for environmental policy are: the Netherlands' 1988 tax on fuels, designed to assist in controlling sulfur dioxide and lead emissions, and France's 1990 taxes on air pollution emissions of sulfur dioxide, nitrous oxides, and hydrochloric acid. The Netherlands also imposed in early 1990 a "carbon tax" based on the carbon-dioxide-generating potential per unit energy of different fuels. Many other countries are currently examining the imposition of carbon taxes (including Canada, the U.S., and Norway). Pesticide and fertilizer taxes have been imposed in Sweden and are being examined in the state of Washington, British Columbia, and Ontario. Effective in 1990, the U.S. imposed a tax on the ozone-depleting factor for a variety of chlorofluorocarbon and halon chemicals. Environmental taxes have also been a part of CERCLA (the Comprehensive Environmental Response, Compensation and Liability Act of 1980), albeit with an objective of financing the Superfund. These have included taxes on crude oil and 42 different industrial petrochemicals and inorganic compounds.

(2) as the increased marginal cost of production due to an environmental restriction (e.g., requiring cleaner fuel) when abatement is possible.[8] The *welfare* consequences of a shift in market structure, however, differ somewhat under the two interpretations, and Section C concentrates on the tax interpretation. A tradable permit system, regulations that fall on fixed costs, and the welfare effects of a regulation falling on marginal costs are discussed in Section E.

Changes in pollution taxes/regulations change the payoffs to a firm in the second-stage game, which in turn alter the location decision in the first stage at critical values of the policy variable. Changes in market structure have four discrete effects under the tax interpretation: they alter the level of pollution, they change product prices and hence consumer surplus, they change the level of government tax revenue, and they change the profits of the local firm (assumed to enter the local income stream). Some of these four effects generally move in opposite directions. Under the regulatory interpretation of the policy variable, there is no tax revenue, but pollution per unit of output varies with regulation. Again, certain welfare effects move in opposite directions.

One feature of the model deserves a brief comment before continuing. Subject to some restrictions, analytical results can be derived specifying market structure as a function of the policy variable. However, the analysis of the welfare effects of market structure shifts proceeds by way of numerical analysis. Market structure makes a discrete change at certain critical values of the environmental and technology parameters. Significant structure is needed to evaluate the combined contributions of conflicting welfare effects at these jumps.

B. A SIMPLE TWO-COUNTRY, TWO-FIRM, THREE-GOOD MODEL WITH POLLUTION

Two regions exist, A and B. Each region is endowed with an identical amount of a homogeneous factor input, L. A homogeneous traded good Z can be produced by each region with its units chosen so that $Z = L_Z$. Z (or L) is the numeraire. There is no pollution associated with the production or consumption of Z.

There is a firm based in Region A that can produce a good X with increasing returns to scale, and there is firm based in Region B that can produce a symmetric substitute good Y.[9] Each firm can either produce in just their own region and export to the other region, have plants in both regions, or not operate. Notionally, let $X^a(Y^a)$ and $X^b(Y^b)$ denote the amount of product $X(Y)$ produced in Regions A and B, respectively. Assume one unit of homogenous pollution is produced in a region for each unit of X or Y produced in that region.[10]

The cost functions for both potential firms (expressed in units of L) are identical where $F \equiv$ firm specific fixed costs, $G \equiv$ plant specific fixed costs, $m \equiv$ constant marginal cost, $s \equiv$ per unit transport costs between the regions, and $t_a \equiv$ the per unit pollution tax in Region A. The firm-specific costs represent joint inputs across plants such as firm-specific knowledge. Multi-plant economies of scale result because the fixed costs of a two-plant firm, $2G + F$, are less than the combined fixed cost of two one-plant firms, $2G + 2F$. Under the regulatory interpretation of the model (discussed

---

[8] Other examples of environmental policies that affect marginal costs of production include: the ban on leaded gasoline in Canada — the additive replacing lead is more expensive, adding to refinery costs; and the substitution of bleaching compounds in the pulp and paper industry from dioxin-creating chlorines to alternative inputs. To get a permit to operate a new plant in British Columbia, the more expensive nonchlorine compounds (and process) must be used. To meet pollution requirements, Volvo may be switching to more expensive water-borne rather than oil-based paints. Scrubbers, an essential part of air pollution control regulations for coal-fired electric power plants in the United States, add both to operating and to capital costs.

[9] Scale economies are assumed sufficiently large relative to demand such that the two regions can support at most one X and one Y firm. Each of these firms therefore has market power.

[10] Under the tax interpretation of the environmental policy variable, the possibility of abatement (reducing pollution per unit of output) is assumed away.

in Section D), t_a is the increased marginal cost of production due to an environmental restrictions such as requiring a less polluting but more expensive input.

Demand for the three products is generated by N consumers in each region, where N is assumed to be a large number. All these individuals have identical preferences which can be represented by the same single quadratic utility function. Specifically, utility for an individual in Region i ($i = a, b$) is

$$U_i = \alpha x_i - (\beta/2)x_i^2 + \alpha y_i - (\beta/2)y_i^2 - \gamma x_i y_i + z_i - \tau(X^i + Y^i) \tag{4.1}$$

where $x_i(y_i)$ is the amount of good $X(Y)$ consumed by each individual in region i, and $(X^i + Y^i)$ is the total amounts of pollution in region i. The parameter τ reflects the constant marginal disutility from pollution. Each individual views the total production of pollution as exogenous. In the absence of a pollution tax, or regulation, this externality, ceteris paribus, will lead to market failure.[11]

Assume profits from the X firm (Y firm) and Region A's revenues from the pollution tax are distributed equally amongst the N individuals in Region A. Given this, the individual budget constraints in Region A and B are, respectively,

$$\left(L + \pi_x + t_a(X^a + Y^a)\right)/N = p_x^a x_a + p_y^a y_a + z_a \tag{4.2a}$$

$$\left(L + \pi_y\right)/N = p_x^b x_b + p_y^b y_b + z_b \tag{4.2b}$$

where $p_x^a(p_y^a)$ is the price of good $X(Y)$ in Region A, and π_x and π_y are the profits of the X and Y firms. Focusing on Region A, the inverse aggregate demand functions are found by maximizing (Equation 4.1) ($i = a$) subject to (Equation 4.2a).

$$p_x^a = \alpha - \beta(X_a/N) - \gamma(Y_a/N), \qquad p_y^a = \alpha - \beta(Y_a/N) - \gamma(X_a/N), \tag{4.3}$$

where $X_a \equiv Nx_a$ and $Y_a \equiv Ny_a$.[12] $x_a(y_a)$ is the demand for $x(y)$ by a representative individual in Region A. The inverse aggregate demand functions for Region B have the same form. Note from (Equation 4.3) that the inverse demand functions do not depend on income and not on tax receipts in particular. We refer to this result again in Section E when discussing the regulatory interpretation of the policy variable t_a.

General equilibrium is characterized by a situation where: (1) each individual is maximizing his/her utility given exogenous prices, profits, and aggregate pollution; (2) each firm is maximizing its profits given the number of plants operated by the other firm; (3) supply = demand for all three goods in each region; and (4) $L = L_X + L_Y + L_Z$ in both regions. Equilibrium social welfare in Region A is the sum of consumer surplus, profits, tax revenue, the disutility of pollution, and labor income. In a short Appendix, we show that this is given by

$$SW_a = \left[\beta(X_a^2 + Y_a^2)/(2N) + \gamma X_a Y_a/N\right] + \pi_x + \left[(t_a - \tau N)(X^a + Y^a)\right] + L. \tag{4.4}$$

The first square-bracketed term is consumer's surplus from X and Y (the marginal utility of Z is constant and hence there is no consumer surplus associated with Z), while the other term in square brackets is tax revenue minus the disutility of pollution. The equilibrium social welfare function

---

[11] Note that the system is also distorted by the market power of the X firm and the Y firm.

[12] Note the distinction between $X_a(Y_a)$ and $X^a(Y^a)$. X_a is the amount of good X consumed in region A and X^a is the amount of good X produced in region A.

for Region B is identical except one substitutes π_y for π_x, and production and consumption levels in B for those in A. Section E discusses the small modification to (4) needed under the regulatory interpretation of t_a.

Equilibrium market structure is determined in a two-step procedure corresponding to a two-stage game. In stage one, X and Y producers make a choice among three options: no production, a plant only in their home region, or a plant in both regions. In stage two, X and Y play a one-shot Cournot game. Moves in each stage are assumed to be simultaneous, and the usual assumptions of full information hold.

The game is solved backward. The maximized value of profits for each firm is determined for the three options listed above, given, in turn, each of the three options for the other firm. Profit levels for the firms in each of these nine cases are then the payoffs for the game in which the strategy space is the number of plants.[13] The Nash equilibrium (or equilibria) of this game in the number of plants determines the equilibrium market structure for the model.

We illustrate the determination of profits by solving for maximum profits in the simple structure where firm X operates one plant in Region A and firm Y does not operate — structure $(1, 0)$.

In the first case, $(1, 0)$, $\pi_y^*(1, 0) = 0$ and

$$\pi_x(1,0) = \left[\alpha - \beta(X_a/N)\right]X_a + \left[\alpha - \beta(X_b/N)\right]X_b - mX_a - (m+s)X_b - t_a(X_a + X_b) - F - G. \quad (4.5)$$

Maximizing, and solving, the profit maximizing levels of sales in the two regions are

$$X_a(1,0) = N(\alpha - m - t_a)/(2\beta) \qquad X_b(1,0) = N(\alpha - m - t_a - s)/(2\beta). \quad (4.6)$$

Substituting Equation (4.6) into (4.5), maximum profits for firm X are

$$\pi_x^*(1,0) = N\left[(\alpha - m - t_a)^2 + (\alpha - m - t_a - s)^2\right]/(4\beta) - F - G. \quad (4.7)$$

Consider now the structure where both firms operate in both regions — $(2, 2)$:

$$\pi_x(2,2) = \left[\alpha - \beta(X_a/N) - \gamma(Y_a/N)\right]X_a + \left[\alpha - \beta(X_b/N) - \gamma(Y_b/N)\right]X_b$$
$$- m(X_a + X_b) - t_aX_a - 2G - F \quad (4.8)$$

Using the related expression for π_y, maximizing and solving, the four supply functions are

$$X_a(2,2) = Y_a(2,2) = N(\alpha - m - t_a)/\delta \qquad \delta \equiv (2\beta + \gamma) \quad (4.9)$$

$$X_b(2,2) = Y_b(2,2) = N(\alpha - m)/\delta. \quad (4.10)$$

Substituting Equations (4.9) and (4.10) into Equation (4.8) and correspondingly for π_y, maximum profits for the two firms in the $(2, 2)$ case are

---

[13] The nine cases are $(0, 0)$, $(1, 0)$, $(0, 1)$, $(2, 0)$, $(0, 2)$, $(2, 1)$, $(1, 2)$, $(1, 1)$, and $(2, 2)$, where the number in the first (second) position is the number of plants by the $X(Y)$ producer. $(2, 0)$ denotes, for example, a market structure in which the X producer has plants in both regions, and Y does not enter. In order to limit the dimensionality of the problem to nine cases, we assume that the $X(Y)$ firm cannot have a single plant in Region $B(A)$: A firm must have a plant in its home region (or none at all). Without this restriction there would be a total of 16 possible cases.

$$\pi_x^*(2,2) = N\left[(\alpha - m - t_a)^2 + (\alpha - m)^2\right]\big/(\beta/\delta^2) - (F + 2G) \qquad i = (x,y). \qquad (4.11)$$

While tedious, maximum profits for all of the other structures can be worked out in similar fashion. Note that while welfare is a decreasing function of τ, profit levels and equilibrium market structure are not a function of τ.

The maximum profits for both firms in each of the nine structures are the payoffs in the second stage of the game. The Nash equilibrium (equilibria) is (are) that (those) market structure(s) such that given the number of plants operated by firm X, firm Y cannot increase its profits by changing its number of plants; and given the number of plants operated by Y, firm X cannot increase its profits by changing its number of plants.

To gain some insights into how different plant locations and market structures can originate, consider four simple example games. All that varies from one game to the other is the magnitudes of F and G (firm- and plant-specific fixed costs). In each game, X and Y are assumed to be imperfect substitutes ($\gamma = \beta/2$), and marginal cost is zero ($m = 0$). Other parameter values are shown at the top of Table 1. These values are chosen to demonstrate how an interesting and empirically relevant sequence of market structure can be generated by varying fixed costs (scale economies).[14]

To isolate on the endogeneity of market structure independent of pollution taxes, initially pollution taxes in Region A are set to zero (pollution taxes in Region B already equal zero).

Table 4.1 reports the profits and equilibrium market structure for the four games. The first (second) number of each pair is the maximum profits for the $X(Y)$ firm in that structure. The Nash equilibrium is denoted with an asterisk.

Roughly speaking, the games are ordered by decreasing F and increasing G, holding the transport cost s constant. In Game 1, with the highest F and lowest G, the multiplant market structure is the unique equilibrium. In Game 2, an increase in G with an equal decrease in F yield two symmetric equilibria, with only a single two-plant firm operating one plant in each region. In Game 3, there are three possible equilibria. In Game 4, with a low F but high G, the unique equilibria is a duopoly between single-plant firms, each firm exporting to the other firm's home market.

The general result is that a multi-plant market structure is more likely with a high F and low G, while a single-plant (for each firm) outcome is more likely with a low F and high G for the given value of s. This is an intuitive result: firms are likely to serve the other market by exports when the fixed costs of a new plant are high relative to the unit transport costs. General cases are found in Horstmann and Markusen (1992).[15]

C. THE IMPACT OF A UNILATERAL POLLUTION TAX (OR A REGULATION RAISING MARGINAL COST) ON EQUILIBRIUM PLANT LOCATION AND MARKET STRUCTURE

Suppose that Region A imposes a unilateral pollution tax $t_a > 0$, while Region B has no tax ($t_b = 0$). Alternatively, t_a could be the additional marginal cost of an emission restriction, perhaps due to the added costs of cleaner fuel. A number of possible sequences of market structures, as a function of t_a, are possible. Rather than present a taxonomy, we choose to present two interesting possibilities: (i) the initial market structure is mutually invading multinationals (2, 2), corresponding to a technology with high firm-specific costs, high transport costs, and low plant-specific costs

---

[14] The (2, 2) market structure, for example, is the case of "mutually invading multinationals" as it is known in the international trade literature. Chemicals and pharmaceuticals are two such industries which also pollute. (1, 1) is an exporting duopoly, a market structure that has been heavily analyzed in the "new" trade literature. Steel, and pulp and paper, are two such industries that also pollute.

[15] Extensive empirical evidence strongly supports this association between the importance of firm-specific costs and the multinationality of an industry. (See, for example, Caves, 1982.)

TABLE 4.1
Equilibrium Market Structure and Plant Location in
Four Different Simple Games

	Number of Plants for Firm Y		
Firm X	2 Plants	1 Plant	0 Plants

Game 1. G = 5,000 and F = 30,000: Nash equilibrium (2, 2), denoted by *

2	(960, 960)*	(2,700, –300)	(24,000, 0)
1	(–300, 2,700)	(1,440, 1,440)	(21,500, 0)
0	(0, 24,000)	(0, 21,500)	(0, 0)

Game 2. G = 6,000 and F = 29,000: Nash equilibria (2, 0) or (0, 2)

2	(–40.0, –40.0)	(1,700, –300)	(23,000, 0)*
1	(–300, 1,700)	(1,440, 1,440)	(21,500, 0)
0	(0, 23,000)*	(0, 21,500)	(0, 0)

Game 3. G = 7,000 and F = 28,000: Nash equilibria (1, 2), (2, 0), or (0, 2)

2	(–1,040, –1,040)	(700, –300)	(22,000, 0)*
1	(–300, 700)	(1,440, 1,440)*	(21,500, 0)
0	(0, 22,000)*	(0, 21,500)	(0, 0)

Game 4. G = 7,000 and F = 27,000: Nash equilibrium (1, 1)

2	(–40.0, –40.0)	(1,700, 700)	(23,000, 0)
1	(700, 1,700)	(2,440, 2,440)*	(22,500, 0)
0	(0, 23,000)	(0, 22,500)	(0, 0)

Note: Parameter values: $\alpha = 16$, $\beta = 2$, $\gamma = 1$, $m = 0$, $s = 2$, $\tau = 0.0035$, $N = 1,000$, $L = 50,000$, and $t_a = t_b = 0$. The first (second) number of each pair is the maximum profits for the $X(Y)$ firm in that structure.

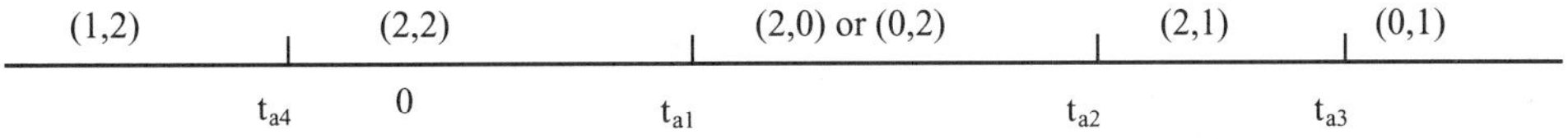

FIGURE 4.1 Assumptions 1A through 1C yield this sequence of market structures as a function of Region A's tax rate.

(e.g., Game 1 of Table 4.1). (ii) The initial market structure is exporting duopoly (1, 1), corresponding to a technology with high plant-specific costs, low firm-specific costs, and low transport costs (e.g., Game 4 of Table 4.1). Consider the following three assumptions where t_{a1} and t_{a2} denote specific values of t_a.

Assumption 1 (High transport costs and high firm-specific costs, low plant-specific costs).

(A) The initial equilibrium market structure is (2, 2) at $t_a = 0$.
(B) At $t_a = t_{a1} > 0$ such that $\pi_x(2,2) = \pi_y(2,2) = 0$, $\pi_y(2,1) < 0$.[16]
(C) At $t_a = t_{a2} > t_{a1}$ such that $\pi_y(2,1) = 0$, $\pi_x(2,1) > 0$.[17]

---

[16] Given a sufficiently large F, only a small t_a is needed to reduce $\pi_x(2,2) = \pi_y(2,2)$ to zero. If G is small and s is large, Y cannot then earn positive profits by dropping its plant in Region A (saving G and t_a) and exporting to A (incurring s) from its single plant in B.

[17] Again, under a sufficiently large F and s, the value of t_a that will set $\pi_y(2,1) = 0$ will be small relative to s. In market structure (2, 1), Y does not pay t_a or a second G, but does pay s and vice versa for X. If G and t_{a2} are small relative to s, then assumption (1C) will be supported.

RESULT 1. *Given Assumptions 1A through 1C, the sequence of market structures as a function of Region A's tax is given schematically by Figure 4.1.*

Proof. Consider first raising t_a from zero. The properties of the model imply that $\pi_y(2,1)$ must increase and $\pi_y(2,2)$ must decrease with an increase in t_a. It follows from (1B) that for $t_a < t_{a1}$, we must have $\pi_y(2,1) < 0$. Assumption 1B thus implies that the Y producer cannot do better by serving Region A from a single plant over the region $0 < t_a < t_{a1}$. If this is true, given the symmetry in the model, the X producer cannot do better by serving Region B from its home plant in A (X would incur more tax *and* transport costs). $(2, 2)$ is therefore the equilibrium over the interval $0 < t_a < t_{a1}$.

Just beyond $t_a = t_{a1}$, one firm must exit completely. Profits for both firms in the $(2, 2)$ market structure are negative. Assumption 1B implies that the transport cost is sufficiently high that the Y producer cannot make profits by deviating to one plant. The symmetry of the model then implies that the X producer would do even worse by deviating to one plant. Thus, one firm must exit. For some values of $t_a > t_{a1}$, the equilibrium must be either $(2, 0)$ or $(0, 2)$. In order to avoid a taxonomy of all possible cases, we assume that the equilibrium is $(2, 0)$ (the home government will not allow its firm to go out of business) but we report welfare values for a $(0, 2)$ outcome in a later figure.

Further increases in t_a reduce the output of X's plant in Region A and, of course, X has no interest in deviating to a single plant in Region A. This reduction in the output of X^a increases the demand for Y. Assumption 1C implies that the Y producer will eventually find it profitable to reenter the market with a single plant serving both A and B. We denote this critical tax level by t_{a2}. Assumption 1C also implies that the profits of the home firm remain positive when Y enters with a single plant (and, again X will not deviate to a single plant), so the new equilibrium at $t_a = t_{a2}$ will be $(2, 1)$.

Increases in t_a beyond t_{a2} continue to reduce the profits of the X producer and increase the profits of the Y producer. Eventually we will reach a value of $t_a = t_{a3}$ such that $\pi_x(2,1) = 0$. The X producer cannot increase his profits by deviation to one plant, and so must exit. The equilibrium for all tax rates $t_a > t_{a3}$ is $(0, 1)$.

Finally, consider reducing t_a below zero, subsidizing the production of X and Y in A. This increases profits from producing in A, so Y will not want to deviate from its initial two plants. But at some point, the subsidy and the reduction in plant-specific costs will outweigh the transport cost, and the X producer will shut his plant in B. We denote the critical value of the tax (subsidy) as $t_{a4} < 0$. The new equilibrium will become $(1, 2)$. This completes the derivation of the sequence of equilibria given by Result 1.

This sequence of equilibria can, in turn, translate into a number of qualitatively different welfare graphs due to conflicting welfare effects. For example in the transition from $(2, 1)$ to $(0, 1)$, Region A experiences a loss of product X and an increase in p_y, but the level of pollution falls. Further, welfare effects depend on whether or not t_a is a tax or the additional marginal cost of an emissions restriction as noted earlier. In what follows we assume that t_a is a tax. An emission restriction will be discussed in a subsequent section.

Two outcomes corresponding to different value of τ, the marginal disutility of pollution, are shown in Figures 4.2 and 4.3. In both cases, the initial parameterization of the model corresponds to that of Game 1 in Table 4.1, and the sequence of market structures corresponds to that of Result 1. Figure 4.2 assumes a relatively low $\tau = 0.0035$. Region A's equilibrium level of social welfare is 59,280 in the absence of a pollution tax. This welfare level is generated by a market structure of $(2, 2)$. For small tax levels, $0 \leq t_a \leq 0.39999$, social welfare gradually increases as the tax is increased and market structure remains at $(2, 2)$. The small increase in welfare is a reflection of the fact that the positive contributions of decreased pollution and increased tax revenue are largely offset by a loss of consumer surplus from the consumption of X and Y (the prices of X and Y increase) and by a loss in the profits of the domestic firm. At a tax rate of $t_a = 0.4$, the market structure switches to either $(2, 0)$ or $(0, 2)$. Since X and Y are symmetric substitutes, the only difference between the two market structures is in the profits of the domestic X producer. In market structure $(2, 0)$, the increased

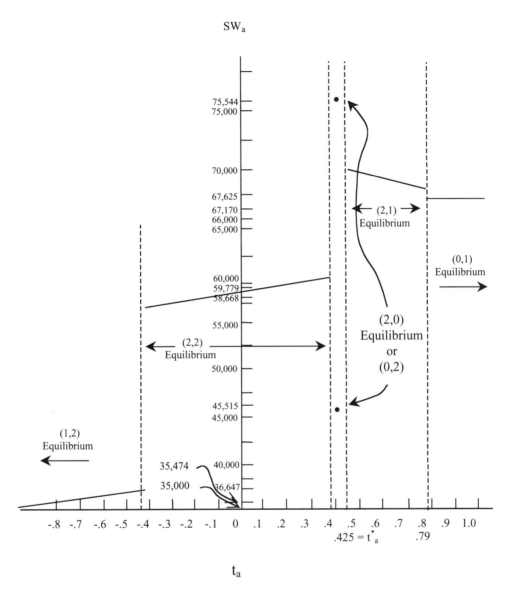

FIGURE 4.2 Equilibrium social welfare in Region A as a function of t_a. Parameter values: $F = 30,000$; $G = 5,000$; $\tau = 0.0035$; $\alpha = 16$; $\beta = 2$; $\gamma = 1$; $m = 0$; $s = 2$; $N = 1,000$; $L = 50,000$; $t_b = 0$.

profits of the local firm outweigh the loss of consumer surplus coming from both the loss of Y and from the increased monopoly price of X. In market structure $(0, 2)$ we have only the latter two effects and no profits for the X producer, so welfare takes a discrete drop.

A further increase in t_a to 0.425 leads to the market structure $(2, 1)$ for the reasons discussed: reduced production of X in A increases the demand for Y to the point where the Y producer reenters the market with a single plant, exporting to A. Welfare is higher than in the $(2, 2)$ market structure due to a combination of several conflicting effects. Pollution is lower, as we indicated above, because Y is imported. Profits of the local firm are higher (for a given tax rate) since X now competes in Region A with higher cost imports. There is a discrete loss of consumer surplus (the price of Y jumps up) and tax revenue decreases at the switch in market structure. Further increases in t_a reduce welfare, suggesting that further reductions in pollution and increases in tax revenue are outweighed by the loss of consumer surplus and profits. At the tax rate of $t_a = 0.79$, the local firm (X) exists,

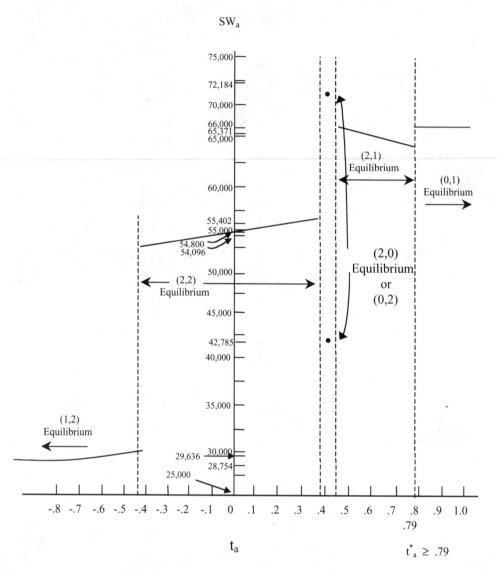

FIGURE 4.3 Equilibrium social welfare in Region A as a function of t_a. Parameter values: $F = 30,000$; $G = 5,000$; $\tau = 0.0042$; $\alpha = 16$; $\beta = 2$; $\gamma = 1$; $m = 0$; $s = 2$; $N = 1,000$; $L = 50,000$; $t_b = 0$.

and Region A suffers a discrete loss of both consumer surplus and profits that outweighs the decrease in pollution to zero. If we assume that Region A cannot engineer outcome $(2, 0)$ (i.e., it cannot avoid the $(2, 0)$, $(0, 2)$ indeterminancy) we thus have $t_a = 0.45$ as the optimal tax.[18]

Figure 4.3 examines what happens if the disutility from pollution is increased by 20% (the sequence of market structures is independent of τ as noted above). In this case, given the greater welfare loss from pollution, it is optimal to impose a tax that drives all the polluters from Region A. A tax of 0.79 is sufficient to do this and there is no cost to imposing a higher tax; i.e., with this higher level of disutility from pollution, the optimal t_a is not unique ($t_a^* \geq 0.79$). The optimal market

---

[18] Note that this tax affects both production and pollution levels in Region A. This tax rate is therefore "best" given that there is only one producer of X and one producer of Y. Given the market power of the two firms, t_a will not, in general, eliminate the inefficiency caused by *both* the pollution distortion and the market-power distortion. Nor should we expect it to, as a single instrument cannot, in general, simultaneously correct two separate distortions.

FIGURE 4.4 Assumptions 2A and 2B yield this sequence of market structures as a function of Region A's tax rate.

structure requires that firm X does not operate, and that firm Y operates in Region B – (0, 1). Increasing t_a beyond 0.79 does not change anything.

We now examine one of a number of possible sequences of market structures as functions of t_a given an initial market structure of (1, 1). The assumptions are as follows:

Assumption 2 (High plant-specific costs, low firm-specific costs, and transport costs).

(A) The initial market structure is (1, 1) at $t_a = 0$, implying $\pi_x(1,1) > \pi_x(2,1)$ at $t_a = 0$.
(B) There exists a $t_a = t_{a1} > 0$ such that $\pi_x(1,1) = \pi_x(2,1) > 0$, and $\pi_y(2,1) > 0$.[19]

RESULT 2. *Given Assumptions 2A and 2B, the sequence of market structures as a function of Region A's tax is given schematically by Figure 4.4.*

Proof. Consider first raising t_a from zero. The X producer will not wish to deviate initially to two plants and the Y producer will never wish to open a plant in A. Neither firm will wish to exit given that profits are still positive. As t_a increases, the properties of the model imply that $\pi_x(1,1)$ falls faster than $\pi_x(2,1)$ because all of the firm's production is taxed in the former market structure. At the level $t_a = t_{a1}$, the tax has reached the level where X is now just indifferent to supplying B by building a branch plant, thereby incurring a fixed cost G but saving transport costs and the pollution tax in A. By Assumption 2B, profits of the Y producer are positive at the (2, 1) market structure, so (2, 1) becomes the new market structure at $t_a = t_{a1}$.

Further increases in t_a beyond $t_a = t_{a1}$ result in a decrease in profits for X and an increase in profits for Y. At some critical value $t_a = t_{a2}$, the X producer exits from the market (recall that X cannot produce only in B). Y will not deviate to two plants, so (0, 1) becomes the market structure for $t_a > t_{a2}$. Finally, consider negative values of t_a. X will not wish to deviate from one plant, but at some critical value $t_a = t_{a3}$, Y will find that the subsidy and the savings on transport costs will outweigh the plant-specific cost, and will open a plant in A. (1, 2) then becomes the market structure for $t_a < t_{a3}$.

Figure 4.5 gives a numerical example of the welfare effects of t_a given initial parameters corresponding to those of Game 4 in Table 4.1. As the tax is increased from zero, social welfare decreases at a gradual rate as long as the market structure remains fixed. This case of an exporting duopoly has been analyzed by Brander and Spencer (1985), among others. What happens (with Cournot behavior) is that the tax puts the domestic firm at a competitive disadvantage such that the loss of its profits reduces welfare. In the case we consider here, this profit effect obviously dominates the positive effect of the reduction in pollution.

When t_a reaches 0.275, the equilibrium market structure switches to (2, 1) and welfare in Region A jumps to its maximum value. The shift by the local firm of its production for Region B to Region B causes a discrete fall in pollution with no adverse consequences for consumer surplus or profits (at the point of switch). There is a fall in tax revenue, but this is obviously outweighed by the discrete drop in pollution. Further increases in the tax reduce welfare for reasons identical to those

---

[19] Assumptions 2A and 2B must hold as s and F approach zero. Consider the case where $s = 0$. (1, 1) must be the market structure if profits are positive. $\pi_y(1,1) = \pi_y(2,2)$ since X's marginal cost of supplying Region B is the same in either case, and hence both firms' supplies in B are the same under either (1, 1) or (2, 1). But $\pi_y(1,1) > \pi_x(1,1)$ since Y pays no tax. At t_{a1}, we therefore have $\pi_y(2,1) = \pi_y(1,1) > \pi_x(1,1) = \pi_x(2,1)$. Finally, $\pi_x(2,1) > 0$ at t_{a1} provided that fixed costs are not too large.

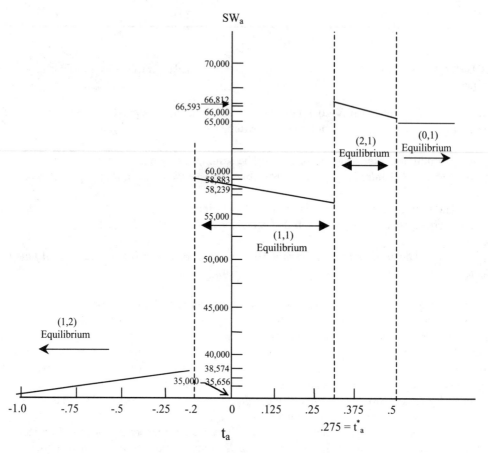

FIGURE 4.5 Equilibrium social welfare in Region A as a function of t_a. Parameter values: $F = 27,000$; $G = 7,000$; $\tau = 0.0035$; $\alpha = 16$; $\beta = 2$; $\gamma = 1$; $m = 0$; $s = 2$; $N = 1,000$; $L = 50,000$; $t_b = 0$.

discussed in Figure 4.2, and as in that case, here the optimal value of the tax is that which is just sufficient to cause the jump in market structure to (2, 1).

D. THE COST OF IGNORING THE ENDOGENEITY OF MARKET STRUCTURE IN THE DETERMINATION OF AN "OPTIMAL" (SECOND-BEST) TAX[20]

Consider how costly it is to ignore the endogeneity of market structure when determining pollution taxes. Assume, as has much of the literature, that the market structure existing in the absence of a pollution tax will not change due to the imposition of such a tax. The exogenous structure is the equilibrium associated with $t_a = 0$. In the context of our model, if one pretends that this market structure is exogenous, one can determine the "optimal exogenous" tax by plotting equilibrium social welfare as a function of t_a, *holding* market structure at its zero tax level. However, this tax rate will not be optimal and will usually result in a suboptimal level of social welfare.

---

[20] Recall that there are two distortions: pollution and market power. The single tax here can only be optimal in a second-best sense, since in general it cannot correct both distortions and produce a full social optimum. Since the pollution tax tends to increase the market power distortion (i.e., we reduce further the output of a good that is already underproduced), there is some presumption that the second-best tax may be smaller than a first-best pollution tax when another policy instrument is available to correct the market power distortion.

In Figure 4.2, the equilibrium market structure when there is no tax is (2, 2). If the regulator incorrectly assumes that the market structure will remain (2, 2) independent of the tax, he or she will determine that the best that can be done is to set $t_a = 3.5$.[21] The regulator anticipates that this tax rate will generate a social welfare level of 61,730. He or she will be wrong. If t_a is set equal to 3.5, equilibrium market structure will switch to (0, 1) as shown in Figure 4.2 where firm X will be driven out of business and firm Y will close its plant in Region A. Equilibrium social welfare will be 66,000, not 61,730. While the outcome is better than expected, it is still less than the welfare level that could have been achieved, 67,625, if the tax had been set at its optimal rate of 0.425.

Consider now Figure 4.3. In this case, the equilibrium market structure in the absence of the tax is again (2, 2). If the regulator incorrectly assumes that market structure will remain at (2, 2) independent of the tax rate, he or she will conclude that the optimal tax is 4.2 and anticipates that a welfare level of 58,320 will result. What a tax of 4.2 will do is generate a (0, 1) equilibrium and a welfare level of 66,000. In this case, there is no cost to ignoring the endogeneity of market structure because imposing any $t_a \geq 0.79$ will drive pollution to zero; its optimal amount when τ is 0.0042.

In the two cases considered so far, imposing the "optimal exogenous" tax was better than doing nothing at all ($t_a = 0$), but this is not always the case. Consider Figure 4.5. The equilibrium market structure when there is no tax is (1, 1) and the corresponding welfare level is 58,556. If the policy maker incorrectly assumes that the market structure will remain (1, 1) independent of the tax, he or she will determine that the best that can be done is to set $t_a = -1.5$ and anticipates that this tax rate will achieve a welfare level of 61,292.[22] The "optimal exogenous" tax is negative for the reasons developed in the strategic trade-policy literature discussed above: artificially holding market structure at (1, 1), the increased rents for the local firm outweigh the increased pollution and loss of tax revenue. However, market structure does not remain at (1, 1), a tax rate of −1.5 shifts the equilibrium to (1, 2) and generates an equilibrium welfare level of 33,483. Imposing the "optimal exogenous" tax results in a 57% decrease in social welfare relative to doing nothing at all. As noted above, if the policy maker had taken the endogeneity of market structure into account, she would have imposed a tax of 0.275 and achieved a welfare level of 66,812.

A second example of when it is better to not tax the pollution than to impose the "optimal exogenous" tax is if $F = 30,000$, $G = 1,000$, and $\tau = 0.0035$ (we have not analyzed this case previously). In this case of very low plant-specific fixed costs, the equilibrium market structure in the absence of the tax is (2, 2) and generates a welfare level of 67,280. If the policy maker incorrectly assumes that the market structure will remain at (2, 2), independent of the tax rate, she will impose a tax of 3.5 with the expectation that it will generate a welfare level of 69,730. Rather it will generate a welfare level of 66,000 and a (0, 1) equilibrium. The optimal tax in this case is 1.7 and generates a welfare level of 73,914, a 10% increase over the no-tax case.

E. ADAPTING THE MODELING FRAMEWORK TO REGULATIONS, PERMITS

As we noted earlier, much of the model can be adapted to consider a variety of regulatory constraints. We mentioned several times that t_a can be interpreted as the added marginal cost of production in Region A due to a regulatory constraint, such as one on the quality of inputs, or indeed any abatement technology that falls exclusively on marginal costs. Because demand is independent of tax revenue, the *positive* analysis of market structure is exactly the same as in the tax interpretation, and Results 1 and 2 apply equally well. However, the welfare implications of these sequences of market structures

---

[21] This value was determined by finding that value of t_a that maximizes SW_a holding market structure constant at (2, 2).

[22] When the equilibrium is constrained to remain at (1, 1), SW_a is a decreasing function of t_a. A tax rate of −1.5 is the smallest tax rate consistent with nonnegative profits in the Y industry, a necessary condition for a (1, 1) equilibrium.

are somewhat different. Referring back to (4), there is no tax revenue term under the regulatory interpretation. On the other hand, the pollution per unit of output is lower when the marginal-cost abatement technology is used.

In order to illustrate the close relationship between the two interpretations, suppose that we choose units such that an abatement technology that raises marginal costs by t_a per unit of output results in a reduction of exactly t_a units of pollution per unit of output. Then (4) continues to give the exact formula for social welfare. The term that is tax revenue, $t_a(X^a + Y^a)$, now becomes simply the (utility) increase from the abatement activity. Of course, negative values of t_a have little meaning under the regulatory interpretation.

Adapting the model to deal with tradable emissions permits is a bit less straightforward due to the imperfect competition (e.g., the monopoly firm would bid zero for the permits and the permits would simply become a binding output restriction). But suppose that we simply take a partial-equilibrium interpretation of the model, and assume that the X and Y producers are only small sectors in a large, competitive economy where many industries pollute. Assume further that X and Y are price-takers in a competitive market for permits, and that abatement is not possible as in our tax interpretation of the model. In this case, t_a becomes the price of a permit per unit of output, and the positive analysis of market structure proceeds as before.

Our analysis also points the way for an analysis of an abatement technology that falls on fixed costs. We can see from the results of Table 4.1 that we will likely get interesting sequences of market structures as G changes much as we do here (general results about market structure as a function of G and F are found in Horstmann and Markusen, 1992). If the regulation raises G in Region A, for example, we could construct a case where the sequence of market structures is (2, 2), (2, 1), (0, 1). Welfare diagrams similar to those of Figures 4.2 and 4.3 would then follow.

F. CONCLUSIONS AND EXTENSIONS

The model presented here is a first attempt at linking pollution policy with a model of endogenous plant location and industrial structure. This topic is currently one of the principal concerns in the formulation of (in particular opposition to) the U.S.–Mexico free-trade agreement. The model demonstrates, in a simple framework, that plant location and market structure can be a function of environmental policy. The model also demonstrates that the cost can be quite high if environmental policy ignores this endogeneity. We also argue that the modeling framework is general enough to encompass tax policies, permit systems, regulations that fall on marginal costs, and regulations that fall on fixed costs.

The model is in the process of being extended (Markusen et al., 1991). Rather than assuming an exogenous pollution tax in Region B, the extension models the simultaneous determination of the pollution tax rates in the two regions as the outcome of a game between the governments of the two regions. Bilateral agreements are also considered, a topic important to many countries including Canada, Mexico, the U.S., and those in the European economic community.

APPENDIX

The purpose of this Appendix is to give the derivation of the social welfare function in (Equation 4.4). Multiplying the utility function in (Equation 4.1) by N and noting that $x_a = X_a/N$ and $y_a = Y_a/N$ aggregate utility of welfare is given by

$$SW_a = NU_a = \alpha X_a - (\beta/2) X_a^2/N + \alpha Y_a - (\beta/2) Y_a^2/N - \gamma X_a Y_a + Z_a - \tau N(X^a + Y^a) \quad (4.12)$$

Multiplying (Equation 4.2a) through by N gives the aggregate budget constraint, which we can rearrange as

$$Z_a = L + \pi_x + t_a(X^a + Y^a) - p_x^a X_a - p_y^a Y_a. \tag{4.13}$$

Using (Equation 4.3) for p_x^a and p_y^a, the last two terms in (Equation 4.13) are

$$p_x^a X_a = \alpha X_a - \beta(X_a^2/N) - \gamma Y_a X_a/N \tag{4.14}$$

$$p_y^a Y_a = \alpha Y_a - \beta(Y_a^2/N) - \gamma Y_a X_a/N. \tag{4.15}$$

Substitute (Equation 4.14) and (Equation 4.15) into (Equation 4.13). Then substitute the right-hand side of (Equation 4.13) for Z_a in (Equation 4.12). SW_a is now given by

$$SW_a = \alpha X_a - (\beta/2) X_a^2/N + \alpha Y_a - (\beta/2)Y_a^2/N - \gamma Y_a X_a/N$$

$$-\alpha X_a + \beta(X_a^2/N) + \gamma X_a Y_a/N - \alpha Y_a + \beta(Y_a^2/N) + \gamma Y_a X_a/N$$

$$+L + \pi_a + (t - \tau N)(X^a + Y^a) \tag{4.16}$$

Canceling and collecting terms yields (Equation 4.4).

REFERENCES

Asako, K., Environmental pollution in an open economy, *Econom. Rec.*, 55, 359, 1979.

Bartik, T. J., The effects of environmental regulation on business location in the United States, *Growth Change*, 19, 22, 1988.

Baumol, W. J. and Oates, W. E., *The Theory of Environmental Policy*, Prentice-Hall, Englewood Cliffs, NJ, 1975.

Besanko, D., Performance versus design standards in the regulation of pollution, *J. Public Econ.*, 34, 19, 1987.

Brander, J. A. and Spencer, B. J., Export subsidies and international market share rivalry, *J. Int. Econ.*, 18, 83, 1985.

Buchanan, J. M., External diseconomies, corrective taxes, and market structure, *Am. Econ. Rev.*, 59, 174, 1969.

Burrows, P., Controlling the monopolistic polluter: nihilism or eclecticism?, *J. Environ. Econ. Manage.*, 8, 372, 1981.

Caves, R. E., *Multinational Enterprise and Economics Analysis*, MIT Press, Cambridge, 1982.

Comoli, P., Pollution control in a simplified general equilibrium model with production externalities, *J. Environ. Econ. Manage.*, 4, 289, 1977.

Dixit, A. K., International trade policy for oligopolistic industries, *Econom. J. Suppl.*, 1, 1984.

Eaton, J. and Grossman, G. M., Optimal trade and industrial policy under oligopoly, *Q. J. Econ.*, 383, 1986.

Forster, B., Pollution control in a two-sector dynamic general equilibrium model, *J. Environ. Econ. Manage.*, 4, 305, 1977.

Forster, B., Environmental and the distribution of income in simple general equilibrium models, in *Economic Perspectives*, Ballabon, M. G., Ed., Harwood, New York, 1981.

Forster, B., Spatial economic theory of pollution control: reflections of a paradox, *J. Environ. Econ. Manage.*, 15, 470, 1987.

Forsund, F. R., Allocation in space and environmental pollution, *Swed. J. Econ.*, 74, 19, 1972.

Gokturk, S. S., A theory of pollution control, location choice, and abatement decisions, *J. Reg. Sci.*, 19, 461, 1979.

Helpman, E., International trade in the presence of product differentiation, economies of scale and monopolistic competition. A Chamberlinian–Heckscher Ohliln approach, *J. Int. Econ.*, 11, 304, 1981.

Horstmann, I. J. and Markusen, J. R., Endogenous market structures in international trade, *J. Int. Econ.*, 32, 109, 1992.

Krugman, P. R., Increasing returns, monopolistic competition, and international trade, *J. Int. Econ.*, 9, 469, 1979.

Laplante, B., Design standard versus performance standard with heterogeneous firms, manuscript, Queen's University, 1989.

Laplante, B., Producer surplus and subsidization of pollution control devices: a non-monotonic relationship, *J. Ind. Econ.*, 39, 15, 1990.

Markusen, J. R., Cooperative control of international pollution and common property resources, *Q. J. Econ.*, 89, 618, 1975.

Markusen, J. R., International externalities and optimal tax structures, *J. Int. Econ.*, 5, 15, 1975.

Markusen, J. R., Trade and gains from trade with imperfect competition, *J. Int. Econ.*, 11, 531, 1981.

Markusen, J. R., Multinationals, multi-plant economies, and the gains from trade, *J. Int. Econ.*, 16, 205; reprinted in Bhagwati, J., Eds., *International Trade: Selected Readings*, 2nd ed., MIT Press, Cambridge, 1984.

Markusen, J. R., Morey, E. R., and Olewiler, N. D., Noncooperative equilibria in regional environmental policies when plant locations are endogenous, Discussion paper, Department of Economics, University of Colorado, Boulder, 1991.

Mathur, V. K., Spatial economic theory of pollution control, *J. Environ. Econ. Manage.*, 3, 16, 1976.

McConnell, V. D. and Schwab, R. B., The impact of environmental regulation on industry location decisions: the motor vehicle industry, *Land Econ.*, 66, 67, 1990.

McGuire, M., Regulation, factor rewards, and international trade, *J. Public Econ.*, 17, 335, 1982.

Meade, J., External economies and diseconomies in a competitive situation, *Econ. J.*, 62, 54, 1952.

Merrifield, J. D., The impact of selected abatement strategies on transnational pollution, the terms of trade, and factor rewards: a general equilibrium approach, *J. Environ. Econ. Manage.*, 15, 259, 1988.

Misiolek, W., Pollution control through price incentives: the role of rent seeking costs in monopoly markets, *J. Environ. Econ. Manage.*, 15, 1, 1988.

Pethig, R., Pollution, welfare, and environmental policy in the theory of comparative advantage, *J. Environ. Econ. Manage.*, 2, 160, 1976.

Pigou, A. C., Divergences between marginal social net products and marginal private net product, in *The Economics of Welfare*, 4th ed., Macmillan, London, 1932, chap. 9.

Siebert, H., Eichberger, J., Gronych, R., and Pethig, R., *Trade and Environment: A Theoretical Enquiry*, Elsevier, Amsterdam, 1980.

Tietenberg, T. H., Spatially differentiated air pollution emission charges: an economic and legal analysis, *Land Econ.*, 54, 265, 1978.

Yohe, G. W., The backward incidence of pollution control — some comparative statics in general equilibrium, *J. Environ. Econ. Manage.*, 6, 187, 1979.

5 On Ecological Dumping

Michael Rauscher

A. INTRODUCTION

Environmental legislation provides a means for the government to influence the competitiveness of an economy or of some of its sectors. According to the standard theorems of international-trade theory, a restrictive policy towards the users of environmental resources will distract internationally mobile factors of production and, inside the economy, cause factor movements from the pollution-intensive sectors of the economy to the cleaner sectors. The comparative advantage of the former is diminished by such a policy. Thus, environmental legislation affects the international division of labor and can be used to achieve trade-policy objectives. This is particularly appealing if the traditional instruments of trade policy, tariffs and quotas, are not available, for instance if the country has signed treaties that prohibit restrictions on imports from other countries. Examples are the EC treaties and the GATT. Thus, the politician may be tempted to use the tools of environmental policies to achieve objectives other than the internalization of the social costs of pollution. The objectives may be terms-of-trade improvements, strategic trade-policy considerations, or the protection of infant or ailing industries or sectors which are supported by powerful lobbies.

The protection of particular sectors or subgroups of a society by means of environmental policies may result in an environmental legislation which is too lax in some sense. Producers obtain hidden subsidies in terms of low pollution abatement requirements and they can dump their products in international markets at prices that do not reflect the true cost of production. This is considered to be a practice of unfair trade. Thus, the term "ecological dumping" is often used in the public discussion to paraphrase the phenomenon. This catchphrase will be taken up for the following investigation. It should be kept in mind, however, that, in contrast to normal dumping, ecological dumping is an activity performed by the government and not by an individual firm. Moreover, it does not affect the price of a tradable commodity, but that of a factor of products which is internationally immobile: nature's capability to provide environmental resources.

This chapter is an attempt to cast some light on the issue of ecological dumping. First, a sensible definition of the subject is sought. What is ecological dumping and how can it be measured? We shall then try to identify economic motives underlying a government's decision to engage in ecological dumping. It will be assumed that the government follows a rational strategy, i.e., it maximizes national welfare or another objective function. Firstly, the terms-of-trade argument will be addressed. It will be seen that the terms of trade of a country are not in general improved by lax environmental policies. Thus, two alternative approaches to explaining ecological dumping are presented. One of them is strategic trade policy: lax environmental standards may be used to improve the market position of a domestic oligopolist and to shift profits from abroad to the domestic economy. The other approach is to model the impact of domestic exporters' lobbies on the political decision-making process. Some final remarks conclude the chapter.

Finally, it may be useful to say what is missing in this chapter. The theoretical framework of the analysis is an international-trade model in which the production of traded commodities harms the environment. Of course, there are also environmental problems connected with the consumption of these goods, and they may have strong implications on international trade (as the Danish bottle case in the EC has vividly demonstrated). Nonetheless, since the term ecological dumping has been used to characterize the regulation of production activities, consumption externalities are of minor

1-56670-530-4/01/$0.00+$.50
© 2001 by CRC Press LLC

interest here. Another omission is the neglect of transfrontier pollution. It has been shown by Merrifield (1988) and Rauscher (1991), that the combination of transfrontier pollution spillovers and international trade create interesting phenomena in the standard model of international trade, but it would be trivial to identify transfrontier externalities as an explanation of too-lax environmental regulation, and the term ecological dumping has not been used in this sense. Therefore, I shall deal with purely national environmental problems. The final restriction of the model is the absence of trade interventions. Many countries have signed international agreements that prohibit the use of these instruments. Therefore, environmental regulation may serve as a second-best alternative to tariffs and quotas in a world of free trade.

B. THREE DEFINITIONS OF ECOLOGICAL DUMPING

Environmental dumping characterizes a situation in which a government uses lax environmental standards to support domestic firms in international markets. Low emission taxes and pollution-abatement requirements enable these firms to dump their goods into foreign markets at relatively low prices. This is viewed as being unfair and should be prevented. It is, however, far from clear what is meant by too-lax environmental standards.

In the public opinion, the term ecological dumping characterizes a situation in which the environmental standards in one country are lower than those in other countries. By undercutting the environmental standards of other countries, a government reduces the production costs of domestic firms. They can produce at lower costs than their foreign competitors and this is often considered unfair. As a consequence, a desirable world of fair trade would be characterized by perfect harmonization of environmental policies: all countries should use the same environmental standards.

There are two objections against applying this concept to international differences in environmental regulations. First, if one believes in factor price equalization, the implicit prices of environmental resources should be the same in all countries if there is trade. Differences in emission taxes across countries would not occur. To detect eco-dumping activities, one would have to employ the autarky prices. The second critique is more important. To a trade theorist, it does not make much sense to postulate that all countries should use the same level of environmental regulation. International differences in the endowments with environmental resources do exist, whether because of differences in physical characteristics of the countries or differences in the tastes of the people. Removing these differences by means of harmonization is equivalent to removing a part of the basis of gains from trade. See Hansson (1990) for instance. Thus, ecological dumping by employing lower environmental standards than the rest of the world can be a good thing.

Nonetheless, I think that ecological dumping is a problem. With another definition of the subject, this becomes obvious at once. The definition is related to the modern view of dumping in commodity markets, which defines the subject of its analysis as pricing at less than marginal cost.[1] See Davies and McGuinness (1982) and Ethier (1982). Correspondingly, ecological dumping can be defined as a policy which prices environmentally harmful activities at less than the marginal cost of environmental degradation, i.e., a policy which does not internalize all environmental externalities. Therefore, firms can dump their output into international markets at prices which do not meet the marginal social cost of production.[2]

---

[1] For the traditional view of dumping as price discrimination between foreign and domestic markets, see Viner (1923).

[2] There may be additional components of the social costs of using environmental resources that are neglected by this definition. The full social cost would also include the effects of environmental regulation on the terms of trade or on other variables that affect welfare. Nonetheless, it appears to be sensible to restrict the definition of ecological dumping to pure production costs, i.e., the private costs of production plus the environmental damage. The same concept has been used by Davies and McGuinness (1982) for their definition of dumping in goods markets. One of the scenarios they consider is that of a monopolist who uses a dumping strategy to deter the entry of a competitor. The definition of dumping as pricing at less than marginal production cost neglects the fact that the entry of a competitor is costly to the incumbent and may be viewed as a part of her cost function.

The problem with this definition is that there are many reasons for too low a level of environmental regulation, and it may be hard to distinguish between ecological dumping and other kinds of under-regulation. For instance, it is often argued that producers' lobbies are more influential in the political decision-making process than the consumers of environmental quality. If this is true, the under-internalization of external effects would be possible even in closed economies, and this can hardly be called ecological dumping. A way out of this dilemma is to compare different sectors of the economy. One may argue that trade-related measures of environmental policy are targeted primarily at the producers of traded goods, who face competition in international markets, but not at the producers of non-traded goods that are not subject to international competition. One could, therefore, take the non-traded goods sector as a point of reference and define ecological dumping as a scenario in which environmental standards are tighter in the non-tradables than in the tradables sector. A prerequisite for this is that the government has the power to use sector-specific instruments of environmental policy. Anecdotal evidence tells us that this is indeed the case. There are even plant-specific differences in pollution abatement requirements, as is vividly attested to by the electricity generation sectors in various countries. In practical applications, however, it may be difficult to compare the tightness of environmental standards across different sectors. Matters are simple if both sectors discharge the same pollutant and cause the same environmental damage at the margin. This will be assumed in the theoretical parts of this chapter, and the emission tax rate is an appropriate measure of the tightness of environmental standards. In the more general case of different pollutants and different marginal damages, one should compare the degrees of internalization of social costs across different sectors of the economy.

Given that public opinion does not provide a sensible definition of ecological dumping, there remain two approaches to investigation. The major difference between the two definitions of ecological dumping is their point of reference. In the first case, it is the full internalization of the social costs of production. In the second case, it is the policy applied to the non-tradables sector (which does not necessarily internalize the external effects of production).

C. THE MODEL

Since one of the definitions of ecological dumping uses the non-traded goods sector as a point of reference, a three-goods model of the economy is needed. There are imported goods, exported goods, and non-tradables, and they are produced with two factors. To reduce complexity, assume that only two goods are produced domestically. Sector 1 of the economy produces non-tradable goods, and sector 2 produces the export good. Commodity 3 is produced only in the foreign country.[3] The two factors of production are capital and an environmental resource, which will also be referred to as emission. Let K^i be the capital stock and E^i be the quantity of the environmental resource employed in the i th sector. The output of this sector, Q^i, then is

$$Q^i = F^i\left(K^i, E^i\right), \qquad i = 1, 2 , \tag{5.1}$$

The production functions are assumed to be well behaved. They are concave and exhibit constant returns to scale. Marginal productivities are positive and declining and the cross derivatives F^i_{KE} are positive. This implies that capital and emissions are imperfect substitutes or, more technically speaking, that their elasticity of substitution is finite and has the normal sign.[4]

Let P^2 and P^3 be the prices of commodities 2 and 3 in terms of the non-traded good which serves as the numéraire. $P = P^2/P^3$ denotes the country's terms of trade, i.e., the price of the export

---

[3] A similar model with traded and non-traded goods has been analyzed by Jones (1974). In this model, however, the exported good is not consumed in the home country.

[4] Algebraically, the condition $f^i_{KE} > 0$ follows from constant returns to scale and the concavity of the production function.

good in terms of the import good. Moreover, let C^i denote the domestic consumption and let $m(P)$ be the foreign country's excess demand function for commodity 2. An equilibrium requires that the excess supply of the home country equals the excess demand of the foreign country

$$Q^2 - C^2 = m(P) \tag{5.2}$$

Moreover, if international trade is balanced, we have

$$P^2(C^2 - Q^2) + P^3 C^3 = 0 \tag{5.3a}$$

or, together with (Equation 5.2)

$$C^3 = Pm(P) \tag{5.3b}$$

The demand for commodities 1, 2 and 3 and the optimal environmental policy can be derived from the national welfare $W(.,.,.,.)$ which has as its arguments the consumption of the three goods and environmental quality, A. For the sake of simplicity, I assume that the welfare function is additively separable in its arguments.[5] Therefore

$$W(C^1, C^2, C^3, A) = U^1(C^1) + U^2(C^2) + U^3(C^3) + V(A) \tag{5.4}$$

with positive but decreasing partial derivatives.[6]

There is a fixed capital stock, K, which can be moved without costs between sectors 1 and 2

$$K = K^1 + K^2 \tag{5.5}$$

For reasons of comparability, let us assume that all sectors discharge the same pollutant. Then, the environmental quality may be defined as

$$A = -E^1 - E^2 \tag{5.6}$$

Inserting equations (5.1), (5.2), (5.3b), (5.5), and (5.6) into the welfare function yields

$$W = U^1\left[F^1(K^1, E^1)\right] + U^2\left[F^2(K - K^1, -A - E^1) - m(P)\right] + U^3\left[Pm(P)\right] + V[A] \tag{5.7}$$

This welfare function is to be maximized by the government, which chooses the optimal state of the environment, A. The firms decide on the allocation of capital to the producing sectors of the economy. The emission levels E^1 and E^2 can be chosen either by the producers or by the government. In one case, the government chooses just the level of environmental policy and then leaves it to the producers to allocate the emissions. This can be done by choosing appropriate emission taxes or by using a tradable permits scheme. On the other hand, the government itself may wish to determine the emissions of the two producing sectors.

---

[5] Even with this restrictive assumption, there will be a large variety of feasible results and a more general welfare function would not add much to the analysis in this respect.

[6] This type of welfare function is known from models with endogenous factor supply. See Kemp and Jones (1962).

If the country is small, P is given and the excess demand of the rest of the world is perfectly elastic. The optimal level of environmental quality, A, is determined by

$$U_C^2 F_E^2 = V' \tag{5.8}$$

where primes and subscripts denote (partial) derivatives of functions. Under the assumption of perfect competition, factors are remunerated according to their marginal productivity. If the government left the allocation decision to the producers, the equilibrium price of environmental resources (equaling the Pigouvian tax rate), T, would be

$$P^2 V'/U_C^2 = P^2 F_E^2 = F_E^2 = T \tag{5.9}$$

If the government itself decides on the allocation issue, it maximizes (Equation 5.7) also with respect to E^1, which implies

$$U_C^2 F_E^1 = U_C^2 F_E^2 \tag{5.10}$$

Noting that utility maximisation of the households implies $U_C^1 = U_C^2/P^2 = U_C^3/P^3$, it follows from (Equation 5.10) that $F_E^1 = P^2 F_E^2$.

This implies that the results of economy-wide and sector-specific environmental regulation are the same. The same tax rate will be applied to all sectors. Thus, it does not pay to discriminate against one sector and favor another if the country under consideration is small. Moreover, the emission tax rate covers the marginal social cost of environmental degradation. Thus, there is no ecological dumping, independently of the definition applied.

D. THE TERMS-OF-TRADE ARGUMENT

Matters may be different if the country under consideration is large. It has an impact on its terms of trade and this affects its welfare. The effect is

$$dW = \left[-U_C^2 m' + U_C^3 (m + Pm') \right] dP$$

Noting that due to utility maximization of the households $U_C^2 = P U_C^3$, this can be rewritten

$$dW = U_C^3 m dP \tag{5.11}$$

As an exporter of good 2, the country will benefit from an increase in the relative price of this good compared to the imported good. One can now determine the impact of the supply of the environmental factor of production on the terms of trade.

In a first step, we shall deal with a scenario in which the government determines merely the level of environmental quality and then imposes a tradable-permits scheme or levies an emission tax, which the firms use to decide on the allocation of the environmental factor of production to the two producing sectors of the economy. For the standard model without non-tradable goods, the optimal environmental policy has been investigated by Markusen (1975) and Rauscher (1991). A country should use its public policy to increase the relative price of the factor with which it is relatively well-endowed to improve its terms of trade. The reduction of the availability of this factor tends to increase its price and the relative price of the commodity that uses this factor intensively in its production. Since this good is the export good, the terms of trade are improved by such a policy. Applying this to a model with environmental resources yields the result that a country well-endowed with environmental resources should employ a particularly restrictive policy towards users of environmental policy. In contrast, countries not so well endowed with environmental resources should reduce their emission tax rates in order to improve their terms of trade.

Matters are different in a case of complete specialization, i.e., if there is no domestic consumption of the imported good. Under normal parameter constellations, it is optimal to reduce the supply of the export good in order to raise its price. This can be done by reducing the supply of the environmental factor of production, regardless of whether the country is well-endowed with resources. In the model considered here, there is an additional complication, since the non-traded goods sector uses environmental resources. The impact of a change in the environmental policy A on the terms of trade, P, is determined by

$$U_C^2 - P^2 U_C^2 = 0 \tag{5.12a}$$

$$U_C^2 - P U_C^3 = 0 \tag{5.12b}$$

$$F_K^1 - P^2 F_K^2 = 0 \tag{5.12c}$$

$$F_E^1 - P^2 F_E^2 = 0 \tag{5.12d}$$

Equations (5.12a, b) are derived from utility maximization by consumers, whereas (5.12c, d) follows from profit-maximisation behavior of firms. Eliminating P^2 yields

$$U_C^2 - P U_C^3 = 0 \tag{5.13a}$$

$$U_C^1 F_K^1 - U_C^2 F_K^2 = 0 \tag{5.13b}$$

$$U_C^1 F_E^1 - U_C^2 F_E^2 = 0 \tag{5.13c}$$

Total differentiation with respect to P, K^1, E^1, and A gives the desired result. Some basic but cumbersome algebraic exercises show that the sign of dP/dA is ambiguous: there are nine positive and two negative terms.[7] This is due to the fact that both sectors compete for the environmental factor of production. A reduction of the availability of these resources reduces the marginal productivity of capital in both sectors since $F_{KE}^i > 0$. If this reduction is larger in the non-tradables than in the tradables sector, capital will be moved from the non-tradables to the tradables sector of the economy. Thus, the output of traded goods may actually rise if the reduction of emissions is dominated by an increase in the capital stock. As a consequence, the terms of trade may be reduced. Under certain circumstances, it may be advisable to relax environmental regulations to improve the terms of trade. It is not optimal to fully internalize the domestic costs of environmental disruption, and this may be called ecological dumping. If, however, the direct effects of environmental policy changes dominate the indirect general-equilibrium effects, it is advisable to use more restrictive policies.

---

[7] An additional ambiguity arises when the determinant of the matrix of the partial derivatives of (Equation 5.13a, b, c) with respect to P, E^1, and E^2 has the "wrong" sign. An explanation of this will be given below.

The type of environmental policy considered here uses one instrument to cope with two distortions. There are the social costs of environmental degradation and the potential to improve the terms of trade. In this case, two policy instruments are better than one. Of course, the optimal combination of policy measures would be an emission tax plus a tariff. However, if tariffs are not available as an instrument of economic policy, the government may wish to enlarge its set of policy measures by imposing sector-specific environmental policies. The question to be considered here is whether such a policy tends to discriminate against the sectors producing non-traded goods. For this purpose it is convenient to rewrite the objective function by substituting for A

$$W = U^1\left[F^1\left(K^1, E^1\right)\right] + U^2\left[F^2\left(K - K^1, E^2\right) - m(P)\right] + U^3\left[Pm(P)\right] + V\left[-E^1 - E^2\right] \quad (5.7')$$

The allocation of capital and the terms of trade are determined by

$$U_C^2 - PU_C^3 = 0 \quad (5.13a)$$

and

$$U_C^1 F_K^1 - U_C^2 F_K^2 = 0 \quad (5.13b)$$

Using (5.13a, b), one can derive the welfare effects of changes in sector-specific environmental policies. Optimal policies are characterized by

$$\frac{dW}{dE^1} = U_C^1 F_E^1 - V' + U_C^3 m \frac{dP}{dE^1} = 0 \quad (5.14a)$$

and

$$\frac{dW}{dE^2} = U_C^2 F_E^2 - V' + U_C^3 m \frac{dP}{dE^2} = 0 \quad (5.14b)$$

Noting that in a competitive economy the marginal productivities of emissions equal the emission tax rates, i.e., $T^1 = F_E^1$ and $T^2 = P^2 F_E^2$ where $P^2 = U_C^1 / U_C^2 F_E^2$, one can easily determine a condition for ecological dumping defined in the sense of discrimination of non-traded goods production. Ecological dumping according to this definition is when the emission tax rate in the export sector is lower than that in the non-traded goods sector or, equivalently

$$T^2 < T^1 \quad \text{iff} \quad \frac{dP}{dE^1} < \frac{dP}{dE^2} \quad (5.15)$$

Ecological dumping is optimal if the terms-of-trade effect of additional emissions in the non-tradables sector is smaller than the corresponding effect in the sector producing traded commodities. If this condition holds, the improvement in the terms of trade by an additional unit of emissions in the tradables sector exceeds the improvement achieved by increasing emissions in the non-tradables sector. Therefore, it is optimal to apply lower environmental standards to the exporting industries.

The terms-of-trade effects of sector-specific environmental policies can be determined by total differentiation of the private sector's optimality conditions (Equations 5.13a, b). This yields

$$
\begin{pmatrix}
-m'U_{CC}^2 - U_C^3 - P(m+Pm')U_{CC}^3 & -U_{CC}^2 F_K^2 \\
m'U_{CC}^2 F_K^2 & U_C^1 F_{KK}^1 + U_{CC}^1\left(F_K^1\right)^2 + U_C^2 F_{KK}^2 + U_{CC}^2\left(F_K^2\right)^2
\end{pmatrix}
$$

(5.16)

$$
\times \begin{pmatrix} dP \\ dK^1 \end{pmatrix} = \begin{pmatrix}
0 & -U_{CC}^2 F_E^2 \\
-U_C^1 F_{KE}^1 - U_{CC}^1 F_K^1 F_E^1 & U_C^2 F_{KE}^2 + U_{CC}^2 F_K^2 F_E^2
\end{pmatrix} \begin{pmatrix} dE^1 \\ dE^2 \end{pmatrix}
$$

The sign of the determinant, D, of the matrix on the left-hand side is ambiguous. This is due to the terms occurring in its first-row first-column element. If the foreign country's import demand is inelastic, then m' is close to zero, implying that this element may be positive. In this case, the determinant is negative. But the opposite scenario is also feasible. In order to give an economic interpretation, imagine a situation in which the supply of good 3 is increased exogenously, e.g., by some manna from heaven. The intuition is that this should improve the terms of trade, since the scarcity of the import good is reduced and its gets cheaper. It can be shown that this happens only if the determinant is positive.[8] If the determinant is negative, general-equilibrium interdependence produce the counter-intuitive result of declining terms of trade. In what follows, let us assume that the intuition is correct and that the determinant has a positive sign.

The terms-of-trade effects of sector-specific environmental policies can be determined by applying Cramer's rule to Equation (5.16). This yields

$$
\frac{dP}{dE^1} = -D^{-1}U_{CC}^2 F_K^2 \left[U_C^1 F_{KE}^1 + U_{CC}^1 F_K^1 F_E^1 \right]
$$

(5.17a)

$$
\frac{dP}{dE^2} = D^{-1} \left[U_{CC}^2 F_K^2 U_C^2 F_{KE}^2 - U_{CC}^2 F_E^2 \left(U_C^1 F_{KK}^1 + U_{CC}^1 \left(F_K^1\right)^2 + U_C^2 F_{KK}^2 \right) \right]
$$

(5.17b)

The terms-of-trade effect of an increase in emissions is negative in the traded commodities sector and ambiguous in the non-tradables sector. This can be explained as follows. An increase in the availability of environmental resources in sector 2 increases the supply of exportable goods and, under normal circumstances, their price will decline. For sector 1, which produces the non-traded good, there are two opposing effects. On one hand, an increase in emissions raises the marginal productivity of capital since $F_{KE}^1 > 0$. If commodity prices are given, capital is moved from sector 2 to sector 1. This reduces the supply of the export good and tends to improve the terms of trade. On the other hand, the increase in the availability of the environmental factor of production in the non-tradables sector increases the supply of this good. Its price relative to that of the other good which is produced at home is reduced. For given productivities, capital tends to move from sector 1 to sector 2. This raises the supply of the exported good and leads to a deterioration of the terms of trade.

Given the terms-of-trade effects of sector-specific environmental policies, the policy implications can easily be derived. Since under normal circumstances the terms-of-trade effect of additional emissions in the export sector is negative, one should attempt to reduce these emissions below those of the reference scenario in which the terms-of-trade effects are not taken into account. The policy implication for the non-tradables sector is ambiguous. If capital and emissions are good substitutes in these sectors (if F_{KE}^1 is large), then there is a positive terms-of-trade effect, and the environmental policy measures applied to the non-tradables sector should be relaxed. In this scenario the policy implication is the opposite of ecological dumping: discriminating against the sector which produces traded goods. It should be noted, however, that a number of different scenarios are

---

[8] Let M be the quality of manna falling from heaven. Then $dp/dM = D^{-1}PU_{CC}^3 (U_C^1 F_{KK}^1 + U_{CC}^1 (F_K^1)^2 + U_C^2 F_{KK}^2 + U_{CC}^2 (F_K^2)^2)$. It follows that sign $(dP/dM) =$ sign (D).

imaginable, some of which can indeed result in discrimination against the non-tradables sector, but such an outcome is not particularly likely.

We can summarize that there is a tendency towards stricter environmental legislation, either in the export sector or for the economy as a whole, if policy makers wish to improve the terms of trade. Due to general-equilibrium considerations, however, the opposite policy recommendation is also possible for some parameter combinations. It should be noted that the tendency towards better protection of the environment is an artifact of the assumptions used in this model. The crucial assumption is the absence of an import-competing industry. If such a sector existed, the policy recommendation would mainly depend upon whether the economy is well-endowed with environmental resources or not.

E. STRATEGIC TRADE POLICY

In the previous sections, it has been assumed that perfect competition prevails on factors and goods markets. Over the last decade, numerous models have been developed which differ from the standard Heckscher–Ohlin–Samuelson framework in this respect and allow the derivation of interesting policy implications. One is the strategic trade policy framework which shows that, under certain circumstances, it may be advisable to subsidize domestic firms in order to increase their market power in international markets.

The basic model is that of Brander and Spencer (1985). They show that the government can turn a firm playing Nash–Cournot into a Stackelberg leader by subsidizing its production. Rents are shifted to the home country and this is welfare-improving. Direct subsidies, however, can easily be detected by foreign competitors. Therefore, it may be better to use indirect methods of subsidization. One possibility is to subsidize domestic research and development (See Spencer and Brander, 1983). Another alternative is to impose relatively modest pollution abatement requirements on the firms competing in international oligopoly markets. There have been some recent attempts to analyze this possibility in theoretical models (See Barrett, 1992; Ulph, 1992; and Conrad, 1993). I will refer mainly to Barrett's paper and try to apply his results to the general-equilibrium framework used above.

Barrett (1992) uses the standard model of strategic trade policy extended by the introduction of costly pollution abatement. There are two firms, one domestic and one foreign. They sell their production in a third country. Each firm takes as given the quantity supplied by the other firm and the environmental policy of its home government. In this situation, the market solution is the Nash equilibrium which is represented by point N in Figure 5.1. Q and q denote the quantities supplied by the domestic and the foreign firm, respectively, and R and r are their reaction curves. Let B and b be the iso-profit curves of the two firms. If a direct subsidy is given to the domestic firm, it increases its output and its curve is shifted to the right. The equilibrium moves along the foreign country's reaction curve. This is beneficial to the home country as long as the net profits (profits minus subsidies) of the domestic industry are raised. The optimum is given by point S, where the foreign firm's reaction curve touches the best-possible domestic iso-profit curve. This is known as the Stackelberg solution.

Matters are slightly different if the subsidy is given indirectly by relaxing environmental standards. This kind of subsidy is not merely a purely redistributive policy instrument, but its application incurs some real cost. The environmental quality deteriorates, and this has to be taken into account by the government. The emission tax rate should be chosen such that the marginal social cost due to environmental degradation and the marginal benefit from the increase in the market position of the domestic firm are equated. The optimal solution is to be found somewhere between the Nash and the Stackelberg points N and S.

Applying these results to the eco-dumping problem yields the following propositions. It can be optimal to favor the export industries if the supply side of the international market is oligopolistic. This implies that the non-tradables sector is discriminated against. Moreover, the environmental

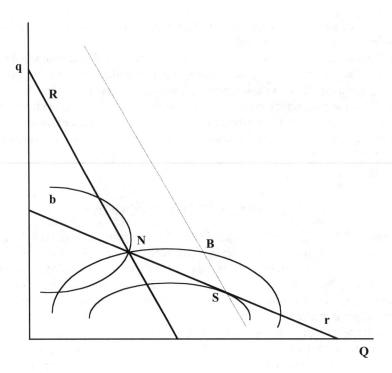

FIGURE 5.1 Strategic environmental policy.

standards are lower than those that internalize the pure social cost of pollution. Therefore, ecological dumping according to both our definitions may be explained by strategic trade policy considerations. Finally, there may be an international competition in environmental regulations which leads to undesirably low levels of environmental quality in many countries.

Matters can be rather different if the general-equilibrium effects are taken into account. As can be seen in Figure 5.1, the objective of strategic trade policy is to raise the production of the domestic export industry for any given quantity supplied by the foreign competitor. In a partial equilibrium, this is achieved by increasing the supply of the environmental factor of production for this sector. In a general equilibrium, the change in the allocation of capital must also be taken into account. We will look at a situation in which the government chooses sector-specific environmental measures. For the rest of this part of the chapter, let us use the assumption often employed in the strategic trade literature that the good is not sold on the domestic market, i.e.,

$$m(P) = F^2\left(K^2, E^2\right) \tag{5.18}$$

Let $m(P)$ now be the inverse demand function faced by the domestic oligopolist for a given level of production of its foreign competitor. If this firm acts as a price taker on domestic factor markets, its profits are passed over to the factor owners. The factors are remunerated at more than marginal cost. This implies

$$F_K^1 = (1+e)P^2 F_K^2 \tag{5.19}$$

where $e = m/(m'P)$, $-1 < e < 0$, is the elasticity of the residual demand function which will be assumed to be constant. Noting that $P^2 = P^3 P$ and $P^3 U_C^1 = U_C^3$, we obtain

$$U_C^1 F_K^1 - (1+e)U_C^3 P F_K^2 = 0 \tag{5.19'}$$

Comparative static results can be derived by totally differentiating Equations 5.18 and 5.19′

$$
\begin{pmatrix}
m' & F_K^2 \\
-F_K^2(1+e)\left[U_C^3 + U_{CC}^3 P(m+Pm')\right] & F_{KK}^1 + U_{CC}^1\left(F_K^1\right)^2 + U_C^3 PF_{KK}^2(1+e)
\end{pmatrix}
$$
$$
\times \begin{pmatrix} dP \\ dK^1 \end{pmatrix} = \begin{pmatrix}
0 & F_E^2 \\
-U_C^1 F_{KE}^1 - U_{CC}^1 F_K^1 F_E^1 & U_C^3 PF_{KE}^2(1+e)
\end{pmatrix} \begin{pmatrix} dE^1 \\ dE^2 \end{pmatrix}
$$

(5.20)

It can be shown that the determinant of the matrix on the left-hand side, D, is positive since the world market demand for 2 is elastic. The impact of changes in environmental policy on the allocation of capital turns out to be

$$
\frac{dK^1}{dE^1} = D^{-1}m'\left(-U_C^1 F_{KE}^1 - U_{CC}^1 F_K^1 F_E^1\right)
$$

(5.21a)

$$
\frac{dK^1}{dE^2} = D^{-1}(1+e)\left(m' U_C^3 PF_{KE}^2 + F_K^2 F_E^2\left[U_C^3 + U_{CC}^3 P(m+Pm')\right]\right)
$$

(5.21b)

In both cases, the effects of environmental policy changes are ambiguous. An increase in emissions in a sector makes capital more productive at the margin since F_{KE}^1 and F_{KE}^2 are positive. Thus, a policy which increases emissions in a sector attracts capital. On the other hand, a good produced in a more pollution-intensive process becomes cheaper and the marginal value product of capital tends to be reduced. This induces a capital movement into the opposite direction. In the case of a change in E^2, however, this capital removal due to price changes is always dominated by the direct output expansion effect of the increase in emissions.[9] Thus, in order to increase the output of the export history, it is advisable to relax the sector-specific environmental policy. It may be optimal to relax the policy for the other sector even more if this induces a very massive capital flow to the export sector. Although this is not particularly likely, it may be optimal to use an environmental policy which is the opposite of ecological dumping.

Matters become more complicated algebraically if an economy-wide policy is sought which increases the output of the export industry. An additional equation for the cross-sector equalization of emission taxes has to be introduced and the effect of a change in environmental policy on the allocation of emissions to the two sectors has to be considered, too. The result is ambiguous: at least one of the sectors will raise its output if pollution abatement requirements are reduced. It may, however, happen that factors move to the non-tradables sector of the economy. In this case, the output of the export industry may be reduced by ecological dumping.

Summarizing the results of this part of the chapter, we can conclude that strategic behavior in international trade does not necessarily lead to eco-dumping, independently of the definition applied. As an additional caveat, one should note that, as Eaton and Grossman (1986) have shown, the policy implications are turned into the opposite direction when the oligopolists play Bertrand instead of Cournot.

F. TOWARDS A PUBLIC-CHOICE EXPLANATION OF ENVIRONMENTAL POLLICIES IN OPEN ECONOMIES

In the previous sections, environmental policies have been investigated which maximize an index of national well-being. It is commonplace, however, that real-world political decision makers do

---

[9] This can be verified by computing $dQ^2/dE^2 = F_E^2 - F_E^2(DK^1/dE^2)$.

not use this kind of welfare criterion. Instead, they try to follow their own objectives. In what follows, I shall consider two models in which these objectives are taken into account. The first one looks at some of the intrinsic driving forces of a public bureaucracy, the second approach considers the impact of the power of sector-specific interest groups on governmental decisions. In order to make matters simpler and to separate the effects from the terms-of-trade effects, we shall consider a small economy for which the world market prices are given.

According to Niskanen (1977) and others, one major characteristic of a governmental bureaucracy is the desire to maximize the budget it has to administer. Taking this into account, the objective function has to be rewritten. Assume that the policy maker maximizes a weighted average of measure of national welfare and the benefits from tax revenues. Let the benefits from tax revenues be measured by an increasing and strictly concave function $G(.)$. The new objective function turns out to be

$$W^G = W + G\left(T^1 E^1 + T^2 E^2\right) \tag{5.22}$$

where W is given by Equations 5.7 or 5.7′.

If $T^1 = T^2$, both sectors face identical environmental regulations. In order to detect ecological dumping, one should compare the solution of the maximization of W^G with a policy that maximizes national welfare. There are two effects of a change in the emission tax rate on the tax revenue. First, for given emissions, the revenue will increase by $(E^1 + E^2)$ times the change in the tax rate. But with an increasing tax rate, emissions should be reduced, which implies that for given levels of the tax rate, the revenue is reduced. Either of these effects may dominate or there may be a bell-shaped Laffer curve. Depending on whether the revenue function is an increasing or decreasing function in the social optimum, the policy maker may wish to raise or reduce the tax rate. Of course, similar arguments apply to a situation in which the government chooses sector-specific environmental policies and it is not clear which sector is discriminated against in an optimum.

A more promising approach to identifying motives of ecological dumping is the consideration of sector-specific interest groups. In a representative democracy, the voter does not decide directly upon all the relevant issues but merely elects a government to represent her will. The government has some discretion in interpreting the will of the electorate. Due to this discretion, interest groups have a chance to influence the government's decisions by lobbying, which may range from spending resources on public relations and financial support of campaigns to bribing and blackmailing. There are good arguments in favor of the hypothesis that the traded-goods industries (cars, chemicals, pharmaceuticals, etc.) have a stronger impact on the government than the producers of non-traded goods (mainly services). If this is true, then the policy maker's objective function is changed by the impact of lobbying.

In what follows, it will be assumed that the lobbyists are concerned about the output of the export industry. This is a sensible approach if there are structural problems with a threat of unemployment and declining profits or if there are sector-specific factors that experience income increases when the output of the sector is increased. If the lobby's impact on political decisions is sufficiently large, the government will maximize an objective function, which is a weighted average of the national welfare function and the utility function of the lobbyists. Thus, we have

$$W^L = W + L\left(Q^2\right) \tag{5.23}$$

where $L(.)$ is increasing and strictly concave and W is defined by Equation (5.7′).

To get an idea of what the major impacts are, the impact of a change in environmental policies on the government's objective function can be determined. In a first step, it is necessary to consider the impacts of the environmental policy on the allocation of the factors of production. This can be done by total differentiation of the first-order optimality conditions of the households and firms

$$U_C^1 F_K^1 - U_C^2 F_K^2 = 0 \tag{5.13b}$$

$$U_C^1 F_E^1 - U_C^2 F_E^2 = 0 \tag{5.13c}$$

Total differentiation with respect to K^1, E^1 and A shows that the impact of a change in environmental policy, A, on the allocation of factors is ambiguous. In the case of capital this may have been expected. But the same is true for emissions: although the overall level of emissions is increased by a relaxation of environmental policy, emissions in one of the two sectors may be reduced. The impact on production is also ambiguous. Ecological dumping may result either in an increase or a reduction of the export industry's output.

Matters are a bit simpler if we consider sector-specific environmental policies. Equation (5.13c) is no longer relevant and total differentiation of Equation (5.13b) yields

$$\frac{dK^1}{dE^1} = \frac{-U_C^1 F_{KE}^1 - U_{CC}^1 F_K^1 F_E^1}{U_C^1 F_{KK}^1 + U_{CC}^1 \left(F_K^1 \right)^2 + U_C^2 F_{KK}^2 + U_{CC}^2 \left(F_K^2 \right)^2} \tag{5.24a}$$

$$\frac{dK^1}{dE^2} = \frac{U_C^2 F_{KE}^2 + U_{CC}^2 F_K^2 F_E^2}{U_C^1 F_{KK}^1 + U_{CC}^1 \left(F_K^1 \right)^2 + U_C^2 F_{KK}^2 + U_{CC}^2 \left(F_K^2 \right)^2} \tag{5.24b}$$

An increase in emissions in sector 1 moves capital into this sector if F_{KE}^1 is large, i.e., if there is a substantial increase in capital productivity, or it $-U_{CC}^1$ is small. The latter condition means that the price elasticity of demand is large and this implies that an increase in supply results only in a small price reduction. Thus the increase in E^1 has only a small effect on the marginal value product of capital for a given level of the physical productivity. The same arguments apply for the other sector of the economy.

Now consider the welfare effects of a change in environmental policy

$$dW^L / dE^1 = dW / dE^1 - L' F_K^2 \left(dK^1 / dE^1 \right) \tag{5.25a}$$

$$dW^L / dE^2 = dW / dE^2 + L' \left(F_E^2 - F_K^2 \left(dK^1 / dE^2 \right) \right) \tag{5.25b}$$

Starting from a situation of a social optimum in which $dW/dE^1 = dW/dE^2 = 0$, ecological dumping by discriminating against the non-traded goods sector is beneficial to the government if

$$\left(F_E^2 - F_K^2 \, dK^1 / dE^2 \right) > -F_K^2 \left(dK^1 / dE^1 \right) \tag{5.26}$$

Inserting Equations (5.24a, b) yields

$$\frac{F_E^2 \left(U_C^1 F_{KK}^1 + U_{CC}^1 \left(F_K^1 \right)^2 + U_C F_{KK}^2 \right) - F_K^2 \left(U_C^1 F_{KE}^1 + U_C^2 F_{KE}^2 + U_{CC}^1 F_K^1 F_E^1 \right)}{U_C^1 F_{KK}^1 + U_{CC}^1 \left(F_K^1 \right)^2 + U_C^2 F_{KK}^2 + U_{CC}^2 \left(F_K^2 \right)^2} > 0 \tag{5.27}$$

The last term in the numerator in Equation (5.27) may exhibit the "wrong" sign. Thus it is possible that the export industry is supported best by applying more restrictive environmental policies there than in the non-tradables industry. This is just the opposite of ecological dumping defined as the preferential treatment of the export industry in terms of low emission tax rates. However, this result requires a very particular parameter constellation, and in the normal case it is advisable to apply lower levels of environmental regulation in the export industry than in the non-traded goods sector.[10]

Summarizing the results of the models presented in this section, one arrives at the conclusion that it is by no means clear that it is in the interest of the exporters' lobby or the government itself to have eco-dumping as an environmental policy.

G. SUMMARY AND CONCLUSION

The preceding investigation was an attempt to give more economic content to the catchword of ecological dumping, which has been frequently (mis)used in the recent public discussion on environmental policy in open economies. In a first step, a definition of ecological dumping was sought. Three definitions were proposed and one, based on an international comparison of environmental policies, was rejected. Considering the remaining two, one uses the optimal internationalization of social costs as its point of reference, the other is based on the comparison of regulations applied to sectors producing traded and non-traded commodities. We then tried to find the economic motives of ecological dumping. Under what circumstances is it optimal to employ a policy that may be termed "ecological dumping" according to one of the two criteria? It was noted that neither terms-of-trade considerations nor maximization of the revenues from emission taxation explain ecological dumping. These objectives may be achieved sometimes by ecological dumping and sometimes by its converse, depending on parameters of the demand and supply side. Strategic trade policy and lobbying activities of the exporters are more successful in explaining ecological dumping. However, there are still a number of ambiguities. The general equilibrium repercussions of changes in environmental policy may turn the results around. Lobbyists might be surprised to know that relaxing environmental standards can make the exporters worse off.

Applied to the real world, the interest group approach is the most promising for an explanation of ecological dumping. If an explanation of economic policy decisions is sought, it is not important what is optimal, but what is thought to be optimal. It is a truism (but not always the truth) that a producer will benefit from relaxed pollution abatement requirements applied to the production processes used, but this conjecture, wrong as it may be, provides incentives to lobby for lower emission tax rates. If the exporters' lobbies are more influential than other lobbies, this may explain ecological dumping. The model presented used a shortcut to translate the interest of lobbyists into political decisions. A complicated process was modeled by simply adding some terms to the government's objective function. It is desirable to model this process more explicitly. A promising approach originates in Hillman and Ursprung (1992) who model the impact of interest groups with motives related to emissions, pollution, and the environment on international trade. If a similar analysis could be undertaken for the opposite problem of how trade-related interests affect environmental policies, this would be another major step towards an understanding of the driving forces of ecological dumping.

Another problem which was neglected in the preceding analysis is retaliation. What happens if the other country also uses environmental policy measures to improve its terms of trade or to support its export industries? Will there be "prisoners' dilemma" situations in which both countries undercut each other's environmental taxes in order to keep their traded-goods industries competitive? This may lead to undesirably low levels of environmental regulation everywhere, and to disastrous

---

[10] It follows from Equation (5.24b) that an increase in emissions in the export industry will always raise its output. At least in this model framework, a situation in which exporters benefit from more rigid environmental policies in their own sector is not possible. This may, however, change if more complicated versions of the model are considered, e.g., if the country is large and if more general utility functions are introduced.

consequences for environmental quality. The question arises under which circumstances this result of an international environmental-policy game is feasible and what kind of institutions are necessary to cope with this prisoners' dilemma. Future research into the problem area is desirable.

ACKNOWLEDGMENTS

Earlier versions of this chapter have been presented in seminars at the Kiel Institute of World Economics, the Universities of Munich and Konstanz, and during the 1992 European Research Workshop in International Trade held at the Universidade Nova of Lisbon. I am indebted to the participants for their helpful comments and suggestions. Moreover, I have benefited from discussing problems of environmental policies in open economics with Kym Anderson, Richard Blackhurst, Scott Barrett, Carlo Carraro, and Henry Tulkens. Last but not least, I received useful comments by one of the editors and a referee. The responsibility for any remaining deficiencies in this chapter is solely mine.

REFERENCES

Barrett, S., Strategic environmental policy and international trade, CSERGE working paper GEC, 92–19, Norwich, 1992.

Brander, J. A. and Spencer, B., Export subsidies and international market share rivalry, *J. Int. Econ.*, 18, 83, 1985.

Conrad, K., Taxes and subsidies for pollution-intensive industries, *J. Environ. Econ. Manage.*, 25, 121, 1993.

Davies, S. W. and McGuinness, A. J., Dumping at less than marginal cost, *J. Int. Econ.*, 12, 169, 1982.

Eaton, J. and Grossman, G. M., Optimal trade and industrial policy under oligopoly, *Q. J. Econ.*, 102, 383, 1986.

Ethier, W. J., Dumping, *J. of Political Economy*, 90, 487, 1982.

Hansson, G., *Harmonization and International Trade*, London, 1990.

Hillman, A. L. and Ursprung, H. W., The influence of environmental concerns on the political determination of international trade policy, in *The Greening of World Trade Issues*, Anderson, K. and Blackhurst, R., Eds., Harvester–Wheatsheaf, 1992.

Jones, R. W., Trade with non-traded goods: the anatomy of interconnected markets, *Economica*, 41, 121, 1974.

Kemp, M. C. and Jones, R. W., Variable labor supply and the theory of international trade, *J. Political Economy*, 70, 30, 1962.

Markusen, J. R., International externalities and optimal tax structures, *J. Int. Economics*, 5, 15, 1975.

Merrifield, J. D., The impact of abatement strategies in transnational pollution, the terms of trade, and factor rewards: a general equilibrium approach, *J. Environ. Econ. Manage.*, 15, 259, 1988.

Niskanen, W., *Bureaucracy and Representative Government*, Chicago, 1977.

Rauscher, M., Foreign trade and the environment, in *Economics and the Environment: The International Dimension*, Siebert, H., Ed., Mohr, Tübingen, 1991.

Spencer, B. and Brander, J. A., International R & D rivalry and industrial strategy, *Rev. Economic Stud.*, 50, 707, 1983.

Ulph, A., The choice of environmental policy instruments and strategic international trade, in *Conflicts and Cooperation in Managing Environmental Resources*, Pethig, R., Ed., Springer, Berlin, 1992.

Viner, J., *Dumping: A Problem in International Trade*, University of Chicago Press, Chicago, 1923.

6 Strategic Environmental Policy and International Trade

Scott Barrett[1]

This chapter demonstrates that governments may have incentives to impose weak environmental standards on industries that compete for business in imperfectly competitive international markets, where "weak" means that the marginal cost of abatement is less than the marginal damage from pollution. However, such an intervention is not as efficient as an export or R & D subsidy in improving competitiveness, and depending on the form of competition and market structure, it may instead be optimal for governments to impose strong environmental standards, where "strong" means that the marginal cost of abatement exceeds the marginal environmental damage.

> The existence of less strict environmental standards in a lower income country ... is not a sufficient basis for claiming that the environmental standards are "too low" or that the country is manipulating its environmental standards in order to improve the competitiveness of its producers. To substantiate such a claim, it would be necessary at the very least to demonstrate that the standards are even lower than would be expected on the basis of such factors as the level of per capita income and the characteristics of the physical environment.
>
> Clearly, that would be very difficult to do. Moreover, the charge might also be aimed at highly developed countries in which stringent environmental standards may have been adopted in some areas but where, because of competitiveness considerations, governments have shied away from high standard in others. [GATT (1992, p. 29)]. © 1994 Elsevier Science B.V. All rights reserved.

A. INTRODUCTION

Concern is growing that environmental policy may not only interfere with free trade, but in some instances may be purposefully designed to confer competitive advantage upon the home country industry at the expense of competitive disadvantage abroad.[2] As an example, an article appearing in the European edition of *The Wall Street Journal* (July 8, 1991), reported that U.S. environmental policy, as devised by the Environmental Protection Agency, has in some instances been modified to accommodate concerns expressed by the White House Council on Competitiveness. Such interventions are potentially worrying, not only because they may worsen environmental quality, but also because they cannot be challenged under the GATT,[3] and hence may inspire offended countries to retaliate by changing their standards or by imposing countervailing duties on imports.[4]

---

[1] I am grateful to Paul Geroski, Michael Hoel, David Pearce, Candice Stevens and two anonymous reviewers for commenting on an earlier version of this chapter.

[2] For discussions of this issue, see Grimmett (1991), Reistein (1991), Whalley (1991), Anderson and Blackhurst (1992) GATT (1992), Low (1992), Pearce (1992), and *The Economist* (May 30, 1992).

[3] In the words of the GATT (1992, p. 23): "Generally speaking, a country can do anything to imports or exports that it does to its own products, and it can do anything it considers necessary to its own production processes."

[4] It is this last possibility that worries the GATT most. Using uncharacteristically strong language, the GATT (1992, p. 30) argues: "To allow each contracting party unilaterally to impose special duties against whatever it objects to among domestic policies of other contracting parties would risk an eventual descent into chaotic trade conditions similar to those that plagued the 1930s."

1-56670-530-4/01/$0.00+$.50
© 2001 by CRC Press LLC

Are the incentives to distort environmental standards for reasons of competitiveness quantitatively important? They may be in some cases,[5] but empirical work is ambivalent on the related question of whether sectoral trade flows are affected by environmental standards. Low and Yeats (1992) find that environmentally "dirty" industries have migrated over the last two decades towards lower income countries where environmental standards are weaker. Lucas et al. (1992) find that stricter regulation of pollution-intensive production in the OECD countries "has led to significant locational displacement, with consequent acceleration of industrial pollution intensity in developing countries." However, Grossman and Krueger (1992) find that the sectoral pattern of trade between the United States and Mexico has not been influenced by pollution abatement costs in U.S. industry. Several earlier studies support the view that the link between trade flows and environmental standards is weak, or nonexistent (Dean 1992).

Theoretical papers consider related questions of strategy.[6] Ulph (1992), using a three-stage model in which governments choose environmental policies (to meet a fixed environmental target) in the first stage and firms choose capital stocks and output levels in the second and third stages, shows that if governments behave strategically, they will want to impose quantity standards rather than pollution taxes (see also Rauscher, 1991). The reason, roughly, is that quantity standards allow firms to commit to producing lower output levels, and hence to earn higher profits.[7] Markusen et al. (1992) also employ a stage game to consider whether governments might want to distort their environmental policies in order to influence the location decision of a single polluting firm. They show that governments may compete by weakening environmental policy in order to attract investment, or by strengthening policy in order to induce the firm to locate elsewhere. Finally, Copeland (1990a), again employing a stage game, shows that countries may wish to "overinvest" or "underinvest" in activities that affect the costs and benefits of exploiting a shared resource (a fishery). For example, if enhancement of fish stocks by one country lowers harvesting costs more for its fishermen than its rivals' fishermen, then the effect of the investment may be to decrease harvesting effort by rivals, to the benefit of the home country. In a related paper, Copeland (1990b) shows that a similar result can be obtained for transboundary pollution when governments choose R & D in pollution control as well as pollution levels.[8]

This chapter considers the strategic role of environmental policy from a different perspective. Unlike Ulph (1992), this chapter is not concerned with the choice of policy instrument, and solves endogenously for the environmental target. Unlike Markusen et al. (1992), this chapter is not concerned with the location decisions of firms, but with the competitiveness of existing industries (as determined by their ability to commit to output and price levels). Unlike Copeland (1990a, b), the interdependence that exists between countries is transmitted not through a shared resource (here, environmental damage is strictly local), but through competition in international markets. Unlike all of the above papers, this chapter also considers alternative market structures and alternative forms of industry competition.

The basic model employed is a stage game involving two governments and their respective industries, which sell all their output in a third market. Governments move first by choosing

---

[5] Grossman and Krueger (1992, p. 22) cite a survey by the U.S. General Accounting Office, "suggesting that a few American furniture manufacturers may have moved their operations to Mexico in response to the State of California's tightening of air pollution control standards for paint coatings and solvents."

[6] The literature on trade and the environment is large, and growing. For recent surveys, see Cropper and Oates (1992) and Dean (1992).

[7] Copeland (1992) provides a richer model of the strategic implications of taxes versus standards under various forms of imperfect competition, but does not consider the international dimension. Markusen et al. (1991) analyze the effect of taxes on the location decisions of firms under imperfect competition.

[8] Oates and Schwab (1988) also find that it may be optimal for governments to set "weak" or "strong" standards, but not for strategic reasons. Employing public choice theory, they show that government officials may want to impose weak environmental standards in order to attract capital investment. Assuming a median-voter rule and a heterogeneous population, environmental standards could be either weak or strong, depending on whether the majority of the population preferred to tax or subsidize capital.

environmental standards for their own firms. Firms take these standards as given and compete by choosing either output levels or prices. It is assumed that environmental damage is local, and that the governments of consumer countries have no means of influencing environmental standards in producer countries. This last assumption is entirely consistent with the GATT (see Pearce, 1992).

The chapter shows that if the domestic industry is a monopoly, the foreign industry is imperfectly competitive, and industrial competition is Cournot, then the domestic government has an incentive to set a weak environmental standard, where "weak" means that the marginal cost of abatement is less than the marginal damage from pollution. However, this conclusion hinges on the inability of the domestic firm to act like a Stackelberg leader, and on the inability of the domestic government to employ industrial policy for the same strategic purpose. What is more, if the domestic industry is an oligopoly, then the incentive to impose weak environmental standards is diminished, and it may even become attractive to impose strong environmental standards, where "strong" means that marginal abatement costs exceed the associated marginal environmental damage. If firms compete in prices rather than quantities, then strategic considerations will favor strong standards, whatever the structure of the domestic industry. The conclusion that strategy may demand either weak or strong environmental standards is reminiscent of the results obtained by Copeland (1990a, b) and Markusen et al. (1992), but arises for quite different reasons.

These theoretical findings are as ambivalent as those appearing in the empirical literature. Taken together, they suggest that the incentives for governments to behave strategically in devising environmental policy are limited. Competitiveness may be more effectively enhanced by improving the efficiency of non-strategic environmental policy, which is exactly what economists have long argued (Cropper and Oates, 1992).

B. UNILATERAL STRATEGIC ENVIRONMENTAL POLICY

Denote the domestic firm's output by q^1 and the foreign firm's output by q^2, and let $C^i(q^i)$ be firm i's production cost function. Denoting derivatives by subscripts, assume $C_q^i > 0$. Total revenues are $R^i(q^1, q^2)$. Assuming that the two outputs are substitutes, $R_j^i < 0$. Abatement costs depend on the level of output and the emission standard, which is set by government. Denote the domestic emission standard by e^1 and the foreign standard by e^2. The abatement cost function for firm i is $A^i(q^i, e^i)$, with $A_q^i \geq 0$, $A_e^i \leq 0$, and $A_{qe}^i \leq 0$. If the emission standards are binding, then all of these inequalities are likely to hold strictly. Assume that this is the case. Each firm chooses its output taking as given both the rival's output and the emission standard set by government. Formally, letting π^i denote firm i's profits, the following optimization problem is solved:

$$\max_{q^i} \pi^i = R^i\left(q^1, q^2\right) - C^i\left(q^i\right) - A^i\left(q^i, e^i\right) \tag{6.1}$$

The Nash equilibrium quantities, given emission standards e^i, are determined by

$$\pi_i^i = R_i^i - C_q^i - A_q^i = 0 \tag{6.2}$$

Second-order conditions for a maximum are

$$\pi_{ii}^i = R_{ii}^i - C_{qq}^i - A_{qq}^i < 0 \tag{6.3}$$

Also assume

$$R_{ij}^i < 0 \tag{6.4}$$

and

$$\pi_{11}^1 \pi_{22}^2 - \pi_{12}^1 \pi_{21}^2 < 0 \tag{6.5}$$

Condition (Equation 6.4) ensures that the reaction functions are downward sloping. Condition (Equation 6.5) ensures stability.

The domestic and foreign governments choose their emission standards e^i so as to maximize their respective net benefits, NB^1 and NB^2. Recalling that firm i's output is not consumed within country i, the maximization problem may be stated formally as

$$\max_{e^i} NB^i = \pi^i\left(q^1, q^2, e^i\right) - D^i\left(e^i\right) \tag{6.6}$$

where $D^i(e^i)$ is the (purely local) environmental damage suffered by country i; $D_e^i \geq 0$. If the emission standards are binding, it is likely that these inequalities will hold strictly. Assume that this is the case. The first-order conditions are

$$NB_e^i = -A_e^i - D_e^i = 0 \tag{6.7}$$

and the corresponding second-order conditions are

$$NB_{ee}^i = -A_{ee}^i - D_{ee}^i < 0 \tag{6.8}$$

Stability requires

$$NB_{ii}^i NB_{ee}^i - NB_{ie}^i NB_{ei}^i > 0 \tag{6.9}$$

$$NB_{ii}^i NB_{je}^j NB_{ej}^j + NB_{ij}^i NB_{ji}^i NB_{ee}^j - NB_{ii}^i NB_{jj}^j NB_{ee}^j > 0 \tag{6.10}$$

Equation (6.7) can be rewritten as $\tilde{e}^i = \phi^i(\theta^i)$. Define these schedules as the *environmentally optimal emission standards* (EOSs). These standards obey the usual optimality rule in environmental economics: pollution is reduced to the level where the marginal damage caused by the pollution (D_e^i) just equals the marginal cost of abatement $(-A_e^i)$. These schedules may vary depending on environmental assimilative capacities, incomes, preferences, and abatement costs.

The domestic firm can earn higher profits if it can commit to producing a higher level of output compared with the Nash equilibrium. While the firm cannot make such a commitment credible, its government may do so on its behalf by departing from the EOS. Suppose, then, that the domestic government solves (Equation 6.6) subject to (Equation 6.2) and (Equation 6.7) for country 2. Then we have the first-order condition

$$\pi_1^1 dq^1/de^1 + \pi_2^1\left(dq^2/dq^1\right)\left(dq^1/de^1\right) + NB_e^1 = 0 \tag{6.11}$$

Upon substitution, (Equation 6.11) can be written as $e^{-1} = \xi^1(q^1)$. Define e^{-1} as country 1's *strategically optimal emission standard* (SOS). We know that the domestic firm will set $\pi_1^1 = 0$. However, the second term in (Equation 6.11) may be positive, in which case the domestic government would want to choose an emission standard (for given q^1) at which the marginal damage of pollution exceeds the marginal cost of abatement. Proposition 1 establishes that this is indeed the case.

Proposition 1. The SOS is weaker than the EOS ($\bar{e}^1 > \tilde{e}^1$ for q^1 given).

Proof. By assumption, $R_2^1 < 0$; hence, $\pi_2^1 < 0$. To sign dq^2/dq^1, totally differentiate (Equation 6.2) for firm 2 and (Equation 6.7) for country 2:

$$NB_{ee}^2 de^2 = -NB_{e2}^2 dq^2 \tag{6.12}$$

$$NB_{21}^2 dq^1 + NB_{22}^2 dq^2 + NB_{2e}^2 de^2 = 0 \tag{6.13}$$

Combining (Equation 6.12) and (Equation 6.13) and rearranging yields

$$dq^2/dq^1 = \left(-NB_{21}^2 NB_{ee}^2\right) / \left(NB_{22}^2 NB_{ee}^2 - NB_{2e}^2 NB_{e2}^2\right) \tag{6.14}$$

By (Equation 6.9), the denominator of (Equation 6.14) is positive. By (Equation 6.4), $B_{21}^2 < 0$, and by (Equation 6.8), $NB_{ee}^2 < 0$. Hence, the numerator of (Equation 6.14) is negative, and $dq^2/dq^1 < 0$. It remains to sign dq^1/de^1. Totally differentiating (Equation 6.2) yields

$$NB_{11}^1 dq^1 + NB_{12}^1 dq^2 + NB_{1e}^1 de^1 = 0 \tag{6.15}$$

Rearranging (Equation 6.15) and substituting (Equation 6.14) yields

$$dq^1/de^1 = \left[NB_{1e}^1\left(NB_{22}^2 NB_{ee}^2 - NB_{2e}^2 NB_{e2}^2\right)\right] / \left(NB_{11}^1 NB_{2e}^2 NB_{e2}^2 + NB_{21}^2 NB_{ee}^2 NB_{12}^1 - NB_{11}^1 NB_{22}^2 NB_{ee}^2\right) \tag{6.16}$$

$NB_{1e}^1 > 0$ if the standard is binding. The term in brackets in the numerator of (Equation 6.16) is positive by (Equation 6.9), and the denominator is positive by (Equation 6.10). Hence $dq^1/de^1 > 0$. Since $\pi_2^1 < 0$ and $dq^2/dq^1 < 0$, (Equation 6.11) implies $NB_e^1 < 0$, or $-A_e^1 < D_e^1$. $\square$

The reason for this result is as follows. If the domestic government increases the domestic standard above $\tilde{e}^1$, then the domestic firm will increase its output, and the foreign firm will decrease its output. Profits will be shifted away from the foreign firm to the domestic firm. This shift in profits compensates the domestic government at the margin for higher environmental damages.[9]

The result is illustrated in Figure 6.1.[10] The action by the domestic government shifts the domestic firm's reaction function to the right, moving the equilibrium from point 1 to point 2. This result is similar to Spencer and Brander's (1983) analysis of R & D subsidies and Brander and Spencer's (1985) analysis of export subsidies. However, in contrast to these analyses of industrial policy, strategic environmental policy does not make the domestic firm a de facto Stackelberg leader — that is, if the domestic firm were a Stackelberg leader, it would produce an even greater level of output than is optimal under the SOS.

---

[9] If the foreign industry is perfectly competitive, then the domestic firm faces a downward sloping demand curve that reflects both third-country demand and the response of foreign firms to price changes, and it is optimal for the domestic government to impose the EOS. Similarly, if the domestic industry faces a horizontal demand curve, the domestic government has no incentive to behave strategically. While this last observation is invariant to the structure of the domestic industry, the former is not; see Section D.

[10] The figure assumes that domestic and foreign output are homogeneous products with linear demand; marginal production costs are constant; the first partial derivatives of $A^i(q^i, e^i)$ are linear; and marginal damages are linear in emission levels.

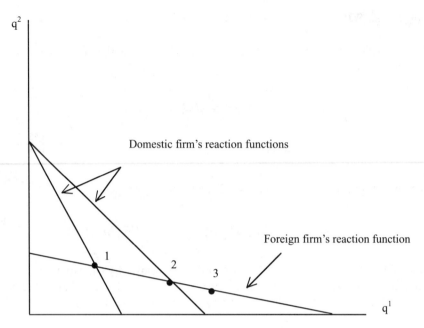

FIGURE 6.1 Strategic environmental policy.

Proposition 2. The SOS does not make the domestic firm a de facto Stackelberg leader. If the domestic firm is a Stackelberg leader, then the domestic government does not have an incentive to behave strategically. Domestic net benefits are higher when the domestic firm is a Stackelberg leader and the domestic government chooses the EOS than when the domestic firm plays Cournot and the domestic government chooses the SOS.

Proof. If the domestic firm is a Stackelberg leader, the foreign firm a follower, and neither government behaves strategically, then the equilibrium is defined by

$$\pi_1^1 + \pi_2^1 dq^2/dq^1 = 0, \qquad \text{with } NB_e^1 = 0 \tag{6.17}$$

plus Equations (6.2) and (6.7) for $i = 2$. If the domestic government imposes the SOS, the foreign government imposes the EOS, and both firms compete Cournot, then the equilibrium is defined by

$$\pi_2^1 dq^2/dq^1 + NB_e^1 de^1/dq^1 = 0, \qquad \text{with } \pi_1^1 = 0 \tag{6.18}$$

plus Equations (6.2) and (6.7) for $i = 2$. The first part of the proposition is proved by noting that (Equation 6.17) and (Equation 6.18) are different.

To prove the second part of the proposition, notice that if the domestic firm were a Stackelberg leader, and the domestic government behaved strategically, the equilibrium would be defined by

$$\left(\pi_1^1 + \pi_2^1 dq^2/dq^1\right) dq^1/de^1 + NB_e^1 = 0, \qquad \text{with } \pi_1^1 + \pi_2^1 dq^2/dq^1 = 0 \tag{6.19}$$

plus Equations (6.2) and (6.7) for $i = 2$. Equation (6.19) implies $NB_e^1 = 0$ which defines the EOS.

To prove the final point of the proposition, suppose the domestic government could choose both q^1 and e^1. Then the equilibrium would be defined by

$$\pi_1^1 + \pi_2^1 dq^2/dq^1, \qquad \text{with } NB_e^1 = 0 \tag{6.20}$$

plus Equations (6.2) and (6.7) for $i = 2$. But (Equation 6.20) is identical to (Equation 6.17), which is different from (Equation 6.18).

In the Spencer and Brander (1983) and Brander and Spencer (1985) models, the optimal export and R & D subsidies move the industry equilibrium to the Stackelberg point: "The government does for the domestic firm what the firm cannot do for itself" (Spencer and Brander 1983, p. 714). Strategic environmental policy moves the industry equilibrium *closer* to the Stackelberg point, but it is not optimal for the domestic government to reach that point. In Figure 6.1, point 2 is the equilibrium where the domestic government imposes the SOS, while point 3 is the equilibrium where the domestic firm is the Stackelberg leader and its government imposes the EOS. The reason for this result is that weaker environmental standards offer an "implicit subsidy," but one that is costly; society suffers because the subsidy worsens domestic pollution. Export and R & D subsidies are not costly to society; they simply transfer resources from the domestic government to domestic industry. Indeed, it can be shown that if the domestic government could subsidize production or exports directly, it would have no incentive to depart from the EOS.[11]

C. BILATERAL STRATEGIC ENVIRONMENTAL POLICY

If the domestic firm's government has an incentive to behave strategically, so would the foreign firm's. Suppose, then, that each government chooses its environmental standard, taking as given the foreign government's environmental standard, but recognizing how this choice will affect the industry equilibrium. Then we have

Proposition 3. The Nash environmental policy equilibrium involves both exporting countries choosing SOSs that are weaker than their respective EOSs.

Proof. The Nash environmental policy equilibrium is defined by (Equation 6.2), (Equation 6.11), which defines the domestic government's SOS, and

$$\pi_1^2 \left(dq^1 / dq^2 \right) / \left(dq^2 / de^2 \right) + NB_e^2 = 0 \tag{6.21}$$

which defines the foreign government's SOS. Totally differentiating (Equation 6.2), setting $de^2 = 0$ and substituting yields

$$dq^2 / dq^1 = -\pi_{21}^2 / \pi_{22}^2 \qquad \text{and} \qquad dq^1 / de^1 = \left(\pi_{22}^2 \pi_{1e}^1 \right) / \left(\pi_{12}^1 \pi_{21}^2 - \pi_{11}^1 \pi_{22}^2 \right) \tag{6.22}$$

Substituting (Equation 6.22) into (Equation 6.11) and repeating the process for the foreign country yields

$$NB_e^i = \left(NB_j^i NB_{ji}^i NB_{ie}^i \right) / \left(NB_{ij}^i NB_{ji}^i - NB_{ii}^i NB_{jj}^j \right) \tag{6.23}$$

Hence, the Nash environmental policy equilibrium is the solution to (Equation 6.23) and (Equation 6.2). It is easy to verify that our assumptions imply $NB_e^i < 0$. Hence, Equation (6.8) implies that in equilibrium the e^i are greater than their EOS levels (for q^i given).

The consequence of bilateral strategic environmental policy is shown in Figure 6.2. Point 1 is the Nash industry equilibrium, assuming governments impose their EOSs. If both governments

---

[11] If the domestic government could choose both e^1 and a domestic export/production subsidy s^1, the optimal values must satisfy $(\pi_1^1 - s + \pi_2^1 dq^2/dq^1)dq^1/ds^1 = 0$ and $(\pi_1^1 - s + \pi_2^1 dq^2/dq^1)dq^1/de^1 + NB_e^1 = 0$. It is easy to show that $dq^1/ds^1 > 0$. Hence, the term in parentheses must equal zero. That means NB_e^1 must equal zero, which defines the EOS.

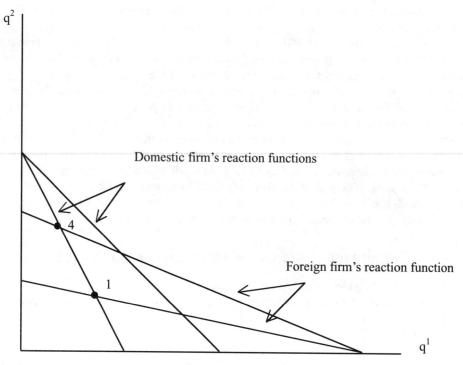

FIGURE 6.2 Nash environmental policy equilibria.

impose their SOSs, the reaction curves tilt outwards, and a new equilibrium is reached at point 4. At this new equilibrium, both countries earn lower net benefits compared with point 1, but neither has an incentive to impose its EOS unilaterally.

D. STRATEGIC ENVIRONMENTAL POLICY FOR OLIGOPOLISTIC INDUSTRIES

Would these results hold for a domestic oligopoly? It seems that they might not. If the foreign industry were perfectly competitive, and the domestic industry oligopolistic, domestic profits could be increased if domestic output were lowered. Hence, the domestic government would have an incentive to impose strong environmental standards unilaterally. When there is imperfectly competitive foreign competition, this incentive must be balanced against the incentive to impose weak environmental standards in order to shift profits away from foreign rivals.

To formalize these points, suppose the domestic industry consists of n producers, the kth of which has profit function

$$\pi^k = R^k\left(\boldsymbol{q}, q^2\right) - C^k\left(q^k\right) - A^k\left(q^k, e^k\right)$$

where $\boldsymbol{q}$ is a vector of n output levels, q^k is k's output and e^k is k's emission standard. For simplicity, it can be assumed that the foreign producer's industry is a monopoly, and produces output q^2 as before. If the foreign industry were oligopolistic, the results would not change substantively; all that matters is that the foreign rival industry respond to changes in the output of the domestic industry.[12]

The domestic government has net benefit function

---

[12] If the foreign industry is perfectly competitive and the domestic industry faces a downward-sloping demand curve, the SOS will be stronger than the EOS. Under these conditions, profits cannot be shifted away from the foreign industry, but strong standards limit domestic production and hence increase domestic industry profits.

$$NB^1 = \sum_{k=1}^{n} \pi^k - D^1(\mathbf{e})$$

where e is a vector of domestic emission levels. The EOS for firm k is given by

$$\frac{-\partial A^k}{\partial e^k} - \frac{\partial D^1}{\partial e^k} = 0$$

The SOS must satisfy

$$\frac{-\partial A^k}{\partial e^k} - \frac{\partial D^1}{\partial e^k} + \sum_{l=1}^{n} \frac{\partial \pi^l}{\partial q^2} \cdot \frac{dq^2}{dq^k} \cdot \frac{dq^k}{de^k} + \sum_{l=1}^{n} \frac{\partial \pi^l}{\partial q^k} \cdot \frac{dq^k}{de^k} = 0 \qquad (6.24)$$

The second term in (Equation 6.24) indicates the contribution to net benefits of the shift in profits from the foreign to the domestic industry. If reaction functions are downward sloping and all goods are substitutes, this term will be positive. What is substantially different about (Equation 6.24) is the last term, which captures the effect of changes in k's emission standard on the profitability of the other domestic firms. If one domestic firm's output is a substitute for another's, this term will be negative. Hence, (Equation 6.24) implies:

Proposition 4. Where the domestic industry is oligopolistic, the SOSs may be either weaker or stronger than the EOSs.
Proposition 4 holds for both unilateral policies and the Nash environmental policy equilibrium.

E. STRATEGIC ENVIRONMENTAL POLICY UNDER PRICE COMPETITION

It is well known that if firms compete in prices rather than quantities, the strategically optimal policy can be reversed (Bulow et al., 1985). In the case of strategic trade policy, if an export subsidy is optimal under quantity competition, an export tax is optimal under price competition (Eaton and Grossman, 1986). A similar qualitative conclusion emerges from the analysis of strategic environmental standards.

Let the prices chosen by the domestic and foreign firms be denoted by p^1 and p^2, respectively. At these prices, firm i sells $S^i(p^1, p^2)$. i's profit function can be written

$$\pi^i = p^i S^i(p^1, p^2) - C^i(S^i(p^1, p^2)) - A^i(S^i(p^1, p^2), e^i)$$

The EOS is unchanged from before; the domestic country's EOS satisfies Equation (6.7) under Bertrand competition. However, the rule for choosing the SOS now becomes (denoting $\partial \pi^1 / \partial \pi^1$ by π_1^1, etc.)

$$\pi_1^1 dp^1 / de^1 + \pi_2^1 (dp^2 / dp^1)(dp^1 / de^1) + NB_e^1 = 0$$

The domestic firm chooses price p^1 such that $\pi_1^1 = 0$. If the two goods are substitutes, $\pi_2^1 > 0$. Assuming that the reaction functions are upward sloping (i.e., $\pi_{12}^1 > 0$), $dp^2 / dp^1 > 0$ if the stability conditions for Bertrand competition are satisfied. These conditions also imply $dp^1 / de^1 < 0$ if the emission standard is binding. Hence, we have $NB_e^1 > 0$, or

Proposition 5. Under Bertrand competition, the unilateral SOS is stronger than the EOS.

Setting a strong emission standard raises marginal costs, thus making it credible for the domestic firm to increase price. Unlike the Cournot case, the foreign firm welcomes this unilateral policy; its profits *rise* as a consequence.

Proposition 2 can be shown to hold for Bertrand competition as well as Cournot competition. However, in contrast to quantity competition, the leader under Bertrand competition may suffer a reduction in profits.[13]

Using the same procedure as in the proof to Proposition 3, it is easy to show that if both countries behave strategically, the Nash equilibrium satisfies

$$NB_e^i = \frac{-NB_j^i NB_{ji}^j NB_{ie}^i}{\left(NB_{ii}^i NB_{jj}^j - NB_{ij}^i NB_{ji}^j \right)} \tag{6.25}$$

If the two goods are substitutes, $NB_j^i > 0$. If the reaction curves are upward sloping, $NB_j^i > 0$. Under our assumptions, $NB_{ie}^i < 0$. The denominators of (Equation 6.25) are positive if the stability conditions are satisfied. Hence, $NB_e^i > 0$.

Proposition 6. Under Bertrand competition, the Nash environmental policy equilibrium involves both countries choosing SOSs that are stronger than their respective EOSs

Finally, note that if the domestic industry is oligopolistic, all the results presented in this section are reinforced. A strong SOS induces domestic firms, as well as foreign rivals, to raise price, thereby increasing domestic industry profits.

F. CONCLUSIONS AND POLICY IMPLICATIONS

This chapter has shown that if the domestic industry consists of one firm, the foreign industry is imperfectly competitive, and competition in international markets is Cournot, then the domestic government has an incentive to impose a weak environmental standard, where "weak" means that the marginal damage from pollution exceeds the marginal cost of abatement. While this result confirms fears that governments may want to weaken standards to improve competitiveness, the chapter shows that environmental policy is inferior to industrial policy as an instrument for improving competitiveness. A production subsidy, for example, would effectively make the domestic firm a Stackelberg leader at no cost to the domestic economy. Hence, governments would only want to interfere in environmental policy if other, more efficient policy interventions could not be made. Furthermore, the conclusion that weak standards improve competitiveness is not robust. If the domestic industry consists of more than one firm, the incentive to weaken standards is reduced, and may even be reversed. If firms compete in prices (Bertrand) rather than quantities (Cournot), countries have an incentive to impose strong standards, where "strong" means that the marginal damage from pollution is less than the marginal cost of abatement, whether the domestic industry is a monopoly or oligopoly. Taken together, the results suggest that incentives to distort environmental standards will typically not be significant and may be of the opposite direction than is usually assumed.

REFERENCES

Anderson, K. and Blackhurst, R., Eds., *The Greening of World Trade Issues*, Harvester Wheatsheaf, London.
Brander, J. A. and Spencer, B. J., Export subsidies and international market share rivalry, *J. Int. Econ.*, 18, 83, 1985.

---

[13] See Tirole (1988, pp. 330–331).

Bulow, J. I., Geanakoplos, J. D., Klemperer, P. D., Multimarket oligopoly: strategic substitutes and complements, *J. Pol. Econ.* 93, 488, 1985.

Copeland, B. R., Strategic enhancement and destruction of fisheries and the environment in the presence of international externalities, *J. Environ. Econ. and Manage.*, 19, 212, 1990a.

Copeland, B. R., Endogenous spillovers, R & D, and transboundary pollution, mimeo, Department of Economics, University of British Columbia, 1990b.

Copeland, B. R., Taxes versus standards to control pollution in imperfectly competitive markets, mimeo, Department of Economics, University of British Columbia, 1992.

Cropper, M. L. and Oates, W. E., Environmental economics: a survey, *J. Econ. Lit.*, 30, 675, 1992.

Dean, J. M., Trade and the environment: A survey of the literature, in *International Trade and the Environment*, Low, P., Ed., The World Bank, Washington, DC, 1992, 15–28.

Eaton, J. and Grossman, G. M., Optimal trade and industrial policy under oligopoly, *Q. J. Econ.*, 101, 383, 1986.

GATT Secretariat, *International Trade 90–91*, GATT, Geneva, 1992.

Grimmett, J. J., Environmental Regulation and the GATT, Report for Congress, Congressional Research Service, 1991.

Grossman, G. M. and Krueger, A. B., Environmental impacts of a North American free trade agreement, Discussion Paper 644, Centre for Economic Policy Research, 1992.

Low, P., International trade and the environment, Discussion Paper 159, The World Bank, Washington, DC, 1992.

Low, P. and Yeats, A., Do "dirty" industries migrate?, in *International Trade and the Environment*, Low, P., Ed., The World Bank, Washington, DC, 1992, 89–103.

Lucas, R. E. B., Wheeler, D., and Hettige, H., Economic development, environmental regulation and the international migration of toxic industrial pollution: 1960–1988, in *International Trade and the Environment*, Low, P., Ed., The World Bank, Washington, DC, 1992, 67–86.

Markusen, J. R., Morey, E. R., and Olewiler, N., Environmental policy when market structure and plant locations are endogenous, Working Paper 3671, National Bureau of Economic Research, 1991.

Markusen, J. R., Morey, E. R., and Olewiler, N., Noncooperative equilibria in regional environmental policies when plant locations are endogenous, Working Paper 4051, National Bureau of Economic Research, 1992.

Oates, W. E., and Schwab, R. M., Economic competition among jurisdictions: efficiency enhancing or distortion inducing?, *J. Pub. Econ.* 35, 333, 1988.

Pearce, D., Should GATT be reformed for environmental reasons?, CSERGE Discussion Paper GEC 92–06, University of East Anglia and University College, London, 1992.

Rauscher, M., Foreign trade and the environment, in *Environmental Scarcity: The International Dimension*, Siebert, H., Ed., J. C. B. Mohr, Tübingen, 1991.

Reinstein, R. A., Trade and environment, mimeo, U.S. State Department, 1991.

Spencer, B. J. and Brander, J. A., International R & D rivalry and industrial strategy, *Rev. Econ. Stud.*, 50, 707, 1983.

Tirole, J., *The Theory of Industrial Organization*, MIT Press, Cambridge, 1988.

Ulph, A., The choice of environmental policy instruments and strategic international trade, in *Conflicts and Cooperation in Managing Environmental Resources*, Pethig, R., Ed., Springer–Verlag, Berlin, 1992.

Whalley, J., The interface between environmental and trade policies, *Econ. J.,* 101, 100, 1991.

7 North–South Trade and the Global Environment

Graciela Chichilnisky[1]

Differences in property rights create a motive for trade among otherwise identical regions. Two regions with identical technologies, endowments, and preferences will trade if one, the South, has ill-defined property rights on environmental resources. Trade with a region with well-defined property rights transmits and enlarges the problem of the commons: the North overconsumes underpriced resource-intensive products imported form the South. This occurs even though trade equalizes all prices, of goods and factors, worldwide. Taxing the use of resources in the South is unreliable as it can lead to more overextraction. Property-rights policies may be more effective. *JEL* classifications: (A13, F10, F02, K11, O10, Q20)

A. INTRODUCTION

Why has the global environment emerged as a North–South issue? There is wide-spread concern about international problems such as acid rain, global warming, biodiversity, and the preservation of the world's remaining rain forests. In June 1992, one hundred nations agreed at Rio de Janeiro to consider a treaty linking environmental policy to economic issues of interest to industrial and developing countries, such as the remission of international sovereign debt and transfer of technology.

In order to develop adequate environmental policies one needs to understand the connection between markets and the environment. Why do developing countries tend to specialize in the production and the export of goods which deplete environmental resources such as rain forests (see Chichilnisky and Geoffrey Heal, 1991)? Do they have a comparative advantage in "dirty industries," and if so, does efficiency dictate that this advantage should be exploited? Is it possible to protect resources without interfering with free markets? Are trade policies based on traditional comparative advantages compatible with environmental preservation?

This chapter proposes answers to these questions. It does so by studying patterns of North–South trade in a world economy where the North has better-defined property rights for environmental resources than the South. Property rights have been neglected in the literature on the economics of environment and trade, although they have been recognized as important in other areas of resource allocation (Coase, 1960; Demsetz, 1967; Cohen and Weitzman, 1975). The chapter analyzes the interactions between property rights and international trade. It considers a trade model with two countries (North and South), two goods, and two factors, which extends that of Chichilnisky (1981, 1985, 1986; Chichilnisky and Heal, 1987). The environment, which is one of the factors of production, is owned as unregulated common property in the South, and as private property in the North.

Section D considers a general completely symmetric case: a world economy consisting of two identical countries, both with the same inputs and outputs, and with the same endowments, technologies, and preferences. The two countries engage in free trade in unregulated and competitive

[1] Research support from the U.S. National Science Foundation Grant No. 92-16028, Stanford Institute for Theoretical Economics (SITE), and Monti di Paschi di Siena is gratefully acknowledged. This chapter appeared as Technical Report No. 31 of SITE, in October 1991. I thank Joshua Gans, Geoffrey Heal, Christopher Stone, Wallace Oates, and anonymous referees for helpful comments and suggestions.

markets. The countries differ only in the pattern of ownership of an environmental resource used as an input to production. I consider this case to demonstrate that lack of property rights alone can create trade, and that trade itself can exacerbate the common-property problem. No trade is necessary for efficiency when the two countries are identical, yet trade occurs when they have different property-rights regimes. In this context, two general propositions are established. First, the country with ill-defined property rights overuses the environment as an input to production, and these ill-defined property rights by themselves create a motive for trade between two otherwise identical countries. Second, for the country with poorly defined property rights, trade with a country with well-defined property rights increases the overuse of resources and makes the misallocation worse, transmitting it to the entire world economy. Trade equalizes the prices of traded goods and of factors worldwide, but this does not improve resource allocation. In the resulting world economy, resources are underpriced; there is overproduction by one country and overconsumption by the other. These results have been extended to a dynamic context in Chichilnisky (1993c).

Section E explores more specific cases. Its purpose is to evaluate policies to check environmental overuse: taxes and property-rights policies. Here, in contrast with Section D, a realistic asymmetry between the North and the South is allowed. The South's resources are produced now either using capital or using labor from a subsistence sector. This labor is not directly traded in the market or employed in other sectors. Subsistence labor is only engaged in the extraction of the environmental resource, which is traded in exchange for capital-intensive goods. In this context it is shown that taxes on the use of environmental resources in the South are generally unreliable at deterring overextraction. Taxes can force lower-income harvesters to work harder and extract more resources to meet their consumption needs. Taxes can therefore lead to more rather than less extraction of the resource. Therefore, property-rights policies may be preferable in the South as a way of correcting environmental overuse. Examples of such policies are discussed.

Property rights can affect market behavior in many ways. Here the focus is on their impact on the supply of resources; these are price-dependent, with their supply curves derived rigorously from micro foundations. It is established that with ill-defined property rights the supply of resources is more price-responsive than it is when property rights are well defined. This price responsiveness is crucial in determining the patterns of international trade; at each price the South offers more resources than the North, leading to apparent comparative advantages in resource-intensive products. This parallels the results of the original North–South model (Chichilnisky, 1981, 1985, 1986), where the price responsiveness of labor supply, called there abundance of labor, played an equally important role in determining the terms of trade and the welfare results from exports of labor-intensive products.

These results offer a new perspective on a current debate, initiated in 1992 by Lawrence Summers, a World Bank economist, about whether developing countries have a comparative advantage in dirty industries (see e.g., *The Economist*, February 8, 1992, Vol. 322, p. 66). If so, the argument goes, is it not efficient that they specialize in dirty industries and environmentally intensive production?

One response to this is that the apparent comparative advantages may not be actual comparative advantages, an issue that this chapter addresses rigorously. They may derive neither from a relative abundance of resources nor from differences in productivity or preferences, not even from lower factor prices, but rather from historical and institutional factors: the lack of property rights for a common-property resource. In this context the South produces and exports environmentally intensive goods to a greater degree than is efficient, and at prices that are below social costs. This happens even if all factor prices are equal across the world, all markets are competitive, and the two regions have identical factor endowments, preferences, and technologies. Under those conditions the trade patterns which emerge are inefficient for the world economy as a whole, and for the developing countries themselves. Developing countries are not made better off by specializing in dirty industries, nor is the world better off if they do.

A. EMPIRICAL MOTIVATIONS: PROPERTY RIGHTS, TRADE PATTERNS, AND TAXES

The problems described in this chapter appear when societies that are still in transition between agricultural and industrialized economies trade with societies already industrialized. Many traditional societies had well developed systems for inducing cooperative outcomes in the use of shared resources. Laws to protect the citizens' property rights in running water were in operation in the U.K. in the Middle Ages. Japan had well-developed systems for the management of traditional communal lands (*Iriaichi*). Other examples are the communal-field agriculture in the Andes and in medieval England, and the successful sea-tenure systems in Bahia, Brazil, before the arrival of outsiders (see Bromley, 1992). These traditional systems, however, appear to lapse in the period of transition between agricultural and industrial economies.

Today many environmental resources are unregulated common property in developing countries. Examples are rain forests, which are used for timber or destroyed to give way to the production and export of cash crops such as coffee, sugar, and palm oil. Other examples include grazing land, fisheries, and aquifers, which by the nature of things must usually be shared property even when the land covering the aquifer is privately owned (see Dasgupta, 1982). These are common-property resources whose ownership is shared with future generations. They are typically used as inputs to the production of goods that are traded internationally.

Recent studies show that 90 percent of all tropical deforestation is for the agricultural use of forests, particularly for the international market (Binkley and Vincent, 1990; Amelung, 1991; Barbier et al., 1991; Hyde and Newman, 1991). The Korup National Park between Cameroon and Nigeria, at 60 million years old one of the oldest rain forests in the world and one of the richest in biodiversity, is exploited as an unregulated common-property resource for the production of palm oil, trapping, and other forest products sold in the international market (Ruitenbeck, 1990). So is the Amazon basin, which is cleared and used as a source of land for the production of cash crops such as soybeans and coffee for the international market.

In the now industrialized countries, communal land was frequently observed prior to industrialization. Industrialization in England was preceded by the "enclosure" (privatization) of common lands (Cohen and Weitzman, 1975). Now, however, industrial countries have much better defined property rights for their resources than do developing countries. The U.S. has property-rights regimes for petroleum. These include laws to prevent the overexploitation of common-property resources such as the Conally "Hot Oil Act" of 1936 and "unitization" laws (McDonald, 1971). Water, however, is still treated as common property in parts of Texas and California, leading to misallocation. Japan is well known for its protection of property rights in environmental resources, including even sunlight. Germany recently initiated a parallel system of national accounts that records the depreciation of environmental assets, effectively treating the accounting of national property on the same basis as that of private property. Indeed, the proposal to update all national accounting systems so that they record the depreciation of environmental assets (see Dasgupta and Heal, 1979) is a move toward treating national property on the same accounting basis as private property.

Although today's differences in property rights between industrial and developing countries are easily observed, only tentative explanations can be offered. It is known that the overexploitation of "the commons" emerges from noncooperative situations, or overuse by "free riders" (Dasgupta and Heal, 1979). Cooperative outcomes in the use of communal property resources, such as shared bodies of water, fisheries, and forests, are achieved through the enforcement of punishment and rewards within a stable group, the members of which share a common land site for many generations. One observes a "repeated game" that enforces cooperative outcomes without the need for formally defined individual property rights (Bromley, 1992).

In the period of industrialization, these age-old institutions cease to work, because small and stable populations become large and transient, so that cooperative solutions are more difficult to enforce. This may be why those societies that have completed the process of industrialization rely instead on formal property rights rather than tradition and custom. The problems described in this chapter appear because the South is still in transition, but it trades with already industrialized economies.

Patterns of trade in environmental resources can defy common wisdom. Small countries such as Honduras, with scarce forest resources, export wood to the U.S., which has some of the largest forests in the world. Honduras's hardwood (e.g., mahogany) forests have been recently nationalized and are treated as unregulated common property, with no accounting of the depreciation of the forest as an asset, while the United States has better-defined property rights. Currently, about two-thirds of Latin American exports are resources, and an even larger proportion holds for Africa. Many of these are produced from common-property sources: Korup National Park and the Amazon are valuable biodiversity reservoirs which are being destroyed in part for the production of export cash crops such as palm oil, soybeans, and coffee. These crops are exported to industrial countries with much higher agricultural productivity and with larger endowments of agricultural land. The puzzle is how to explain such patterns of trade. The exporters do not have a technical advantage in the sense of Ricardo, since their labor is not more productive, and it is unlikely that they have a comparative advantage in the sense of Heckscher–Ohlin, either.

These issues are analyzed in Section D, which establishes that, due to differences in property rights, the countries will trade following the empirically observed patterns of trade just discussed. Since under these conditions markets do not allocate resources efficiently, Section E examines policies to check environmental overuse: taxes and changes in property rights. In Section E, a realistic difference between the North and the South is introduced. In the North, resources are extracted using capital. In the South, resources are extracted either by using capital as in the North, or by using labor from a subsistence sector consisting of workers for whom there is no formal labor market. These workers extract environmental resources from the common-property pool such as a forest. Their labor is not traded directly: they use it to extract resources which are traded for capital-intensive goods at market prices. Workers employed in this fashion exist today in parts of Paraguay, Argentina, and Brazil; they are employed in a so-called "piecemeal" fashion, usually in environmentally related sectors such as timber and agriculture, and are unemployable in other sectors because of their lack of formal education. Workers who extract wood and trappings from common-property land, such as the Korup rain forest in Nigeria, appear to conform to this pattern (see Ruitenbeck, 1990), and so do those who extract rubber in the Amazon. Such patterns also appear within firms: much of the Amazon's land is treated as unregulated common property by firms which buy forest products from their employees in a piecemeal fashion, at market prices. This specification also has some points in common with the formulation of abundant labor in Chichilnisky (1981, 1986), and it fits well with the results of empirical studies of subsistence labor in agriculture in Latin American countries, such as Schweigert's (1993) research on Guatemalan agriculture.

The first policy considered in Section E is a tax that the South levies on the use of environmental resources, the revenues of which are spent on the capital-intensive good. The intention is to raise the producer's costs and to deter its use of environmental resources. In this context I show that taxes will not always work as intended. As taxes decrease the demand for resources, the harvesters receive lower prices for their products. This can force the harvesters to work harder and lead them to extract more rather than less of the resources, a possibility that is explored rigorously in this chapter (Proposition 2; see also World Bank, 1992). For this reason, taxes in the South are generally ineffective and can be unreliable: when the workers' demand for marketed goods is relatively inelastic, taxes can lead to more extraction of resources, defeating their original purpose (Proposition 3).

C. THE NORTH–SOUTH MODEL

Following Chichilnisky (1981, 1985, 1986), consider a model with two goods, two inputs, and two countries, similar to that of Heckscher–Ohlin, except that here the supplies of inputs are price-dependent. For example, the supply of resources is $E^s = E^s(p_E)$, where p_E is the price of the resource E. The model for one region is as follows. Capital (K) and the environmental resource (E) are used to produce two goods, A and B: one good, B, is more intensive in the use of the environmental resources than the other, A, which is more capital-intensive.[2] Production exhibits constant returns to scale; a concave utility function $U(A, B)$ is postulated. Both E and K are supplied by continuous supply functions $E = E(p_E)$ and $K = K(r)$ where p_E is the price of the environmental resource, and r is the rental of capital. E is extracted from a resource pool using an input x which has opportunity cost q; this cost q is endogenously derived in Proposition 2 and Appendix C, as are all other prices and quantities, in a general-equilibrium fashion. To show that the results are robust under different model specifications, the input used to extract E is taken to be quite general in this model: it can be capital (as in Section D, Theorem 1), subsistence labor, or either of the two (as in Section E, Propositions 2 and 3). The supply of the environmental resource E is formally derived in Proposition 1, where it is shown to vary systematically with the property rights of the resource pool. Since endowments, technologies, and preferences are defined, all ingredients of a general-equilibrium model are provided. A one-region competitive equilibrium is a price vector $\mathbf{p}^*$ at which each of the four markets (for A, B, E, and K) clears.

The two-region model (North–South) is defined as usual by considering two one-region models together and, to allow for international trade, by relaxing the hypothesis that each commodity market (for A and B) clears in each region to require instead that the world markets for A and B must clear. Input markets clear in each country because, as is standard in international trade, the factors of production (K and E) are only traded domestically. Appendix B provides a set of equations and computes an equilibrium of the North–South model, which is shown to be unique in Appendix C.

The model differs from earlier versions of the North–South model (Chichilnisky, 1981, 1985, 1986) in that in those versions all goods were privately owned, and the supply functions of the endowments were exogenously given. Here, instead, the supply of input E is derived explicitly from microeconomic behavior and is shown to depend on the property-rights regime. There is, however, an unexpected similarity with the original North–South model: the crucial role played by the responsiveness of resource supplies to their price. With common-property regimes, more is supplied at any given price than is supplied with private-property regimes (Proposition 1). Since at each price the quantity supplied under common property exceeds that supplied under private property, resources appear to be more "abundant" with common property. A similar responsiveness of supplies, but of labor instead of resources, appears as a crucial parameter in Chichilnisky (1981, 1986): it measures "labor abundance" in the South and determines whether or not the country will benefit from increasing its exports of labor-intensive goods.

C. PROPERTY RIGHTS AND THE INTERNATIONAL SUPPLY OF RESOURCES

Property rights act on the market in many complex ways. A simple but critical way in which property rights enter in a general-equilibrium trade model is identified, and a transparent and a general explanation of how they determine the patterns of trade and the welfare of the traders is proposed. *The two regions are assumed to differ solely in their property rights for a pool of resources from which one input to production is extracted.* For example, the property rights on forests from which wood and pulp are produced are different in the North and the South.

How do property rights affect trade? Tracing the impact of property rights is nothing more and nothing less than a comparative-statics exercise: the comparison of a world equilibrium in which both regions have well-defined property rights with a second world equilibrium in which the South does not.[3] In principle, this could be a complex undertaking. However, when property rights involve

---

[2] This is a standard assumption in trade theory: it means that the production of good B requires a higher ratio of input K than does the production of good I. This is formalized in Appendix B.

[3] At the second world equilibrium prices are distorted, since one of the traders, the South, does not satisfy the marginal optimality conditions (Proposition 1 below and Appendix A); therefore although the first world equilibrium is a competitive equilibrium, the second is a general equilibrium but is not a competitive equilibrium.

solely an input of production as they do here, their effects can be summarized in a simple fashion: by their impact on resources supplied. Indeed, Proposition 1 below establishes that, for each price of the resource, more is supplied under unregulated common property than under private property. Therefore, the comparative-statics exercise need only compare a world where the two countries have the same supply curve for resources with another world where the South has a different supply curve than the North: it supplies more at each price.

In practice, therefore, one compares the equilibrium of a market with two identical traders to the equilibrium of another market in which the traders are identical *except* for their supply of resources. In the first equilibrium both countries have the identical *private-property supply curve* $E^p(p_E)$. In the second, the South has, instead, a *common-property supply curve* $E^C(p_E)$, which prevails with ill-defined property rights. The North has one supply curve throughout, which arises with well-defined property rights. Why are the two curves different in the South?

Under unregulated[4] common-property regimes the harvester's cost of extracting an additional unit of E is relatively low. It merely reflects the opportunity costs of the inputs used to extract the resources: in the case of a fishery, the costs incurred in catching; and in the case of an aquifer, the costs incurred in obtaining the water. These costs do not reflect the full impact on other users of one individual's use of the resource pool. In the case of the fishery or a forest, each unit extracted decreases the stock available to others and increases their extraction costs. This could eventually lead to the depletion of the stock. In a private-property regime matters are different. Externalities are fully internalized so that the cost to a harvester of extracting the resource reflects fully the costs this imposes on the extraction by others and could increase rapidly with the level of extraction (Dasgupta and Heal, 1979, Chapter 3).

In its simplest possible form, the argument is that with private property the efficient harvester equates the relative prices of inputs and outputs to the marginal productivity. Instead, with common property and with many producers, the relative prices are often equated to *average* productivity. If production has decreasing returns to scale, then average productivity is always larger than marginal productivity, so that at the same prices there is always an incentive to produce more under common-property regimes. This argument is actually valid much more generally: it is true for any finite number of harvesters and without invoking marginal and average costs; under general conditions it leads to the following proposition, which is rigorously established in Appendix A within a Nash equilibrium framework.

PROPOSITION 1: *The common-property supply curve for the resource lies below the private-property supply curve, so that under common-property regimes, more is supplied at any given price. Both supply curves are increasing functions of resource prices.*
(Proposition 1 is illustrated in Figure 7.1.)

It is worth emphasizing that this proposition does not depend on what input is used for extracting the resource: for example, the input used could be either capital or labor. Indeed, both inputs are considered in the chapter. The difference of supplies with private and common property regimes has substantial practical implications. It leads by itself to different concepts of comparative advantages and gains from trade. These are crucial to current policy because they clarify whether developing countries have a comparative advantage in environmentally intensive exports and, if so, whether such advantage should be exploited.

D. PROPERTY RIGHTS, COMPARATIVE ADVANTAGES, AND GAINS FROM TRADE

The completely symmetric case of the North–South model is utilized here because it shows very clearly the role of property rights in determining patterns of trade. The comparative advantages of

---

[4] I consider here unregulated common property, which is also called "open access"; it must be differentiated from regulated common property.

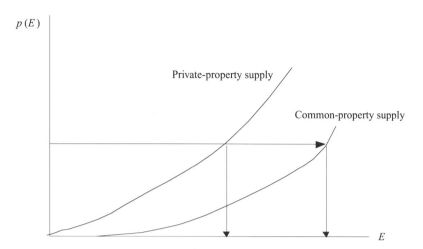

FIGURE 7.1 At each price, the common-property supply exceeds the private-property supply.

nations with price-dependent endowments are defined as follows: region S is said to have a *comparative advantage* in the production of good B which is intensive in the use of the input E, when for each price p_E the supply of E relative to that of K in region S is larger than the corresponding supplies in region N. *Actual comparative advantages* rely on the private-property supply curve for E, which internalizes all the externalities that each unit of extraction has on others. This is the supply curve corresponding to points in the economy's efficient frontier.

It is worth noting that this definition of comparative advantage, while rather natural, differs from classical ones. This must be the case because Ricardian comparative advantages cannot exist in a world where countries have identical technologies, as considered here. Nor can Heckscher–Ohlin comparative advantages be well defined here, since the endowments of factors are price-dependent, so at different prices the countries can exhibit different Heckscher–Ohlin advantages, measured as the relative endowments of production.

A correct understanding of the comparative advantages of developing countries is, as already pointed out, a substantial policy concern. To examine the matter it will prove useful to consider *apparent comparative advantages*. These are defined in the same manner as actual comparative advantages, but using for each price the supply curve corresponding to the prevailing property regimes in each region, which in the South corresponds to the case in which some externalities are not internalized. Apparent comparative advantages of the South reflect its institutional constraints, in this case inadequate property rights, which may hinder the attainment of full optimality. A market response by the South to these institutional constraints leads it to behave as if it has a comparative advantage even if it does not. When institutional constraints are binding, apparent comparative advantages prevail; observers could then attribute to developing countries a comparative advantage in exporting environmentally intensive products, which they do not actually have.

It must be noted that here the two regions are identical except for property rights, so that neither region has a comparative advantage over the other. However, Proposition 1 established that the South, with ill-defined property rights, exhibits an *apparent* comparative advantage in the production of the resource-intensive good. This is because at each price the South supplies more resources than does the North, so its resources appear to be "more abundant."

Theorem 1 below uses another useful concept, *gains from trade*, which are measured as usual by the increase of utility $U(A, B)$ associated with a move from autarky (i.e., where each country maximizes utility within its own production possibility set) to a world equilibrium. Again it is useful to differentiate between apparent and actual gains from trade in the South. *Apparent gains from trade* are computed by comparing welfare in a model where the production possibility sets emerge from common-property supply curves. *Actual gains from trade* are defined, instead, by using the private-property supply curve. The weaker the property rights in the South, namely, the less the production externalities are internalized, the larger will be the divergence between its common-property and private-property supply curves, and between the apparent and actual production possibilities.[5] Thus, the weaker the property rights in the South, the larger will be the divergence between its apparent and its actual gains from trade.

The following theorem shows that trade by a region with ill-defined property rights with another with well-defined rights leads to apparent comparative advantages when none exist, and to apparent gains but actual losses from trade. It is established for two identical regions with the same technologies and preferences, both using the same technology to extract resources E using capital K, $E = F(K)$; the initial endowment of capital $\overline{K}$ is the same in the two regions. The theorem compares two equilibria: one in which both countries have the same property rights (private property) over the resource pool, with another in which the South has common property. The model is formally defined in Appendix B; the theorem and its corollary are proved in Appendix C.

THEOREM 1: *Consider the North–South model in which both regions have identical technologies, the same homothetic preferences, and the same endowment of inputs $\overline{K}$, (a) The model as defined in Appendix B has at most one competitive equilibrium. (b) If the pool from which the environmental resource is extracted is unregulated common property in the South, then the South exhibits apparent comparative advantages in environmentally intensive goods even though neither region has any (actual) comparative advantage over the other. (c) At a world equilibrium the two regions trade, and the South exports environmentally intensive goods at a price that is below social cost. The equilibrium is not Pareto efficient. (d) Trade makes things worse, in the sense that the overuse of the resource increases as the South moves from autarky to trade. Furthermore, trade leads to lower resource prices worldwide. (e) The South shows apparent gains from trade, even though it has no actual gains from trade. It extracts more environmental resources, and it produces and exports more environmentally intensive goods than is Pareto efficient.*

Note that the environmental overuse described in Theorem 1 is induced by a competitive market response. There is no regulation in either country, and all markets are free and competitive.

COROLLARY 1: *Trade between the North and the South leads to the equalization of the prices of all traded goods and of all inputs of production, and in particular to the equalization of the price of environmental inputs in the two regions. However, prices for environmentally intensive goods are below social costs. At a competitive equilibrium the South uses more environmental resources and produces and exports more environmentally intensive goods than the North, and more than is Pareto optimal.*

This corollary calls attention to the possibility that the overexporting of environmentally intensive goods from developing countries, such as mineral by-products, wood products, or cash crops, may not be due to the fact that the production costs are lower in the South than in the North, as is sometimes thought. For example, under unregulated common-property regimes for environmental resources, a country such as Mexico would export environmentally intensive agricultural products, or products produced from dirty industries, even if its forest land, clean air, and machinery were as expensive as they are in the industrial countries and even if it had the same technologies and preferences. Equalizing factor prices through the international market does not resolve the problem of overuse of environmental resources.

E. ENVIRONMENTAL POLICIES

Since unregulated competitive markets do not lead to efficient allocations with ill-defined property rights, the task of this section of the chapter is to analyze policies that could correct the problem

---

[5] The North–South model has one parameter (α, a real number), which can be used to represent the degree of internalization of property rights (see Appendices A and B).

of environmental overuse, focusing first on taxes. Taxes will be considered in a general-equilibrium framework, tracing their full incidence on all prices and quantities worldwide. This approach captures the impact of taxes in an economy where income effects prevail, a case particularly relevant to developing countries.

A realistic asymmetry is introduced between the North and the South in this section. In the South the environmental resource E is extracted either using capital as in Section D above or, alternatively, by workers from a subsistence sector. These workers do not sell their labor directly in the marketplace. Instead, they use their labor to extract the resource, which they exchange for capital-intensive goods in the marketplace. This specification is less symmetric than that of Section D. It captures a stylized fact in developing countries where many markets, including labor markets, are less developed than in industrial countries; empirical motivation for this was provided in Section A. Proposition 3 will contrast the effects of taxes within these two alternative specifications of the process for extracting E in the North and the South. The economy of the North remains as in Section D.

1. Lower Resource Prices Can Lead to More Extraction: The Opportunity Cost of Subsistence Labor

Does the South extract more or less resources after taxes? This is studied rigorously in the next subsection and in Appendix C. Taxes alter extraction because they induce price changes. Here it is traced how this occurs. In particular, it is shown how resource extraction changes with the opportunity costs of the inputs that are used to extract the resources.

No matter what input is used (labor or capital), and no matter what the property-rights regime, in this model the supply of the resource E always increases when the relative price of outputs to inputs increases (Appendix A). However, in a market economy the cost of an input is its opportunity cost, and this typically changes along with the price of the resource itself. The relevant issue is therefore whether the opportunity cost of the inputs increases relative to that of the resource.

What are the opportunity costs of inputs? How are they affected by resource prices? There is a simple answer when capital is employed to extract the resource: the opportunity cost then equals the rental on capital r, and r varies across the equilibria of the model in a well-specified fashion: it is inversely related to the price of the resource [Appendix B, Equation (7.3)].

A difficulty arises when E is extracted instead by subsistence workers. Since by definition no formal market exists for subsistence labor, market wages, which would be the natural proxy for the opportunity cost of labor, are not available. It is nevertheless possible to deduce implicitly the opportunity cost q of subsistence workers' time, to explore how this changes with the prices of the produced goods, and to see how this affects the extraction of resources. Proposition 2 shows how. Its formulation opens up a whole range of issues, which are unobservable in a partial-equilibrium formulation, and which are useful for understanding the behavior of resource markets with low-income populations. It formalizes the intuition that a drop in the price of a resource can force lower-income harvesters to work harder and extract more resources in order to meet their consumption needs. Discussions about this phenomenon appear in the World Bank's 1992 *World Development Report*, without any formal analysis.

Assume that the harvester's endowments consist solely of their labor input ($\bar{x} = 24$ hours a day). A harvester trades his harvest of resources E for the good A at competitive market prices. Each harvester has strictly concave increasing utility function $u_i(A, \bar{x} - x_i)$, which increases in leisure ($\bar{x} - x_i$) and in consumption (A). The harvester's problem is

$$\max_{x_i}\left[u_i\left(A, \bar{x} - x_i\right)\right] \tag{7.1}$$

subject to

$$p_A A = p_E E^C(x_i).$$

For each p_A and p_E, the solution $E^C(x_i)$ is the optimal quantity extracted by the harvester, where p_A and p_E are determined by the market in a general-equilibrium fashion. Proposition 1 (proved in Appendix A) explains the quantity of resource extracted in a different manner: it shows that it depends on resource prices p_E and on the opportunity cost of labor, q. In a general-equilibrium world these two ways of explaining extraction must tally: this observation allows one to derive rigorously the opportunity cost q of a worker's labor (which was previously treated as an exogenous parameter) endogenously, as a function of the prices p_A and p_E. This derivation proves that, as is intuitively obvious, q equals the rate of substitution between leisure and consumption of A [see Appendix C, Equation (7.20)]. Recall that the "terms of trade" between leisure and consumption are determined by the competitive market, because the market determines the price of the good A and the price of the good E, which is exchanged for A by the harvester. Therefore, the market determines implicitly the opportunity cost of subsistence labor, q.

Consider now a North–South model as defined in section B but with the addition indicated above. E is supplied by the type of harvesters just defined: they own only labor, extract and trade the resource E for the good A, and their utilities have elasticity of substitution $\sigma < 1$. This is called *the North–South model with a subsistence sector*. The following proposition studies the consequences of a drop in resource prices; its proof is in Appendix C.

PROPOSITION 2: *Under the assumptions made above, a drop in the price of resources leads to a drop in the opportunity cost of labor of the subsistence worker in the South. At the new, lower resource prices, more effort is applied by the harvester, and more resources are extracted.* (Proposition 2 is illustrated in Appendix C, Figure 7.3.)

Why do lower resource prices lead to more extraction? One explanation is that harvesters have relatively inelastic demand for the marketed good A. If so, as the relative price of the resource falls, the proportion of total expenditure spent on A rises, and the quantity of resources extracted must increase, because each unit of the resource brings now lower revenue. Here, this inelastic behavior of demand is not assumed; it is proven that it arises naturally from utility maximization on the part of the harvester when the harvester's utility u_i has an elasticity of substitution σ less than 1 (Appendix C). This result is useful in practice because the condition $\sigma < 1$ is plausible and rather general: for example, with constant-elasticity-of-substitution (CES) utilities, $\sigma < 1$ is always satisfied when the indifferences of u_i do not intersect the axis (see Varian, 1992, pp. 19–20), which is a standard case. Note, however, that the proof of Proposition 2 applies to a general class of concave utilities with $\sigma < 1$ and does not require that u_i be CES or any other functional form.

The result in Proposition 2 has useful implications under more general conditions as well. It highlights the weakness of any policy that leads to a drop in resource prices in lower-income countries: as prices drop, by continuity, harvesters with relatively low elasticity of substitution do not decrease their extraction very much. Lower resource prices are therefore ineffective at deterring overextraction.

The proposition also highlights the crucial role of the opportunity costs of labor in determining the extraction of resources. This is of practical interest because the opportunity cost of subsistence labor q could be regarded as a policy variable: this opportunity cost could be increased by providing productive job opportunities for subsistence workers outside the environmental sector. The general point is that any policy that increases the opportunity cost of subsistence labor could have beneficial environmental effects. Figure 7.2 illustrates how: at each resource price, a lower opportunity cost q leads to a shift in the supply curve for resources as illustrated (and established rigorously in Appendix C). Under the conditions, lower resource prices lead to an even lower opportunity cost of labor, and the ratio of output to input prices q/p_E actually *increases*. Because of this, at lower resource prices the subsistence harvesters extract more resources than they did before. In Figure 7.2 this is shown as a move from point C to D.

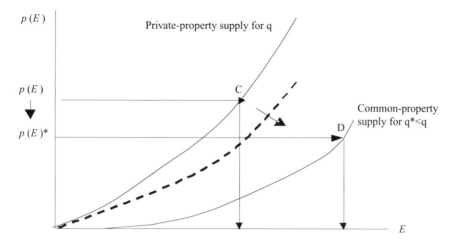

FIGURE 7.2 The common-property supply curve shifts with drops in the opportunity cost of the input used in extracting the resource, q.

Note: As established in Proposition 2, the equilibrium level of output of E increases as $p(E)$ drops to $p(E)^*$, because the corresponding opportunity cost drops to q^*.

In summary, in lower-income regions a drop in the price of resources can lead to more poverty, and this in turn can lead to more extraction. A policy that induces lower resource prices does not provide an effective way of checking environmental overuse. This provides a key to the following results on taxes.

2. TAXES IN THE SOUTH

The next proposition shows that the results of taxing the use of the resource could be just the opposite of what is expected if all prices are allowed to adjust to their new market-clearing levels: taxes could lead to more extraction. Intuitively this occurs because the resource is extracted by lower income subsistence workers whose demand for consumption goods is somewhat inelastic. The trade literature has not covered this case, which is implicit in the discussions of the World Bank's 1992 *World Development Report* with respect to the effects of poverty on the environment. Exploring this possibility requires tracing all impacts of taxes on prices, quantities, and income, a somewhat complex task.[6] Most of the international trade literature on taxation uses a partial-equilibrium framework. However, some of the most important effects of taxes in developing countries are income effects, and these cannot be examined properly in a partial-equilibrium.

For example, corrective tariffs and subsidies in a partial-equilibrium framework have been studied in Dixit and Norman (1980). However, as in all partial-equilibrium analysis, these results occur in a world where the terms of trade and all factor prices remain fixed throughout. Consider for example a standard policy of taxing the use of the common-property resource in order to restrict its use. When prices remain fixed, such a tax has the effect of deterring use. The tax increases the relative price of the resource, and this leads to less use of it as an input.

It is not always realistic to assume that prices remain fixed after taxes. Taxes alter income and therefore alter demand and hence prices. The classical works of Lerner (1936) and Metzler (1949) analyzed tariffs on traded goods which have general-equilibrium effects on the terms of trade of

---

[6] In their *Theory of International Trade* (Section 2, Policy Responses to Distortions and Constraints), Dixit and Norman (1980, p. 176) state: "Having examined the terms of trade and interpersonal distribution at some length, we now put them aside by assuming a *small one-consumer economy* [i.e., partial-equilibrium analysis]. This is done in order to highlight the new questions that are raised by the presence of further constraints on policy. It is not in principle difficult to combine everything into a grand model, but its algebraic complexity obscures all economic understanding."

the trading regions. We shall consider here, instead, a tariff which is levied domestically on the use of the resource E, which is a (nontraded) input to production.

The complexity of tracing the effects of taxes on all prices and quantities is alleviated by an unusual feature of the North–South model which makes it particularly useful for this purpose: it has an explicit analytic solution, which is obtained by solving a second-order equation in the terms of trade, from which all other endogenous variables are determined explicitly. This procedure was introduced in Chichilnisky (1981, 1986) and is the methodology that is used in establishing Proposition 3 below.

The relative price of exports and imports of the South, p_B/p_A ($= p_B$ since $p_A = 1$) are called the South's *terms of trade*.

PROPOSITION 3: *Consider the North–South model with a subsistence sector in the South which extracts the resource E from a common-property pool. The North has well defined property rights. Assume further that a unit tax T is imposed on the producers in the South for their use of the resource E as an input of production: all producers pay p_E + T per unit of the input E. Furthermore, assume that the tax revenues increase the domestic demand for the product that does not use the environmental resource E intensively, namely, for the good A. Then the tax always leads to:*

   (i) *a decrease in the price of the resource E,*
   (ii) *a drop in the South's terms of trade p_B,*
   (iii) *an increase in the extraction of the resource E in the South, and*
   (iv) *an increase in its exports of the environmentally intensive good B for some parameter values.*[7]

If the South produces E using capital, the tax leads also to (i), (ii), and (iv), but instead of (iii) to (iii)′ a drop in the extraction of the resource E in the South.
(See Appendix C for the proof.)

An intuitive explanation for this proposition is as follows. A unit tax T on the use of the resource E is collected from the producers who use E as an input. The tax proceeds TE are allocated to increasing the demand of the nonresource-intensive sector, A, giving the tax the best opportunity to work in the intended direction, namely, reducing the demand for resource-intensive products. The relative price of the environmentally intensive good B then drops; this is (ii). The drop in demand for B leads to a decreased demand, and a lower price, for the resource E; this is (i). Therefore, after taxes, the price of resources, E, is lower.

Consider first the case when the harvesters use subsistence labor as an input. If the resource E is extracted by the subsistence sector, then after taxes the price of E is lower and the supply of E is higher (this follows from Proposition 2); this is (iii).

The second case is when E is extracted using capital as an input. Since p_E is lower after taxes, the cost of capital r is higher: these two variables are always inversely related across equilibria [see Equations (7.3) in Appendix B]. Since the opportunity cost of the harvesters' labor is higher and output prices are lower after taxes, the supply of the environmental resource decreases; this is (iii)′.

Finally the exports of B increase after the increase in p_E if and only if price effects dominate; when income effects dominate then exports decrease (see also Chichilnisky, 1981, 1986); this is (iv).

F. CONCLUSIONS: PROPERTY RIGHTS AND NORTH–SOUTH TRADE

Differences in property-rights regimes for environmental resources can account for some puzzling aspects of the patterns of trade between the North and the South. The problems analyzed in this chapter arise when societies, which are in transition from agricultural to industrial economies, such

---

[7] Formally, exports increase if and only if the condition $c_2/D < 2\,p_E p_B$ is satisfied, a condition that is shown to be satisfied precisely when "income effects dominate" (see the penultimate paragraph of Appendix C).

as the South today, trade with already industrialized societies. The South exports environmentally intensive goods even if it is not well endowed with them. The South overproduces, and the North overconsumes, even if trade equalizes all traded goods and all factor prices worldwide.

Several environmental policies have been discussed. Section E examined tax policies in the South and established that they are not always reliable. When environmental resources are extracted using subsistence labor, taxes can lead to more extraction. It has been shown that anything that decreases the price of the resource can increase poverty and lead to more extraction, an observation which was made in the World Bank's 1992 *World Development Report*. Taxes can have this effect because they can decrease the demand for environmental goods, which are a main source of income for the subsistence sector.

A main argument in favor of property-rights policies is that once these have been implemented, no market intervention is needed. Consider, for example, any policy which improves the property of Amazonian small farmers, such as rubber-tappers. This will change the supply function of Amazonian resources as shown in Proposition 1, reducing output at each price. In turn this will change the computation of comparative advantages and of gains from trade from agricultural exports based on deforestation of the Amazon. Production patterns shift, and export patterns will reflect more fully the social cost of deforesting the Amazon.

Examples of such property-rights approaches are provided by recent agreements involving debt-for-nature swaps (Ruitenbeck, 1990). Another example is provided by recent agreements between the U.S. pharmaceutical industry and Costa Rica. The spearhead of this project is a pair of ingenious efforts to exploit the forests to obtain medicinal products.[8] A Costa Rican research institute (INBIO) is prospecting for promising plants, microorganisms, and insects to be screened for medical uses by Merck and Company, the world's largest drug company. Merck, in turn, is supporting the prospecting efforts financially and will share any resulting profits with Costa Rica. Thus, Costa Rica has acquired property rights over the "intellectual property" embodied in the genetic information within its forests. A similar initiative was taken by a small California company, Shaman Pharmaceuticals, which is tapping the expertise of traditional healers, "shamans," or medicine people, in various parts of the tropics (see Chichilnisky, 1993a). The company intends to promote the conservation of the forests by channeling some of its profits back to the localities whose medicine people provided the key plants. The theory behind both ventures is that everybody wins: the world gets new drugs, the pharmaceutical companies earn profits, and people in the localities are justly compensated for their "intellectual property" and their conservation and collection efforts.[9]

Similar examples hold for land resources. Recently, the government of Ecuador allocated a piece of the Amazon the size of the state of Connecticut to its Indian population, a clear property-rights policy.[10] Under the conditions examined here, this policy should lead to a better use of the forest's resources and to a more balanced pattern of trade between Ecuador and the U.S.[11]

However, property rights change slowly because they require expensive legal infrastructure and enforcement. Poor countries may be unable to implement such policies quickly. The improvement of property rights of indigenous populations in developing countries, which house most of the

---

[8] The plans were described at a symposium held in January 1992 at Rockefeller University, organized jointly by the Rain Forest Alliance, and the New York Botanical Garden's Institute of Economic Botany (see e.g., the report in *Science Times*, science supplement to *The New York Times* 28 (January, 1992), C1).

[9] Examples of successful medical discoveries from rain forests and other natural sources include widely used medicines such as aspirin, morphine, quinine, curare, the rosy periwinkle used to treat childhood leukemia and Hodgkin's disease, and (more recently) taxol.

[10] Indian groups will gain title to land in Pastaza Province, a traditional homelands area covering 4,305 square miles in eastern Ecuador. Ecuador's move is part of a wider trend in the Amazon Basin. Achuar, Shiwiar, and Quiche Indians will soon administer an area where population density averages five people per square mile.

[11] In fact, forest and other environmental assets have public-good aspects, which have not been covered here. Markets with public goods have different behavior and require a different treatment. For example, Lindahl prices or particular distributions of endowments are needed for efficiency (see Chichilnisky et al., 1993).

world's population, should nevertheless be considered a major policy goal.[12] There is a small but apparently growing trend of this type in Brazil, Bolivia, Colombia, Ecuador, French Guyana, and Venezuela.[13] Property-rights policies, either through government action or preferably through private enterprise, as in the examples offered here, provide a hopeful foundation for resolving some North–South environmental issues. Improving property rights should also lead to better, more balanced income patterns, since one of he most direct causes of poverty in the developing countries is the lack of entitlements for land and environmental resources such as clean water (Dasgupta, 1982). Similarly, as shown here, poverty can prevent environmental policies based on taxation from having their intended effects. Poverty and environmental overuse have a common root, and both are at the core of the North–South environmental dilemma.

Inexpensive environmental resources are a main source of environmental overuse. The statement that resources are overconsumed is practically equivalent to the statement that they are underpriced. Environmental overuse does not occur solely because the locals overconsume their resources, but because they export these resources to a rich international market at prices that are below social costs. This is why the global environmental issue is inextricably connected with North–South trade. The South overproduces, and the North overconsumes. The international market transmits and enlarges the externalities of the global commons. No policy that ignores this connection can work.

APPENDIX A: RESOURCE EXTRACTION AND PROPERTY RIGHTS

The Nash Equilibrium of the Harvester Extraction Problem with Unregulated Common-Property Resources — The extraction of the input E is carried out by N "harvesters" of an unregulated common-property pool, indexed $i = 1, \ldots, N$. Let x_i be the input of harvester i, and let $x = \sum_{i=1}^{n} x_i$. All harvesters are identical and interchangeable, so that the total harvest can be expressed as a function $E = F(x)$ of the total input. Assume also that all harvesters are symmetric, so that each harvester obtains as its output a fraction of the total output equal to the fraction that it supplies of the total input, formally, $E_i = F(x)(x_i/x)$. This insures uniqueness of the solutions. These are all standard assumptions (Dasgupta and Heal, 1979). Each harvester chooses its input level x_i to maximize the value of its share of outputs net of costs, $p_E E_i(x_i) - q x_i$, taking as given p_E, q, and the input levels of others, x_j for $j \neq i$. Here p_E is the market-induced price of the resource, and q is the "opportunity cost" of the input x_i; both q and p_E are endogenously determined in Appendices B and C along with all other prices in a general-equilibrium fashion. $F(0) = 0$, $F'(x) > 0$, and $F(x)$ is strictly concave, so there are strictly diminishing returns, arising perhaps from the application of increasing amounts of variable input x to a fixed body of land or water, and this insures existence

---

[12] Judge Alvaro Eduardo Junqueira declared that a shipload of mahogany sold by C & C Industria e Comercio was illegally felled and stolen from Indian reserves in the Amazon region, and he ordered the immediate seizure of thousands of cubic meters of mahogany bound to London. Tradelink, the London-based timber agent, stated that they will cease purchasing wood from C & C Industria e Comercio if the theft is confirmed (see *Financial Times*, 29 October 1993, 1, 34). The matter is causing embarrassment for Britain's Timber Trade Federation: Britain imports about half of Brazil's entire mahogany exports.

[13] In the last three years, the Governments of Ecuador, Colombia, and Venezuela have restricted most of their Amazon areas as national parks or Indian reserves, as have Brazil and Bolivia, and France has made plans to protect a third of French Guyana. Last year, a coalition of Amazon Indians and foreign and local environmentalists confronted U.S. oil companies to induce them to abandon plans for extracting oil in Ecuador's Amazon (see Kane, 1993). Ecuador, one of South America's poorest countries, draws currently about $50 million yearly in revenues from oil exports. In the highlands of Ecuador, Indian groups have expressed similar resistance to export-oriented farming. Jose Maria Cabascango, leader of Indigenous Nationalities of Ecuador, which is said to represent the nation's estimated two million Indians, states: "We should only produce food for our own consumption. The Amazon region has a very fragmented ecology and to continue colonization would destroy it" (see e.g., *New York Times*, 6 September 1992, L10). Similar concerns were expressed by Antonio Macedo, Coordinator of the National Council of Rubber–Tappers of the Amazon, of Cruzeiro do Sul, Acre, Brazil, in a recent interview at Columbia University, 7 December 1992). A recent article in *The New Yorker* (Kane, 1993) highlights the practical issues involved.

of solutions. This models a Nash equilibrium pattern of use of an unregulated common-property resource, which is unique under the symmetry conditions.

PROOF OF PROPOSITION 1: The marginal product of input is $F'(x)$. The average product is $F(x)/x$, and by strict concavity $F(x)/x > F'(x)$. Look first at marginal products. The private-property marginal product of the input is denoted MP^P, and the common-property marginal product is MP^C. Now, $MP_i^C = d/dx_i[x_iF(x)]$ equals $d/dx_i[x_iF(x_i + x_{-i})]/(x_i + x_{-i})$, where $x_{-i} = \Sigma_{j\neq i}x_j$. Hence, under the assumptions,

$$MP_i^C = F(x)/x + x_i\left\{\left[xF'(x) - F(x)\right]/x^2\right\}$$

$$= F(x)/x$$

$$+\left(x_i/x\right)\left[F'(x) - F(x)/x\right].$$

Note that the analysis provided here is independent of the number of harvesters as long as there is more than one ($N > 1$). As $N \to \infty$, $x_i/x \to 0$, and the common-property marginal product becomes the average product. In this limiting case we recover the well-known result that harvesters equate input prices to average return rather than to marginal product, the basis of the "tragedy of the commons." Since $MP_i^P = F'(x)$,

$$MP_i^P - MP_i^C = F'(x) - F(x)/x$$

$$-\left(x_i/x\right)\left[F'(x) - F(x)/x\right]$$

$$= \left[F'(x) - F(x)/x\right]\left[1 - x_i/x\right]$$

$$< 0$$

Therefore, the common-property marginal product is lower than the private-property one. The private-property supply curve, E^P, is obtained by equalizing the opportunity cost q with the value of MP_i^P, $q = p_E MP_i^P$, and the common-property supply curve E^C is obtained by equating $q = MP_i^C p_E$: both supply functions of E as a function of p_E are parameterized by the opportunity cost q of the input x_i.

The last task is to show that, with both private- and common-property regimes, both supply curves increase with the price of the resource E at any given q. For simplicity in the rest of this proof, I use E_i to indicate either the common-property supply curve or the private-property supply curve, because the argument that follows applies to both. For each p_E, the harvester's objective is to find x_i, which optimizes

$$p_E E_i\left(x_i\right) - qx_i \qquad (7.2)$$

For each fixed q, the solution to this problem, denoted $x_i(p_E, q)$ is an increasing function of p_E: as the market price of E increases, the marginal productivity of x_i, which maximizes the objective function of the harvester must satisfy $p_E(\partial E_i/\partial x_i) = q$. For each p_E and q, this maximization problem defines $E_i = E_i(x_i(p_E, q))$. Note that, for a given q, as p_E increases (2) implies that $\partial E_i/\partial x_i$ must decrease. The concavity of E_i implies then that the optimal level of input $x_i(p_E, q)$ increases with p_E for any given q (i.e., $\partial x_i/\partial p_E > 0$). Therefore, for any given q, *the supply curve $E_i(x_i)$ is always increasing in the resource's price p_E.*

APPENDIX B: FORMULATION OF THE NORTH–SOUTH MODEL

Consider a Heckscher–Ohlin formulation of two competitive economies trading with each other, except that factor endowments in the two countries are not fixed, but variable, depending on factor prices. This follows Chichilnisky (1981, 1985, 1986, 1990, 1993b) but differs from these works because they considered privately owned goods only. Variable factors are crucial in the consideration of property rights: Proposition 1 showed that at each price the supply of resources under common-property regimes is larger than with private property, a characterization that requires price-dependent endowments. This chapter considers simple production functions with constant returns to scale; the model and its results have been extended to a variety of utility demand specifications, to Cobb–Douglas and CES production functions, and to economies with increasing returns (Chichilnisky, 1993b). Specify first one economy: the South. It produces goods A and B using two inputs, E and K. We consider a fixed-proportions technology in each sector, although there is substitution of factors at the aggregate level. This is because changes in factor prices lead to changes in factor endowments and so to changes in the composition of output and the factor intensity of final production. This could not happen in a Heckscher–Ohlin model, but *can* happen here because factor supplies vary with prices (see also Chichilnisky, 1986).

Production is specified by $B^S = E^B/a_1 = K^B/c_1$, and $A^S = E^A/a_2 = K^A/c_2$, where the superscript S denotes supply and superscripts A and B denote sectors, and where a_1, a_2, c_1, and c_2 are input-output coefficients. $E^S = E^B + E^A$ varies with prices, and so does $K^S = K^B + K^A$. The good B is more resource-intensive than A so that $D = (a_1c_2 - a_2c_1) > 0$. The following equations identify an equilibrium. First, with zero profits in the production of A and B,

$$p_A = a_1 p_E + c_1 r \qquad (7.3)$$

$$p_B = a_2 p_E + c_2 r$$

or

$$p_E = \frac{p_B c_2 - c_1}{D} \qquad r = \frac{a_1 - p_B a_2}{D}$$

where p_A and p_B are the prices of A and B, respectively, p_E is the price of E, and r is the rental on capital, K. Consider next the exogenously given supply functions for the inputs E and K. It was established in Proposition 1 (see Appendix A) that the supplies of the environmental resource E increase with p_E/q, where q is the opportunity cost of the input used in extracting E. To simplify the computation of solutions consider a simple form of the supply function:

$$E^S = \alpha p_E/q + E^0 \qquad \alpha > 0 \qquad (7.4)$$

The variable q in (Equation 7.4) takes different values depending on the input used to extract E: q is r when E is produced from capital K; when E is produced from subsistence labor, p_B is an appropriate proxy for q. Note that when the supply function has the form in (Equation 7.4) the slope α represents the property-rights regimes for the pool from which E is extracted: as established in Proposition 1 (proof in Appendix A), the supply of E has a larger slope when resources are common property and a smaller one in the case of private property, quite independently of which input is used to extract E. The assumptions made on the property-rights regimes of the economies of the North and the South made in Section C translate therefore into: $\alpha(N) < \alpha(S)$ and $\alpha(S)$ large, where the letters in parentheses denote the North and the South. The supply for capital is similarly

$$K^S = \beta r + \overline{K} \qquad (7.5)$$

where $\beta \geq 0$; everything that follows applies for $\beta = 0$ as well (i.e., when K^S is a constant $\overline{K}$). In equilibrium all markets clear:

$$E^S = E^D \qquad K^S = K^D \tag{7.6}$$

$$E^D = E^B + E^A = B^S a_1 + A^S a_2 \tag{7.7}$$

$$K^D = K^B + K^A = B^S c_1 + A^S c_2$$

where superscript D denotes demand. When the extraction of E uses capital, $K^D = K^B + K^A + K^E$,

$$B^S = B^D + X_B^S \tag{7.8}$$

$$A^S = A^D + X_A^D$$

where X_B^S and X_A^D denote exports of B and imports of A, respectively, and

$$p_B X_B^S = p_A X_A^D \tag{7.9}$$

(i.e., the value of exports equals the value of imports). The North is specified by a set of equations similar to Equations (7.3) to (7.9) with the same technology parameters and the same capital supply functions, but with different α's (i.e., different supply functions for environmental resources E), as explained in the paragraph following (Equation 7.4), and discussed in Section C. The two supply functions are now denoted $E^S(S)$ and $E^S(N)$. In a world equilibrium, the prices of the traded goods (A and B) are equal, and exports match imports:

$$p_A(N) = p_A(S) \tag{7.10}$$

$$p_B(N) = p_B(S)$$

$$X_A^S(N) = X_A^D(S)$$

$$X_B^S(S) = X_B^D(N)$$

where the terms in parentheses, (S) and (N), indicate the North and South, respectively. Since the economies are identical except for property rights, in the two regions there are nine exogenous parameters: α_1, α_2, c_1, c_2, β, $\overline{K}$, E^0, and $\alpha(N)$ and $\alpha(S)$. We add a price-normalization condition $p_A = 1$ and obtain a total of 26 independent Equations, (7.3)–(7.9) for the North and for the South, plus (Equation 7.10) and $p_A = 1$.[14] There are in total 28 endogenous variables, 14 for each region: p_A, p_B, p_E, r, E^S, E^D, K^S, K^D, A^S, A^D, B^S, B^D, X_B^S, and X_A^D, so the system is underdetermined so far up to two variables, which reflects the fact that demand has not been specified yet. We consider a demand specification that leads to simple analytics (more general utility functions can be considered at the cost of more computation). In each region consider the utility function:

---

[14] Walras's law assures market-clearing in one of the markets whenever all other markets clear, thus reducing the number of equations presented here by one.

$$U(A, B) = B + k \qquad \text{if } A \geq A^{D^*} \tag{7.11}$$

where $k > 0$, and

$$U(A, B) = B + \gamma A \qquad \text{otherwise}$$

where $\gamma = k/A^{D^*} > 0$. For $p_B > 1/\gamma$, agents demand A^{D^*}, so by choosing k and γ in U appropriately, one may assume:

$$A^D(N) = A^{D^*}(N) \tag{7.12}$$

$$A^D(S) = A^{D^*}(S)$$

Thus, we have a system of 28 equations on 28 variables, depending on nine exogenous parameters, and the model is now fully specified. The economies of the two regions are identical except for the parameters $\alpha(N)$ and $\alpha(S)$ which depend on the property rights for the common-property resource in each region. We shall say that property rights are better defined in the North when $\alpha(N) < \alpha(S)$; both countries have the same property right when $\alpha(N) = \alpha(S)$. Note that by inverting equations $E^S = a_1 B^S + a_2 A^S$ and $K^S = c_1 B^S + c_2 A^S$ one obtains

$$B^S = (c_2 E - a_2 K)/D \tag{7.13}$$

$$A^S = (a_1 K - c_1 E)/D$$

where as before $D = a_1 c_2 - a_2 c_1 > 0$. When the extraction of E uses capital, K in the two equations in (Equation 7.13) is replaced by $K - K^E$.

APPENDIX C: PROOFS OF THE RESULTS

Proof of uniqueness of a competitive equilibrium

The North–South model defined in Appendix B has, at most, one competitive equilibrium for any parameters $\alpha(S)$ and $\alpha(N)$ representing the structure of property rights in the two regions. In the following, the supply function of resources in (Equation 7.4) is $E = \alpha p_E/p_B + E^0$ which, as discussed in the paragraph following (Equation 7.4), corresponds to the case in which E is extracted using subsistence labor, as in section E. Similar computations obtain when E is extracted using capital (K) as an input; in this latter case, q is replaced by r in (4) and $K^D = K^S = K^A + K^B + K^E$, where $F(K^E) = E$ (see Proposition 1). From (Equation 7.10) and (Equation 7.12),

$$A^{D^*}(S) - A^S(N) = A^S(N) - A^{D^*}(N). \tag{7.14}$$

Now rewrite (Equation 7.14) as a function of one variable only, p_B. Substituting Equations (7.3)–(7.10) and (Equation 7.13) into (Equation 7.14), one obtains:

$$p_B^2[\psi(S) + \psi(N)] + p_B[A^{D^*}(S) + A^{D^*}(N) + \Gamma(S) + \Gamma(N)] - [\rho(S) + \rho(N)] = 0 \tag{7.15}$$

where

$$\psi = \beta a_1 a_2 / D \qquad \rho = \alpha c_1^2 / D^2$$

and

$$\Gamma = (1/D)\left[c_1 E^0 - \alpha_1 \overline{K} + (1/D) \times \left(\alpha c_1 c_2 - \beta a_1 a_2\right)\right]$$

Equation (7.15) is a quadratic equation in p_B, which has at most one positive root, because the constant term is negative. Therefore, there is at most one equilibrium price p_B^*. From p_B^* one obtains in each country the equilibrium levels of all other variables: p_E^* and r^* from (Equation 7.3), E^{S*} and K^{S*} from (Equation 7.4) and (Equation 7.5), B^{S*} and A^{S*} from (Equation 7.13), X_A^{D*} from (Equation 7.12), and A^{D*}, A^{S*}, and X_B^{S*} from (Equation 7.9) and (Equation 7.10), so the (unique) full equilibrium of the North–South model is computed.

PROOF OF THEOREM 1: The completely symmetric case of the North–South model is utilized here because it shows very clearly the role of property rights in determining patterns of trade. Here, the two regions are identical except for property rights: same inputs K and E, same produced goods A and B, same production functions for A and B, and same preferences. In the two regions, E is produced by harvesters using an exogenous, fixed endowment of capital $\overline{K}$ as an input, where $\overline{K}$ is the same in the two regions, and using the same production technology $f: K \to E$ (see Proposition 1). The South extracts E from an unregulated common-property pool, and the North does so under private-property regimes. The market-clearing condition for K is $K^D = K^A + K^B = K^E + \overline{K}$, where $F(K^E) = E$ as derived in Proposition 1. In sum, the two regions are identical, but the South's supply of E is given by the common-property supply curve $E^C(p_E)$, while the North's is given by the private-property supply curve $E^P(p_E)$ (see Appendix A). At each price vector, the supply of E in the South exceeds that of the North (see Proposition 1; proof in Appendix A).

Consider now a world equilibrium commodity price vector $\mathbf{p}_w^*$ at the equilibrium, factor prices p_E^* and r^* are the same in the two regions because the two regions have the same technologies (i.e., the same a_1, a_2, c_1, and c_2), and by the zero-profit conditions on the production of A and B given in (Equation 7.3). However, the South supplies more environmental resources than the North by Proposition 1. By inverting the two linear equations, $E^S = a_1 A^S + a_2 B^S$ and $K^S - K^E = c_1 B^S + c_2 A^S$, one obtains

$$B^S = \frac{c_2 E - a_2\left(K - K^E\right)}{D}$$

which increases with E because $D > 0$. It follows that at $\mathbf{p}_w^*$, the South produces a larger amount of the traded good B than does the North; intuitively this is a consequence of the fact that B is intensive in the input E, which is more abundant in the South.

When the two regions have the same homothetic utilities, since at $\mathbf{p}_w^*$ the two regions face the same relative prices for goods A and B, the North and the South demand goods for A and B in the same proportions. Since in equilibrium the supply of B in the South is proportionately larger, when the international markets clear, the South must export B, and the North must import B; that is, the South is an exporter of environmentally intensive goods at the world equilibrium.

Next, we establish formally the intuitively obvious proposition that in the move from autarky to trade, the South produces more resources. First, we establish that from autarky to trade the relative price of the environmentally intensive good B must increase from the autarchic level in the South. To establish this, we show that, in autarky, the relative price p_B/p_A (recall $p_A = 1$), must be larger in the North than it is in the South, that is, in autarky,

$$\left(p_B/p_A\right)(\text{N}) > \left(p_B/p_A\right)(\text{S}) . \tag{7.16}$$

This is also intuitively obvious because preferences are the same in both countries, and the South has an institutional comparative advantage in B. Formally, it is seen as follows. It is known that *at any given price* the South produces more B than the North because of common-property regimes (Proposition 1). If, in contradiction with (Equation 7.16) $(p_B/p_A)(N) < (p_B/p_A)(S)$, then this effect is emphasized according to Proposition 1 and (Equation 7.3), and therefore in autarky the South supplies more B than the North. But preferences are assumed in this theorem to be homothetic, so that if $(p_B/p_A)(N) < (p_B/p_A)(S)$, in autarky the North consumes proportionately more B than the South. This contradicts the assumption that the two countries are both in an autarchic equilibrium: they are identical, but the South produces proportionately more B than the North, and it consumes proportionately less than the North. Therefore, in one of these countries the market for B must fail to clear, a contradiction. Since the contradiction arises from negating (Equation 7.16), inequality (Equation 7.16) must hold.

Now it is straightforward to show that a move from autarky to trade leads to a world price of B that is higher than the autarky price was in the South (i.e., that $p_B^* > (p_B/p_A)(S)$). Assume not; then, by the above,

$$\left(p_B/p_A\right)(\text{N}) > \left(p_B/p_A\right)(\text{S}) > p_B^* . \tag{7.17}$$

Since preferences are homothetic, (Equation 7.17) implies that the proportion of goods B/A consumed in both countries increases after trade. However, the proportion of goods B/A produced in both countries decreases, because as p_B decreases, so does p_E, while r increases (Equation 7.3), so that the harvesters extract less E, and the B good, which is E-intensive, is produced in smaller quantities, while the A good is produced in higher quantities.

As shown above, (Equation 7.17) implies that at a trade equilibrium the proportion B/A in which goods are produced decreases in both countries with respect to autarky, while the proportion B/A in which goods are consumed increases in both countries. This implies that markets cannot clear at the trade equilibrium when (Equation 7.17) is satisfied, a contradiction. Therefore, at a world equilibrium it must be true that $p_B^* > (p_B/p_A)(S)$ as I wished to prove. Similarly, I have proven that p_B^* is lower than the autarky price in the North, $p_B/p_A(N)$.

Having proven that p_B/p_A increases in the South from autarky to trade, it follows immediately that the extraction of E increases in the South after trade. As the price of the environmentally intensive good B increases from autarky to trade, the price E must increase, and the rental on capital r, which moves in the opposite direction to p_E, decreases (see Equation 7.3). Therefore, after trade, the harvesters face lower opportunity costs for their input, K, and higher prices for the output, E. They therefore harvest more (see the proof of Proposition 1 in Appendix A), and the extraction of E increases.

Since the two countries are identical except for property rights, when the two have well-defined property rights they do not trade (i.e., the result is autarky). By the first theorem of welfare economics, when $K^S = \overline{K}$ the private-property competitive equilibrium is Pareto efficient, and as proven above in this appendix, it is unique. Since as seen in the previous paragraph the South produces more E when it moves from autarky to trade, at a trade equilibrium (with common property) its production of E is larger than is Pareto efficient. Its equilibrium price is lower than in autarky (i.e., when both countries have well-defined property rights), since $p_B^* < (p_B/p_A)(N)$.

PROOF OF COROLLARY 1: The fact that factor prices in the two regions are equalized at a competitive equilibrium follows immediately from Equations (7.3), which shows that the factor prices are determined by the traded-goods prices p_A and p_B and by the technical coefficients a_1, c_1, a_2, and c_2, which are identical in equilibrium in the two regions.

PROOF OF PROPOSITION 2: The resource E is supplied to the market by harvesters as specified in Appendix A. The harvesters own only their input $\bar{x}$ of labor, they exchange their harvest of E for A, and they have utilities $u_i(A, x - x_i)$, which depend on consumption of A and on leisure, with elasticity of substitution between leisure and consumption, $\sigma < 1$. Consider the harvester's optimization problem:

$$\max\left[u_i\left(A, \bar{x} - x_i\right)\right] \tag{7.18}$$

subject to

$$p_A A = p_E E\left(x_i\right)$$

where $E(x_i) = E^C(x_i)$ is as defined in Proposition 1. At each relative market price p_E/p_A the harvester chooses x_i so that

$$\left(p_E/p_A\right)E_i'\left(x_i\right)=\left[\partial u_i/\partial(\bar{x} - x_i)\right]\big/\left[\partial u_i/\partial A\right]. \tag{7.19}$$

The solution to (Equation 7.19) is denoted $x_i(p_E, p_A)$. The opportunity cost of the input q must satisfy

$$\left(p_E/p_A\right)E_i'\left(x_i\right)= q \tag{7.20}$$

so that by (Equation 7.19),

$$q\left(p_E, p_A\right)=\left[\partial u_i/\partial(\bar{x} - x_i)\right]\big/\left[\partial u_i/\partial A\right].$$

That is, *the opportunity cost $q(p_E, p_A)$ of x_i is the ratio of the marginal utility of leisure and the marginal utility of consumption*, as stated in Section D.

In (Equation 7.20), q appears explicitly as a function of p_A and p_E. Using (Equation 7.20) one can now explore how the extraction of the resource E changes with its price p_E, and how the opportunity cost q of the input x_i varies across equilibria. First recall that the workers' elasticity of substitution between leisure and consumption, which measures precisely how substitutable one good (A) is for the other (leisure, $\bar{x} - x_i$), is assumed to be less than 1, that is

$$\sigma = \frac{\partial\left(\dfrac{A}{\bar{x} - x_i}\right)}{\partial\left(\dfrac{p_A}{q}\right)}\left[\frac{\dfrac{p_A}{q}}{\dfrac{A}{\bar{x} - x_i}}\right]<1$$

(Varian, 1992, p. 44). This implies immediately, by definition, that as the relative price of A increases, leisure decreases. This in turn implies that extraction has increased. Figure 7.3 illustrates this fact:

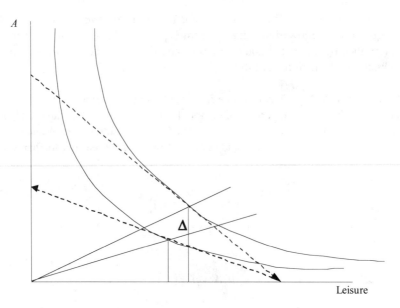

FIGURE 7.3 The harvester maximization problem.

Notes: The curves represent indifferences of the harvester between consumption of A and leisure. The two dashed lines represent budget lines at different prices. They converge to a point on the horizontal axis that is the harvester's initial endowment of labor. With harvesters owning only labor, and with elasticity of substitution less than 1, as the relative price of resources increases (the price of A increases), leisure decreases, and more is harvested.

as the relative price of A increases, the proportion of consumption of A and of leisure must go down by a smaller proportion, because $\sigma < 1$. The vertical segment between the two indifference surfaces in Figure 7.3 represents the change in prices, while the part of this segment indicated with Δ, which indicates the change in the proportion of consumption of A and of leisure, is smaller, leading to less leisure after the price of A increases.[15] More effort x_i is applied, and therefore there is more extraction of E as the price p_E drops. Note that the utilities used here are quite general, requiring only continuity, strict concavity, and $\sigma < 1$. For example, with CES utility functions (which are not required here),[16] the assumption $\sigma < 1$ simply means that the indifference surfaces of a worker's utility do not cross the axis, which is a standard assumption.

The last task is to explain Figure 7.2, which indicates how a decrease in opportunity cost of labor increases extraction. First note that the opportunity cost of the input x_i decreases with p_E: as p_E/p_A decreases, the quantity of $x_i(p_E, p_A)$ supplied increases (as in Figure 7.3), so that the supply of the resource $E = E(x_i(p_E, p_A))$ also increases. By the concavity of the production function for E, this implies that as p_E/p_A drops, the derivative $E_i'(x_i(p_E, p_A))$ must also drop. Since p_E/p_A decreases and $E_i'(x_i)$ has not increased, the opportunity cost q defined in (Equation 7.20), $q(p_E, p_A) = (p_E/p_A)E_i'(x_i)$ must have decreased as well. As the relative price of the resource p_E decreases, the opportunity cost of labor q drops, and the supply curve of resources shifts as was indicated in Figure 7.2.

PROOF OF PROPOSITION 3: Consider a unit tax T on the use of the resource E in the South, which increases the value of demand for A, $A^{D*}(S)$, by TE^* (recall that $p_A = 1$). I shall now consider the cross-equilibria relationship between p_B and $A^{D*}(S)$ to determine the changes in prices of the

---

[15] For example, if utilities are Cobb–Douglas, one obtains the limiting case of constant supply of the resource E as p_E/p_A decreases.

[16] CES utilities are of the form $u(A, \bar{x} - x) = [dA^\rho + e(\bar{x} - x)^\rho]^{1/\rho}$, and their elasticity of substitution is $\sigma = 1/(1 - \rho)$ (see Varian, 1992).

resource-intensive good A in a new equilibrium, following the tax. From (Equation 7.15), by the implicit-function theorem, across equilibria,

$$\frac{\partial p_{B^*}}{\partial A^{D^*}(S)} = \frac{-p_{B^*}}{2p_B\left[\Psi(S)+\Psi(N)\right]+\Gamma(S)+\Gamma(N)+A^{D^*}(S)+A^{D^*}(N)} \tag{7.21}$$

which, when $\alpha(S)$ is sufficiently large, is always negative (see Equation 7.15 and also Chichilnisky, 1986, p. 15). Now, by assumption, the tax proceeds lead to a higher level of demand for the capital-intensive good A in equilibrium, A^{D^*}. Therefore, by (Equation 7.21) the tax leads to a drop in the equilibrium price p_B^* of the resource-intensive good, B. Finally, when $c_2/D < 2\,p_E/p_B$ and α is large, the equilibrium level of exports of B increases as their price p_B drops since $X_B^S(S) = B^S - B^D = B^S - (p_E E + rK - A^{D^*})/p_B$ so that

$$X_B^S(S) = \frac{c_2 E - a_2 K}{D} - \frac{p_E E + rK - A^{D^*}}{p_B} = \left(\frac{\alpha c_1}{D^2 p_B}\right)\frac{c_2 - c_1}{p_B} + \left(\frac{\beta a_1}{D^2}\right)\frac{a_2 - a_1}{p_B}$$

$$+ \frac{c_1 E^0 - a_1 \overline{K}}{Dp_B} + \frac{A^{D^*}(S)}{p_B} \tag{7.22}$$

and thus,

$$\frac{\partial X_B^S}{\partial p_B} = \frac{\alpha c_1}{D^2 p_B^2}\left(\frac{2c_1}{p_B} - c_2\right) + \left(\frac{\beta a_1}{D^2 p_B^2}\right) + \frac{a_1 \overline{K} - c_1 E^0}{p_B^2} - \frac{A^{D^*}(S)}{p_B^2} \tag{7.23}$$

(see also Chichilnisky, 1981, 1986). When α is large, (Equation 7.23) implies that $\partial X_B^S/\partial p_B$ has the sign of $2c_1/p_B - c_2$, which is equal to that of $c_2/D - 2p_E/p_B$ by (Equation 7.3). Since by assumption $c_2/D < 2p_E/p_B$, (Equation 7.23) is negative; therefore taxes lead to a lower p_B and to an increase in exports $X_B^S(S)$. Notice that when α is large, c_2/D is the term representing the supply (B^S) response to the change in terms of trade p_B, while $2p_E/p_B$ is the demand response to p_B. Therefore, the inequality $c_2/D < 2p_E/p_B$ implies that an increase in the price of B leads to a larger increase in the demand for B than in supply of B, a situation which is described by stating that "income effects dominate" in the market for B. The rest of the proof follows from Proposition 2.

To simplify computations, in this proof we have taken utility functions which effectively make the demand for A in each region an exogenously chosen parameter at an equilibrium (see Equation 7.12). This follows Chichilnisky (1981, 1986, 1993b), where it is also shown that the results generalize to more general utilities and demand functions. This procedure allows one to explore the effect of a tax that unequivocally increases the demand for nonresource-intensive goods, the most favorable conditions for the tax to work in the intended direction. Proving the results under such assumptions gives a stronger result than if demand for the resource-intensive goods also increased after taxes.

REFERENCES

Amelung, T., Tropical deforestation as an international economic problem, paper presented at the Egon–Sohmen Foundation Conference on Economic Evolution and Environmental Concerns, Linz, Austria, August 30–31, 1991.

Barbier, E. B., Burgess, J. C., and Markandya, A., The economics of tropical deforestation, *AMBIO*, 20, 55, 1991.

Binkley, C. S. and Vincent, J. R., Forest-based industrialization: a dynamic perspective, Forest Policy Issues Paper, World Bank, Washington, DC, 1990.

Bromley, D.W., Ed., *Making the Commons Work*, ICS Press, San Francisco, 1992.

Chichilnisky, G., Terms of trade and domestic distribution: export-led growth with abundant labour, *J. Dev. Econ.*, 8, 163, April 1981.

Chichilnisky, G., International trade in resources: a general equilibrium analysis, in *Proc. of the Am. Math. Society* 32, 1985, 75; reprinted in McElvey, R., Ed., *Environmental and Natural Resource Mathematics*, American Mathematical Society, Providence, RI, 1985, 75.

Chichilnisky, G., A general equilibrium theory of North–South trade, in *Essays in Honor of Kenneth Arrow, Vol. 2: Equilibrium Analysis*, Heller, W. P., Starr, R. M., and Starrett, D. A., Eds., Cambridge University Press, Cambridge, 1986, 3.

Chichilnisky, G., On the mathematical foundations of political economy, *Contributions to Political Economy* 9, 25, 1990.

Chichilnisky, G., Biodiversity and property rights in the pharmaceutical industry, Case study, Columbia University School of Business, 1993a.

Chichilnisky, G., Traditional comparative advantages vs. external economies of scale, Report to the Economic Commission for Latin America and the Caribbean, Project on Trade Liberalization in the Americas, United Nations, New York, 1993b; *J. Int. Comp. Econ.* (forthcoming).

Chichilnisky, G., North–South trade and the dynamics of renewable resources, *Structural Change and Econ. Dynamics* 4, 219, 1993c.

Chichilnisky, G. and Heal, G. M., *The Evolving International Economy*, Cambridge University Press, Cambridge, 1987.

Chichilnisky, G. and Heal, G. M., *Oil in the International Economy*, Clarendon Press, Oxford, 1991.

Chichilnisky, G., Heal, G. M., and Starrett, D., International markets with emission permits: equity and efficiency, working paper, Stanford Institute for Theoretical Economics, December 1993.

Coase, R. H., The problem of social cost, *J. Law Econ.*, 3, 1, 1960.

Cohen, J. S., and Weitzman, M. L., A Marxian model of enclosures, *J. Dev. Econ.*, 1, 287, 1975.

Dasgupta, P., *The Control of Resources*, Harvard University Press, Cambridge, MA, 1982.

Dasgupta, P. and Heal, G. M., *Economic Theory and Exhaustible Resources*, Cambridge University Press, Cambridge, 1979.

Demsetz, H., Towards a theory of property rights, *Am. Econ. Rev. (Pap. and Proc.)* 57, 347, 1967.

Dixit, A. and Norman, V., *Theory of International Trade*, Cambridge University Press, Cambridge, 1980.

Hyde, W. F. and Newman, D. H., Forest economics in brief — with summary observations for policy analysis, unpublished manuscript (draft report), Agricultural and Rural Development, World Bank, Washington, DC, 1991.

Kane, J., Letter from the Amazon, *New Yorker*, 69, 54, 27 September, 1993.

Lerner, A. P., The symmetry between import and export taxes, *Economica*, 3, 306, 1936.

McDonald, S. L., *Petroleum Conservation in the United States: An Economic Analysis*, Johns Hopkins University Press, Baltimore, 1971.

Metzler, L., Tariffs, the terms of trade, and the distribution of national income, *J. Political Economy*, 57, 1, 1949.

Ruitenbeck, H. J., The rainforest supply price: A step towards estimating a cost curve for rainforest conservation, Development Research Programme Working Paper No. 29, London School of Economics, 1990.

Schweigert, T., Commercial sector wages and subsistence sector labor productivity in Guatemalan agriculture, *World Dev.*, 21, 81, 1993.

Varian, H., *Microeconomic Analysis*, 3rd ed., Norton, New York, 1992.

World Bank, *World Development Report: Development and the Environment*, Oxford University Press, Oxford, 1992.

8 Trade and Transboundary Pollution[1]

Brian R. Copeland and M. Scott Taylor

This chapter examines how national income and trading opportunities interact to determine the level and incidence of world pollution. We find that (i) free trade raises world pollution if incomes differ substantially across countries; (ii) if trade equalizes factor prices, human-capital-abundant countries lose from trade, while human-capital-scarce countries gain; (iii) international trade in pollution permits can lower world pollution even when governments' supply of permits is unrestricted; (iv) international income transfers may not affect world pollution or welfare; and (v) attempts to manipulate the terms of trade with pollution policy leave world pollution unaffected. (JEL F10, H41, Q28)

A. INTRODUCTION

There is growing concern over the effects of international trade on the global environment. While traditional opposition to free trade focused on potential job losses and wage reductions, environmentalists have recently turned the pro-free-trade case on its head by arguing that, even if trade liberalization succeeds in raising incomes and consumption, this will only lead to more pollution.

Economists have responded by pointing out that the national income gains brought about by freer trade will increase the demand for environmental quality, make new investment in pollution abatement affordable, and generate much-needed government revenues for enforcement of environmental regulations. In fact, recent empirical work by Gene M. Grossman and Alan B. Krueger (1993) suggests that income gains can have a significant effect on some types of pollution emissions.

It is clear that both sides may be at least partially right. Trade may tend to increase world pollution by raising the scale of economic policy activity and providing added incentives for polluting industries to locate in countries with low environmental standards. Conversely, the income gains created by trade may increase the pressure for tougher environmental regulation and enforcement. In this chapter we develop a very simple model to investigate these issues. We assume that global environmental quality is a pure public good whose supply responds endogenously to trade-induced changes in relative prices and incomes, and we use the model to explore how welfare and pollution levels are affected by free trade in goods and pollution permits, by international income transfers, and by international agreements limiting or reducing pollution emissions. Our model is designed to highlight income effects, since these are central to the arguments put forward on both sides of the debate.

The model builds on our earlier work (Copeland and Taylor, 1994), where we studied trade in a world where environmental quality is a *local* public good (pollution damage is confined to the emitting country). In that paper, we developed a static two-country general-equilibrium model with a continuum of goods differing in their pollution intensity of production. Countries differed only

[1] Copeland's portion of this research was funded by a grant from SSHRC. Earlier versions of this paper were presented at the Midwest International Economics Meetings, the Third Canadian Conference on Resource and Environmental Economics, the winter meetings of the Econometric Society, the NBER 1994 Summer Institute, and in seminars at UBC and the University of Victoria. We thank participants for their comments. We are also grateful for helpful comments from Jim Brander, Satya Das, Mick Devereux, David Donaldson, and three anonymous referees.

in their endowment of the one primary factor (human capital); and we studied only the case in which this difference was large. Governments set pollution policy endogenously, and because environmental quality is a normal good, the higher-income country had tougher environmental regulations. We found that free trade shifted pollution-intensive production to the human-capital-scarce country and raised world pollution. Nevertheless, there was no market failure because pollution stayed within the country of origin and governments regulated pollution optimally. Consequently, trade always increased welfare.

In the present chapter, global environmental quality is a pure public good (or equivalently, pollution is a pure public bad): all countries are equally exposed to a given unit of pollution, regardless of its source. The pollutants we have in mind feature prominently in much of the debate over global warming, depletion of the ozone layer, and biodiversity. The welfare effects of trade in this case are fundamentally different. If pollution is a pure public bad, the relocation of pollution-intensive industries to countries with less stringent environmental protection may increase the exposure of home residents to pollution, and this works against standard gains from trade. Because pollution crosses borders, uncoordinated regulation of pollution at the national level does not eliminate all market failures, and consequently free trade need not raise welfare.

In addition to allowing for transboundary pollution, we depart from our earlier work in several other significant ways. We allow for an arbitrary number of countries, and we consider two cases: one with a large number of countries (where no country perceives a terms-of-trade effect from changes in its environmental policy) and one with a small number of countries (where terms-of-trade motivations for environmental policy cannot be ignored). This allows us to isolate the effects of terms-of-trade motivations for pollution policy from purely environmental motives. Finally, we study equilibria where factor prices are equalized by trade (FPE equilibria), as well as specialized equilibria. This allows us to examine how inequalities in the international distribution of income influence the effects of trade on the environment.

The bulk of the chapter focuses on the case with a large number of countries. Each government sets a national pollution quota and implements it with marketable permits, treating the rest of the world's pollution as given. We find a Nash equilibrium in pollution levels, examine the effects of liberalizing trade in goods, and analyze various proposals for reducing world pollution. Within this context we obtain several interesting results.

First, if human-capital levels differ substantially across countries, then a movement from autarky to free trade raises world pollution. In contrast, if all countries have similar human-capital levels, then world pollution does not rise with trade. When countries are very different, trade does not equalize factor prices, and consequently pollution permit prices are lower in human-capital-scarce countries. "Pollution havens" are created by trade as the most pollution-intensive industries shift to countries with weak environmental regulations. This tends to raise world pollution above autarky levels. On the other hand, if countries are sufficiently similar, then trade equalizes pollution-permit prices, and the pollution-haven effect is eliminated.

Second, we find that the human-capital-abundant countries lose from trade and the human-capital-scarce countries gain from trade when regions have similar (but still different) income levels in autarky. As we show, lower-income countries have a strategic advantage when setting pollution levels in a free-trade regime, and this allows them to increase their income and pollution at the expense of richer countries.

Third, we find that when free trade in goods raises world pollution, allowing for international trade in pollution permits can lower global pollution. This result holds even when countries are not restricted in the number of permits they issue. This is because free trade in permits equalizes pollution-permit prices and eliminates the pollution-haven effect.

Fourth, we show that untied international transfers of income lower the recipient's pollution but raise the donor's. In an FPE equilibrium, the donor's welfare may be no lower after the transfer than before. This result underscores the potential importance of income effects in analyzing global pollution reform.

Finally, when we move to the case with a small number of countries, we find that the flavor of most of our results carries over; although terms-of-trade motivations for pollution policy reduce the strategic advantage of low-income countries, increase the pollution produced by the high-income North, and render income transfers from North to South welfare-reducing for the donor and welfare-enhancing for the recipient.

While there is an extensive literature on transboundary pollution, there is little work examining the interaction among pollution, income levels, and the pattern of trade in a general-equilibrium setting (see Dean, 1992, for a survey). Optimal unilateral and multilateral approaches to transboundary pollution have been addressed in a number of papers (see in particular, Markusen, 1975a, b), but this literature is mainly concerned with how to regulate pollution and not with the interaction among income levels, pollution policy, and trade. Markusen (1975b) also considered a noncooperative Nash equilibrium in pollution levels between governments, but the pattern of trade was exogenous. Rodney D. Ludema and Ian Wooton (1994) have recently extended this work, but confine their attention to strategic trade policy issues. Michael Rauscher's (1991) work is closest in spirit to ours: he uses a two-country general-equilibrium model with no goods trade, but capital is mobile and responds to differences in environmental regulations. He finds that increased capital mobility leads to a pollution reduction in at least one country but has ambiguous welfare effects. In contrast, we adopt a multi-good, many-country trading model. This allows us to analyze linkages between trade patterns and pollution levels, and also to contrast FPE equilibria with specialized equilibria and to compare pure goods trade with trade in pollution permits.

The remainder of the chapter proceeds as follows. In the first six sections we assume that the number of countries is large, and hence terms of trade motivations for pollution policy can safely be ignored. Section B and C detail the model's assumptions and derive the equilibrium conditions. Section D examines autarky, and Section E considers the effects of trade on welfare and pollution levels. We examine various approaches to pollution policy reform in Section F and study the effects of untied income transfers in Section G. Section H reconsiders our earlier results when the number of countries is small, and Section I presents our conclusions.

B. THE MODEL

We consider a world economy consisting of two regions (North and South), each composed of many countries: n in the North and n^* in the South. All countries within a region are identical. Countries differ across regions only in their endowments of the one primary factor, effective labor, which is supplied inelastically. Effective labor can be thought of as the product of the number of workers and an efficiency index determined by the level of human capital. Since the number of individuals per country plays no independent role in our analysis, we normalize the population of each country to 1.[2] Consequently, international differences in effective labor endowments reflect differences in human capital. Each Northern country has L units of effective labor, and each Southern country has L^* units. We assume $L > L^*$ so that Northern countries are human-capital-abundant relative to Southern countries.

There is a continuum of private consumption goods, indexed by $z \in [0,1]$. Pollution is produced jointly with consumption goods, but the output (y) of a consumption good z can be written as a function of pollution emission (e) and labor input (ℓ).[3] To keep the model simple, we adopt the following functional form:

---

[2] In Section 7 of Copeland and Taylor (1994) we consider an extension of the model in which factors such as country size, population density, and the environment's absorptive capacity allow these two components of effective labor to have different effects. We have not adopted this extension here, given the complications of transboundary pollution.

[3] This requires that the technology satisfy certain regularity conditions. In Copeland and Taylor (1994), we show how (Equation 8.1) can be derived from a joint production technology.

$$y = f(e, \ell; z) = \begin{cases} \ell^{1-\alpha(z)} e^{\alpha(z)} & \text{if } e/\ell \leq \lambda \\ 0 & \text{if } e/\ell > \lambda \end{cases} \qquad (8.1)$$

where $\lambda > 0$, and $\alpha(z) \in (0,1)$ is a parameter varying across goods. Because pollution is a by-product of production, output must be bounded above for any given labor input. This constraint is most easily captured by the requirement $e \leq \lambda\ell$ since this ensures that $y \leq \lambda^\alpha\ell$.

Private firms must obtain permits to emit pollution. Letting τ denote the price of a permit, and w the return to a unit of effective labor, the unit-cost function corresponding to (Equation 8.1) is

$$c(w, \tau; z) = k(z)\tau^{\alpha(z)} w^{1-\alpha(z)} \qquad (8.2)$$

where $k(z) \equiv \alpha^{-\alpha}(1 - \alpha)^{-(1-\alpha)}$ is an industry-specific constant.[4] Since the share of pollution charges in the total cost of producing good z is $\alpha(z)$, we can easily rank goods in terms of pollution intensities so that $\alpha'(z) > 0$. High-z goods are more pollution-intensive than low-z goods at all factor prices.

Northern and Southern consumers have identical utility functions defined over consumption goods and aggregate world pollution. Pollution is a pure public bad: consumers in all countries are harmed by the pollution released from any one country. To simplify matters we assume that pollution affects only the level of utility and plays no role in determining consumer choice among goods. For tractability, we follow Rudiger Dornbusch et al. (1980) and impose constant budget shares; hence,

$$U = \int_0^1 b(z) \ln[x(z)] dz - \beta \left(\sum_{i=1}^{n+n^*} E_i \right)^\gamma / \gamma \qquad (8.3)$$

where β and γ are positive constants, E_i is the total amount of pollution emitted by country i, $x(z)$ is consumption of good z, and $b(z)$ is the budget-share function satisfying $\int_0^1 b(z) = 1$. We assume that $\gamma \geq 1$, to ensure that the marginal willingness to pay for pollution reduction is a nondecreasing function of pollution levels.

C. POLLUTION SUPPLY

There are three types of decision-makers: governments, producers, and consumers. We abstract from all income distributional issues and assume that the government chooses policy to maximize the utility of the representative consumer. Governments move first and set national pollution quotas. Next, consumers and producers maximize utility and profits, treating prices and pollution as given. Finally, markets clear.

We consider a noncooperative Nash equilibrium with each government treating the rest of the world's pollution as given when choosing its own pollution quota E_i. Pollution targets are implemented with a marketable permit system: the government of country i issues E_i pollution permits, each of which allows a (local) firm to emit one unit of pollution. The permits are auctioned off to firms, and all revenue is given to consumers via lump-sum transfers.[5]

We begin with the production sector. Given goods prices $p(z)$, and the government's allotment of pollution permits, profit-maximizing firms maximize national income and, hence, implicitly solve

---

[4] We assume an interior solution, but one always obtains if effective labor endowments are not too small. See Copeland and Taylor (1994) for further details.

[5] This is equivalent to assuming that the aggregate pollution target is implemented with a pollution tax whose revenues are rebated to consumers.

$$G(p, E_i, L_i) = \max_{(\ell(z), e(z))} \left\{ \int_0^1 p(z) f[e(z), \ell(z); z] dz \right\}$$

subject to

$$\int_0^1 e(z) dz = E_i$$

$$\int_0^1 \ell(z) dz = L_i.$$

For given $p(z)$, the market price of a pollution permit in country i can be obtained as

$$\tau_i = \partial G / \partial E_i$$

which measures the marginal cost to the economy of reducing pollution.

Consumers maximize utility, given prices and pollution levels. Let I_i denote national income of country (in equilibrium $I_i = G(p, E_i, L_i)$). Then the indirect utility function corresponding to (Equation 8.3) for the representative consumer in country i is given by

$$V = \int_0^1 b(z) \ln[b(z)] dz - \int_0^1 b(z) \ln[p(z)] dz + \ln(I_i) - \beta \left(\sum_{j=1}^{n+n^*} E_j \right)^\gamma / \gamma. \qquad (8.4)$$

Each government chooses its pollution target E_i to maximize the utility of its representative consumer, treating the pollution level of all other countries as fixed. For country i the first-order condition implies

$$\tau_i = -V_E / V_I + \int_0^1 m_i(z) \frac{dp(z)}{dE_i} dz \qquad (8.5)$$

where $m_i(z)$ is net imports of good z. Equation (8.5) tells us that the pollution target should be chosen so that the equilibrium permit price (the marginal cost of pollution abatement) is equal to the marginal benefit from lower pollution, measured by marginal damage (V_E / V_I) plus an indirect terms-of-trade effect.

When the number of countries is large, no individual government can have a significant effect on its terms of trade. Consequently, the final term in (Equation 8.5) tends to zero as n and n^* increase. In contrast, each country retains an incentive to control its pollution even as the number of countries grows large. The intuition for this result is as follows. Start from an existing equilibrium with $n + n^*$ countries each contributing to global pollution, and add another country. On impact, this extra country adds to the stock of world pollution. Since the marginal damage from pollution is rising in global pollution levels, this extra pollution *increases* the incentive of each existing country to control its own pollution. Hence as n and n^* grow large, each country retains an incentive to limit its contribution to global pollution.[6]

---

[6] For a proof of this assertion see footnote 25 in Section G. This result, which applies to public bads, is in fact just the mirror image of the typical (voluntary provision) public-goods problem. In the public-goods case, the addition of further agents increases the quantity of the public good on impact and reduces the incentive of each agent to provide the public good. The key difference between public bads and goods is that the impact effect of another agent in the public-bads case is to *lower* the utility of all others, thus raising the marginal benefit of controlling the public bad; whereas in the public-goods case, adding an extra agent *raises* the utility of others, thus reducing the marginal benefit of contributing to the public good.

Until Section G, we assume that the number of countries is large enough so that no government will perceive any terms-of-trade benefit from manipulating pollution policy. With this assumption, each government chooses its target so that the equilibrium permit price is equal to marginal damage:

$$\tau_i = -V_E/V_I = \beta \left(\sum_{j=1}^{n+n^*} E_j \right)^{\gamma-1} I_i . \tag{8.6}$$

Permit prices are increasing in income since environmental quality is a normal good,[7] and nondecreasing in the aggregate pollution level since the marginal willingness to pay for pollution reduction is nondecreasing.

To generate a relationship between relative factor prices (τ_i/w_i) and pollution supply, note that national income is the sum of labor income and returns from pollution permits:

$$I_i = w_i L_i + \tau_i E_i . \tag{8.7}$$

Combining (Equation 8.6) with (Equation 8.7), letting $E^w \equiv \Sigma_j E_j$ denote world pollution, yields country i's inverse pollution supply curve:

$$\rho_i \equiv \frac{\tau_i}{w_i} = \frac{\beta L_i \left(E^w \right)^{\gamma-1}}{1 - \beta E_i \left(E^w \right)^{\gamma-1}} . \tag{8.8}$$

Since all countries within a region are identical, it is clear from (Equation 8.8) that their pollution supply curves are identical. In equilibrium (autarky or free trade), all countries within a region attain the same level of utility and emit the same amount of pollution.

D. AUTARKY EQUILIBRIUM

The (derived) demand for pollution in autarky arises from the demand for goods whose production creates pollution. Recalling that $\alpha(z)$ is the share of pollution charges in the cost of production, we have

$$\tau_i e_i(z) = \alpha(z) p_i(z) y_i(z) = \alpha(z) p_i(z) x_i(z) = \alpha(z) b(z) I_i \tag{8.9}$$

where $e_i(z)$ is the number of pollution permits needed to produce autarky consumption $x_i(z)$.

Integrating over all goods and using (Equation 8.7) yields the derived demand for pollution:

$$\rho_i \equiv \frac{\tau_i}{w_i} = \frac{\bar{\theta} L_i}{E_i \left[1 - \bar{\theta} \right]} \tag{8.10}$$

where $\bar{\theta} \equiv \int_0^1 \alpha(z) b(z) dz$ is the share of pollution-permit revenue in national income. Equating country i's pollution demand (Equation 8.10) and supply (Equation 8.8) yields its best response to foreign pollution:

---

[7] The optimal permit supply problem depends on aggregate income, reflecting our decision to abstract from income distributional issues by assuming a representative agent. If instead pollution permits were given to firms (rather than auctioned), and if the government placed different weights on labor income and pollution-permit rents collected by firms, then the supply of pollution would depend on distributional weights.

$$E_i \left(E^w \right)^{\gamma-1} = \bar{\theta}/\beta \,. \tag{8.11}$$

Solving the system (Equation 8.11) simultaneously for all i yields autarky pollution levels:

$$E_i^a = \left(\frac{\bar{\theta}}{\beta \left(n + n^* \right)^{\gamma-1}} \right)^{1/\gamma} \equiv E^a \,. \tag{8.12}$$

World pollution in autarky is obtained by summing (Equation 8.12) over all countries:

$$E^{aw} = \left(\frac{\left(n + n^* \right) \bar{\theta}}{\beta} \right)^{1/\gamma} \tag{8.13}$$

Autarky pollution is independent of the level of human capital, and hence all countries generate the same amount of pollution. A larger production capacity created by higher human-capital levels increases the demand for pollution permits (a scale effect), but the ensuing higher income reduces the amount of pollution the population is willing to supply, leading to a higher pollution-permit price and cleaner techniques of production. As in Copeland and Taylor (1994), these scale and technique effects exactly offset each other in autarky, leaving pollution independent of the level of human capital.

On the other hand, changes in the number of countries do affect pollution. On impact, an increase in the number of countries raises world pollution. This increases marginal damage and raises the marginal benefit of controlling pollution. In response, each country cuts back its pollution (from Equation 8.12, an individual country's pollution is declining in $n + n^*$), but not by enough to prevent global pollution from rising (from Equation 8.13, world pollution is increasing in $n + n^*$).

While pollution levels are the same across countries, the relative price of pollution permits differs: since $L > L^*$, we have $E^a/L < E^a/L^*$, and hence pollution permits are relatively scarce (and expensive) in the North. This is illustrated in Figure 8.1, which plots a Northern and a Southern country's pollution demand (D) and supply (S) (using Equations 8.8 and 8.10), parameterized by the equilibrium autarky pollution level of the rest of the world. Because of its higher income, Northern demand for pollution is higher, and its supply lower, than in the South. Consequently, $\rho^a > \rho^{a*}$, and this provides the basis for trade.

E. TRADING EQUILIBRIUM

If countries have sufficiently similar effective labor endowments, factor prices will be equalized by trade; conversely, if endowments are sufficiently different, trade will not equalize factor prices.[8] Since these two types of equilibria have very different implications, we consider both cases.

1. FACTOR PRICES EQUALIZED

First consider the FPE case. With equal factor prices the exact pattern of commodity trade is indeterminate,[9] but we can obtain results on the pattern of trade in factor services. Moreover,

---

[8] The conditions which generate each type of equilibrium are discussed in the Appendix.

[9] This is a standard feature of trade models in which the number of goods exceeds the number of factors. See, for example, Dornbusch et al. (1980).

because supply curves of countries within a region are identical, all countries within a region will produce the same amount of pollution and attain the same income level in free trade. Consequently, we omit individual country subscripts, except in cases where there may be some ambiguity. Thus, for example, E is the amount of pollution produced by a typical Northern country. Asterisks denote corresponding Southern-country variables.

Since $\tau = \tau^*$ in a FPE equilibrium, we conclude from (Equation 8.6) that $I = I^*$, and hence $wL + \tau E = wL^* + \tau E^*$. Rearranging yields

$$L - L^* = \rho\left(E^* - E\right) \tag{8.14}$$

where $\rho \equiv \tau/w$. By definition we have $L > L^*$, and hence $E^* > E$; more pollution is generated by Southern countries. Moreover, since $I = I^*$ and preferences over goods are homothetic, each country consumes a fraction $1/(n + n^*)$ of the world's (embodied) pollution and effective labor services. Hence each Northern country is a net exporter of $n^*(L - L^*)/(n + n^*)$ units of effective labor services, and each Southern country is a net exporter of $n^*(E^* - E)/(n + n^*)$ units of pollution services.

Equilibrium pollution levels can be obtained by equating world demand and supply for pollution services.[10] Denote world magnitudes with a superscript "w", and use the same argument that led to (Equation 8.10) to obtain the world demand for pollution:

$$\rho = \frac{\overline{\theta} L^w}{\left(1 - \overline{\theta}\right) E^w}. \tag{8.15}$$

Next, invert (Equation 8.8), sum, and rearrange to obtain world supply:

$$\rho = \frac{\beta L^w \left(E^w\right)^{\gamma - 1}}{n + n^* - \beta\left(E^w\right)^{\gamma}}. \tag{8.16}$$

Equating demand and supply yields world pollution in the FPE equilibrium:

$$E^w = \left(\frac{\left(n + n^*\right)\overline{\theta}}{\beta}\right)^{1/\gamma}. \tag{8.17}$$

Comparing (Equation 8.17) with (Equation 8.13), and using (Equation 8.12) and (Equation 8.14), we have shown the following.

PROPOSITION 1: *In an FPE equilibrium, trade raises the level of pollution generated by each Southern country, lowers the pollution level generated by each Northern country, and leaves world pollution unaffected.*

Figure 8.1 illustrates the effects of trade on pollution. In Figure 8.1A and 8.1B, we depict the autarky ("a") and trading ("t") equilibria for typical Northern and Southern countries. Figure 8.1C illustrates the trading equilibrium with aggregate world demand and supply. Trade eliminates the

---

[10] This aggregation is possible because factor prices are equalized by trade and tastes are homothetic over private consumption goods.

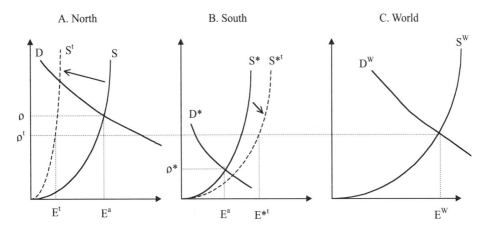

FIGURE 8.1 Pollution demand and supply.

gap between Northern and Southern pollution permit prices. Northern countries move down their supply curves as the relative price of a pollution permit drops, while Southern countries move up their supply curves. In addition, each Northern country's supply curve shifts inward in response to the net increase in foreign pollution,[11] while each Southern supply curve shifts outward in response to the net decrease in pollution from the rest of the world. Consequently, pollution must fall in the North and rise in the South. As is apparent from the diagram, the elastic supply of pollution and the pollution spillovers tend to reinforce the production shifts that trade creates in a standard Heckscher–Ohlin model.

Aggregate world pollution is unaffected by trade for essentially the same reason that pollution is independent of the level of human capital in autarky. Trade leads to real income changes which generate offsetting scale and technique effects. In addition, trade generates a composition effect, as the relatively pollution-intensive industries on average shift to the South from the North. Nevertheless, when factor prices are equalized by trade, the techniques of production are identical across countries, and hence shifting production across regions has no effect on pollution emissions.

Although trade does not affect the level of global pollution, it nevertheless has interesting welfare effects. These can be investigated with the aid of Figure 8.2. Since the world pollution level is unaffected by trade, we can draw an Edgeworth box with dimensions equal to the world supply of pollution E^w and effective labor L^w. Let L^N denote North's aggregate endowment of effective labor. Then the autarky factor supply point is at A, since North produces a share $n/(n + n^*)$ of pollution in autarky (recall that autarky pollution levels are identical across countries). Now consider an *integrated equilibrium* where factor supplies are fixed but freely mobile across countries. This yields some equilibrium $\rho = \tau/w$, and a production locus $y(z)$. If we now divide the world factor supplies between the two regions and allow free trade in goods but not in factors, then, as in Avinash Dixit and Victor Norman (1980), we can find the set of factor allocations where trade in goods alone can replicate the integrated equilibrium. This set is the interior and boundary of the area $O^N a O^S b O^N$. Outside of this area, the full-employment conditions for countries in at least one of the regions cannot be satisfied at the factor price and output vector of the integrated equilibrium.[12]

---

[11] The net result of Southern pollution increases and other Northern reductions leaves a typical Northern country facing more foreign pollution. The shift in the supply curve can be obtained from (8): we have $\partial \rho i/\partial E_j > 0$ for $j \neq i$ since an increase in foreign pollution raises the marginal damage from domestic pollution.

[12] In Dixit and Norman's analysis, the boundaries of the integrated equilibrium region are piecewise linear, with the slope of each piece being the factor-input ratio of each good. With a continuum of goods, the boundary is smooth, and the slope at each point is the factor-input ratio for some good z.

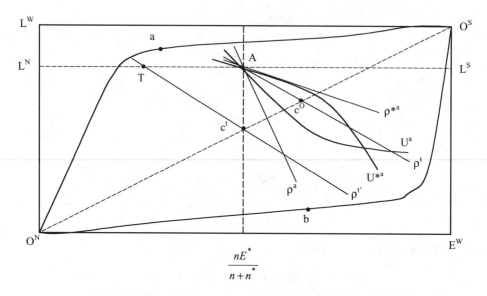

FIGURE 8.2 Gains and losses from trade.

Because of constant returns to scale, we can think of each consumer as buying a bundle of factor services and using the bundle to produce consumption goods.[13] Hence we can draw indifference curves with respect to factor services. Moreover, we can aggregate preferences because all individuals within a region have identical incomes and homothetic preferences over goods. This allows us to draw an indifference curve in Figure 8.2 which represents "Northern utility." In autarky, Northern consumers collectively face the budget constraint labeled ρ^a, and their indifference curve is tangent to this line at point A. Southern consumers face the budget constraint ρ^{*a} and must also be at A. Potential gains from trade lie within the lens-shaped area bounded by the two indifference curves. If pollution were unaffected by trade, the equilibrium free-trade factor-price ratio would generate a budget constraint through point A (labeled ρ^t), and because preferences over goods are identical and homothetic, the equilibrium consumption point would be at c^0. In this hypothetical case, all countries necessarily gain from trade.

---

[13] If consumers have a factor service bundle of (E,L), utility is given by

$$\bar{U}(E,L) = \max\left\{ \int_0^1 b(z)\ln\left(f[e(z),\ell(z);z]\right)dz - \beta\left(E^w\right)^\gamma / \gamma \right\}$$

subject to

$$\int_0^1 e(z)dz = E$$

$$\int_0^1 \ell(z)dz = L$$

where consumers treat world pollution E^w as fixed. Solving, we obtain

$$\bar{U}(E,L) = K + \bar{\theta}\ln E + \left(1-\bar{\theta}\right)\ln L - \beta\left(E^w\right)^\gamma / \gamma$$

where K is a constant. This analysis (i.e., the equivalence of buying goods and factor services) is valid as long as all goods are produced by firms facing the same factor prices.

Although global pollution is not affected by trade, pollution levels within each country do respond to trade. Southern countries increase their pollution, and Northern countries reduce their pollution. This corresponds to a leftward movement along the line $L^N L^S$ to point T. Because pollution permits generate income, this change in the *distribution* of pollution-generating activities has the same effect on the North as a transfer of some of its earning potential to the South. The free-trade budget constraint is thus $\rho^{t'}$, and the free-trade consumption point is at c^{t}.[14]

Several results are immediately apparent from this analysis. First, Southern countries must always gain from trade. World pollution is not affected by trade, and so welfare is affected only by the change in real income. If factor supplies stayed constant before and after trade, then the standard gains-from-trade results would apply. Since the South also generates a greater share of the world's pollution after trade, it realizes additional income gains from its expanded supply of pollution permits.

PROPOSITION 2: *In the FPE equilibrium, Southern countries always gain from trade.*

Whether or not the North gains from trade depends on how much the South increases its pollution as a result of trade. If pollution supplies did not change, the North would gain from access to the South's relatively cheaper (in autarky) pollution-intensive production. However, the North is harmed by South's increase in pollution. As Northern countries cut back on their pollution in response to increased pollution from the South, their collective budget constraint shifts down. In fact, as shown, the North's free-trade budget constraint, $\rho^{t'}$, must lie outside the lens-shaped area, and hence Northern countries lose from trade.

PROPOSITION 3: *Each Northern country is worse off in free trade than in autarky.*
(See Appendix A for the proof.)

The North loses from trade because without any global agreements, Southern countries have a strategic advantage: their lower income allows them to commit to higher pollution levels with the opening of trade. Northern countries can respond either by accepting more global pollution or by cutting back on their own pollution and allowing their incomes to fall. In our model, the reduction in pollution by Northern countries is exactly enough to maintain global pollution at its autarky level. In a more general model, Northern countries may choose a different income/pollution trade-off. But the central point remains: the increased pollution emanating from the South reduces the gains from trade for the North, and this effect can be strong enough to make the North lose from trade.

This suggests that there are strong incentives for the North to link environmental agreements to free-trade agreements. As noted above, if the South were prevented from increasing its pollution, then the North would always gain from trade. Thus, the North has an incentive to link a free-trade agreement with an agreement that freezes pollution at its pre-trade levels, whereas the South has an incentive to oppose this.

PROPOSITION 4: *As compared with unconstrained free trade, each Northern country would gain from an agreement that freezes pollution in each country at its pre-trade levels, and each Southern country would lose from such an agreement.*

Note that global pollution levels are unaffected by such an agreement. The issue here is a conflict over who generates the pollution. For a given level of global pollution, the right to pollute is a valuable asset, and each country can gain if it obtains a greater share of this asset. Since free trade allows the South to exploit its strategic advantage, the North has an incentive to place restrictions on its ability to do so, either by opposing free trade or by insisting that a trade agreement be linked to an environmental agreement.

---

[14] The slopes of $\rho^{t'}$ and ρ^{t} are equal because A and T are both within the FPE region. The equilibrium consumption point must be at c^{t} because with identical homothetic preferences and identical incomes, the North consumes a fraction $n/(n + n^*)$ of the world's factor services.

2. FACTOR PRICES NOT EQUALIZED

If factor prices are not equalized by trade, Northern countries specialize in human-capital-intensive goods, and Southern countries specialize in pollution-intensive goods. Given our ordering on $\rho(z)$ there will exist some $\bar{z}$ such that goods on the interval $[0,\bar{z})$ are produced in the North, and goods on the interval $(\bar{z},1]$ are produced in the South. To determine the equilibrium, we follow Dornbusch et al. (1980), and solve for the marginal good. As in the FPE case, all countries within a region will attain the same outcome, and we omit country subscripts. Aggregate Northern and Southern variables are denoted with superscripts N and S; hence, for example, $L^N = nL$.

Since the marginal good is produced at equal cost in both regions; we have $c(w,\tau,h,\bar{z}) = c(w^*,\tau^*;\bar{z})$, or equivalently, using (Equation 8.2),

$$\omega \equiv \frac{w}{w^*} = \left(\frac{\tau^*}{\tau}\right)^{\alpha(\bar{z})/[1-\alpha(\bar{z})]} \tag{8.18}$$

To eliminate τ^*/τ from the above, note that North's share of world income equals the share of world spending on Northern goods:

$$I^N = \varphi(\bar{z})\left(I^N + I^S\right) \tag{8.19}$$

where $\varphi(\bar{z}) \equiv \int_0^{\bar{z}} b(z)dz$ is the North's share of world income. Using (Equation 8.6) and (Equation 8.19), we have $\tau^*/\tau = n\varphi^*(\bar{z})/n^*\varphi(\bar{z})$ where $\varphi^*(\bar{z})$ is the South's share of world income. Substituting into (Equation 8.18) yields the following:[15]

$$\omega = \left(\frac{n\varphi^*(\bar{z})}{n^*\varphi(\bar{z})}\right)^{\alpha(\bar{z})/[1-\alpha(\bar{z})]} \equiv T(\bar{z}). \tag{8.20}$$

To obtain another equation linking ω and $\bar{z}$, note that income is the sum of wages and pollution-permit revenue, and pollution-permit revenue is the sum of fees paid by all producers. In the North, we have

$$\tau E^N = \int_0^{\bar{z}} \alpha(z)b(z)I^w dz = \frac{I^N \theta(\bar{z})}{\varphi(\bar{z})} \tag{8.21}$$

where the last step uses (Equation 8.19) and where $\theta(\bar{z}) \equiv \int_0^{\bar{z}} \alpha(z)b(z)dz$ is the share of Northern pollution charges in world income. Now use (Equation 8.21) to eliminate pollution charges in (Equation 8.7), do the same for the South, substitute into (Equation 8.19), and simplify to obtain:

$$\omega = \frac{n^*L^*}{nL}\left[\frac{\int_0^{\bar{z}} b(z)[1-\alpha(z)]dz}{\int_{\bar{z}}^1 b(z)[1-\alpha(z)]dz}\right] \equiv B(\bar{z}). \tag{8.22}$$

---

[15] This analysis is valid only if $\tau > \tau^*$, since otherwise we are in the FPE case. Thus we must have $n\varphi^*(\bar{z}) < n^*\varphi(\bar{z})$; or letting $\varphi(\hat{z}) \equiv n/(n - n^*)$, we require $z > \hat{z}$.

Jointly solving (Equation 8.22) and (Equation 8.20) determines the equilibrium.

PROPOSITION 5: *If trade does not equalize factor prices, then (i) global pollution is higher in free trade than in autarky; (ii) pollution in the North falls with trade; and (iii) pollution in the South rises with trade.*
(See Appendix A for the proof.)

In contrast to the FPE case, free trade increases global pollution in a specialized equilibrium.[16] As before, the supply response to the factor-price movements created by trade leads to reduced pollution in the North and increased pollution in the South. However, since the gap between factor prices is not fully eliminated, the South has relatively lower pollution-permit prices than the North. Consequently, the marginal good is produced with more pollution-intensive techniques in the South than in the North. Since the most pollution-intensive industries shift to the country with the lowest pollution-permit price, the increase in pollution generated by the South is less than the fall in the North.[17]

3. TRADE IN POLLUTION PERMITS

As in the Heckscher–Ohlin model, trade in goods in our model is an indirect way of allowing countries to trade factor services. We now consider the effects of allowing direct trade in pollution services when governments agree that a permit issued by one country can be used by a firm that wants to emit pollution in any country.

In the FPE equilibrium, trade in goods and trade in factors are perfect substitutes: allowing trade in pollution permits has no effect on production, incomes, pollution, or welfare. This is a standard result in factor-proportions models. The only difference is that, in our case, the supply of the tradable factor is endogenous. Despite the fact that pollution permits are potential revenue-generators for governments, opening up international trade in pollution permits does not create an incentive to increase their supply beyond the levels of the pure goods-trading equilibrium. This is because pollution is a pure public bad. Since the harm suffered from pollution by the permit-issuing country is independent of the location of production, the income/pollution trade-off is not affected by making permits tradable when factor prices are equalized by goods trade alone.

In the specialized production equilibrium, trade in goods and trade in permits are no longer perfect substitutes. Consider an equilibrium with free trade in goods and then allow pollution permits to be freely traded internationally. Arbitrage equalizes the price of pollution permits across countries, and the zero-profit conditions ensure that τ/w is equalized. Consequently, the FPE equilibrium conditions apply. Since opening up the pollution permit market to international trade induces an FPE equilibrium, pollution must fall.

PROPOSITION 6: *Suppose factor prices are not equalized by goods trade. Then if countries allow pollution permits to be tradable internationally (without any global agreement to restrict their supply), global pollution will fall (relative to the pure goods trade level).*

When pollution-permit prices differ across countries and goods are freely tradable, there is an incentive to shift the most pollution-intensive production to countries with the lowest pollution-permit prices. This composition effect tends to increase world pollution because production techniques are dirtier in countries with low permit prices. Allowing trade in permits equalizes permit prices, ensures that production techniques are identical across countries, and thus eliminates the "pollution haven" effect.

---

[16] The intuition is similar to that for proposition 2 of Copeland and Taylor (1994), where we provide a detailed explanation based on the scale, technique, and composition effects.

[17] A general welfare result for the non-FPE case corresponding to Propositions 2 and 3 has thus far eluded us. It is, however, possible to derive some limited results. For example, at the borderline between FPE and non-FPE, Propositions 2 and 3 hold; as well, if $n^* = 1$, the South must always gain in the non-FPE case.

F. REFORM OF POLLUTION POLICY

In autarky, there are two distortions: relative goods prices differ across countries, and the international externality leads to excessive global pollution. Thus far, we have considered the effects of eliminating the trade distortion, while leaving the pollution distortion unresolved. We now consider the effects of various proposals to control world pollution.

1. GLOBAL REFORM

We begin by briefly discussing the first best. Since environmental quality is an international public good, it is clear that global pollution levels are Pareto inefficient in a Nash equilibrium. The Samuelson condition for efficient public-goods provision requires that permit prices (the marginal pollution-control cost) be equalized across countries, and set equal to the global sum of marginal damage; that is, for any country j,

$$\tau_j = \tau = -\sum_{i=1}^{n+n^*} \left(V_{E_i}^i / V_{I_i}^i \right) = \beta \left(E^w \right)^{\gamma-1} I^w .$$ (8.23)

Combining (Equation 8.23) with (Equation 8.15) yields the optimal global pollution level:[18]

$$\tilde{E}^w = \left(\frac{\bar{\theta}}{\beta} \right)^{1/\gamma}$$ (8.24)

which is less than autarky and free-trade global pollution levels, as one would expect.

Implementation of the first best is straightforward in theory, but difficult in practice. For example, governments could agree to issue a fixed number of internationally tradable pollution permits. Moreover, since permits generate income, negotiations to divide the initial allocation of permits across countries can take the place of explicit international side payments. Alternatively, the permits need not be tradable, but since efficiency requires that their price be equalized, they must be allocated across countries in a manner that is consistent with factor-price equalization. In this case, explicit international transfer payments may be required to support some points on the Pareto frontier.

In practice, large changes in pollution may not be politically feasible. Instead, international agreements often require only small reductions in the target variable. In our model, it is easy to show that equiproportionate reductions in pollution by all countries will be Pareto-improving as long as global pollution exceeds the optimum, and that repeated application of the eqiuproportionate reduction rule will eventually implement a point on the Pareto frontier. In a more general model, the path to the Pareto frontier may not be so simple. But the main point remains: application of standard results from public finance implies that either radical or gradual multilateral reductions in pollution, perhaps combined with international transfers of income, can achieve a first-best allocation.

2. REGIONAL REFORMS

In reality, any multilateral agreement requiring pollution reductions by *all* countries may be difficult to achieve. Some pollution agreements, such as the Montreal protocol, require pollution reductions by only the major polluting countries. To examine the effect of these limited agreements, suppose

---

[18] Note that the optimal level of global pollution is independent of any distributional weights on country utilities. This follows from our assumption that pollution is strongly separable from consumption in the (homothetic over goods) utility function.

all countries in the North agree to reduce their pollution by the same small amount while the South commits to freezing its pollution at current levels. For simplicity, we focus on the FPE case.

First consider the effect on the welfare of a typical Northern country whose indirect utility is given by (Equation 8.4). Using the zero-profit conditions and (Equation 8.2) we have $p(z) = c(z) = k(z)\rho^{\alpha(z)}w$, and using (Equation 8.7) we have an expression for income. Substituting these into (Equation 8.4) and rearranging, we obtain the following expression for Northern country i's indirect utility:

$$V^i = K + \ln(L + \rho E) - \bar{\theta}\ln(\rho) - \beta(E^w)^\gamma / \gamma \qquad (8.25)$$

where K is a constant. Totally differentiating, and noting that each individual country i has chosen its pollution level so that $dV^i / dE_i = 0$, yields

$$dV^i = \left[\frac{\rho E}{L + \rho E} - \bar{\theta}\right]\hat{\rho} - \beta(E^w)^{\gamma-1}\sum_{j \neq i} dE_j \ . \qquad (8.26)$$

To simplify, let $M = E^w/(n + n^*) - E$ denote a Northern country's net imports of embodied pollution services in the FPE equilibrium. Then, after some manipulation, we obtain:[19]

$$dV^i = \frac{\rho E}{L + \rho E}\left[\frac{nM}{E^w} - (n - 1)\right]\hat{E} \ . \qquad (8.27)$$

As (Equation 8.27) illustrates, a cut in pollution by the North has two effects on Northern welfare. First, since world pollution is reduced, the relative price of pollution permits rises ($\hat{\rho} > 0$). Since each Northern country is a net importer of pollution services ($M > 0$), this worsens Northern countries' terms of trade. Although no individual country perceives a terms-of-trade effect from cutting its own pollution, when the coalition of all Northern countries cuts pollution, the terms-of-trade effect is multiplied n-fold and is therefore significant.

If there were only one Northern country ($n = 1$), then the terms-of-trade deterioration would be the only effect of a unilateral pollution cut,[20] and that country would have an incentive to increase its level of pollution in order to improve its terms of trade. With n countries, however, each country benefits from the reduction in pollution by the other pollution-cutting countries. This is the second term in (Equation 8.27), and it provides a counter-balance to the terms-of-trade effect. To determine the net effect, substitute for M in (Equation 8.27), Northern countries gain from a cut in Northern pollution provided that

$$\left[\frac{n}{n + n^*} - (n - 1)\right]E^w < nE$$

---

[19] Since $I = I^*$ in the FPE equilibrium, we have $(L^w + \rho E^w) = (n + n^*)(L + \rho E)$. Combining this with (Equation 8.15) yields $\bar{\theta} = \rho E^w/[(n + n^*)(L + \rho E)]$. Substituting this into (Equation 8.26), using (Equation 8.17) to eliminate E^w, and noting that $dE_j = dE$ for all Northern j, and $dE_j = 0$ for Southern j, yields

$$dV^i = \frac{\rho}{L + \rho E}\left[-M\hat{\rho} - (n - 1)dE\right] \cdot$$

Finally, noting from (Equation 8.15) that $\hat{\rho} = -\hat{E}^w = -(nE/E^w)\hat{E}$, we obtain (Equation 8.27).

[20] Since we are starting at a Nash equilibrium, where each country sets its pollution level optimally, there is no first-order benefit to a country from its own pollution reduction.

which is true for $n \geq 2$. Once there are at least two Northern countries, the benefits of partially correcting the international pollution externally more than offset the terms-of-trade deterioration, and the North has a collective incentive to cut its pollution.

Let us now consider the effects of the North's cut in pollution on the South. Since Southern countries are net exporters of pollution services, they stand to reap a terms-of-trade gain as the relative price of pollution-intensive goods rises. Moreover, the South benefits from a cleaner environment. Hence, the South must gain from a cut in pollution by the North.

For similar reasons, a small cut in pollution by a coalition of all the Southern countries must also benefit the South: their terms of trade improve, and each Southern country benefits from the pollution cuts carried out by the other Southern countries. The North also gains from a pollution reduction by the South, despite a terms-of-trade deterioration. Totally differentiating (Equation 8.25), and noting from (Equation 8.16) that $\hat{\rho} = -\hat{E}^w$ (where $\hat{E}^w$ is the change in world pollution due to the South's cut), the effect on a typical Northern consumer in country i is

$$dV^i = \left[-\frac{\rho E}{L + \rho E} + \bar{\theta} - \beta \left(E^w \right)^\gamma \right] \hat{E}^w .$$

As long as world pollution is above the global optimum (i.e., as long as $E^w > [\bar{\theta}/\beta]^{1/\gamma} = \tilde{E}^w$), Northerners must gain from a fall in Southern pollution.

These results have several interesting implications for policy. First, since one region's reduction in pollution always benefits the other, each region has an incentive to offer a cash payment to the other in return for a reduction in pollution; however, net exporters of pollution services have a stronger incentive to do this. Second, a net exporter of pollution services stands to gain from its own region's cut in pollution. This means that selfish interests of pollution-intensive countries need not necessarily be in conflict with global environmental quality. Finally, the use of pollution policy by a coalition of countries to improve its terms of trade can be Pareto improving in some cases. If the pollution-intensive countries tighten their pollution regulations, they benefit through improved terms of trade, and their trading partners benefit from improved environmental quality.

It is, however, important to keep in mind that these results require that those countries not cutting pollution must commit to freezing pollution at its current level. Moreover, each country would prefer to be in the group subject to the freeze rather than in the group actually reducing emissions. Hence while the cuts are "unilateral," a multilateral treaty must be in force to restrict opportunistic behavior.

G. TRANSFERS

Because global pollution levels are inefficient at the Nash equilibrium, coalitions of countries have incentives to pay their trading partners to reduce pollution. Therefore a transfer tied to pollution reduction can lead to a Pareto improvement. In this section, we consider the effects of *untied* transfers. If there is no externality, then in a standard two-country trade model with identical homothetic tastes, a transfer leaves world prices unaffected and must harm the donor country. This result requires that, on impact, the transfer not directly affect the supply side of the model. In our case, however, the supply side of the model is affected, since pollution supplies respond to income changes. Despite this complication, we will show that a transfer has no real effects even if tastes are neither identical nor homothetic.

We consider only the FPE case. Let labor be the numeraire, and suppose a Northern country gives a transfer T to a Southern country.[21] Each country chooses its pollution level based on

---

[21] The same analysis applies if a coalition of Northern countries gives transfers to a group of Southern countries.

(Equation 8.6), where income is interpreted as being net of the transfer. Using a derivation similar to that which led to (Equation 8.8), the pollution supply of country i is given by

$$E_i = \frac{\rho - \beta (E^w)^{\gamma-1}(L_i + T_i)}{\rho\beta(E^w)^{\gamma-1}} \tag{8.28}$$

where $T_i = -T$ for the donor, $T_i = T$ for the recipient, and $T_i = 0$ for other countries.

Summing over all countries, we find that world pollution supply is unaffected by the transfer.[22] With FPE, both countries face the same income/pollution trade-off at the margin. Consequently, one country's marginal increase in pollution exactly off-sets the other's reduction.

Perhaps more surprisingly, incomes are also unaffected by the transfer. The donor issues enough new pollution permits to bring its income back to where it was prior to the transfer. Similarly, the recipient's cut in pollution reduces its net income back to its pre-transfer level. From (Equation 8.28), we have $\Delta E_i = -\Delta T_i/\rho$, and since $I_i = L_i + \rho E_i + T_i$, we have $\Delta I_i = 0$.

Since global pollution and each country's income are unaffected by the transfer, goods prices are also unaffected, and there is no effect on welfare in either country. The only effect of a transfer is to change the location of pollution emissions: the donor becomes more pollution-intensive, and the recipient becomes less pollution-intensive. To summarize, we have shown the following.

PROPOSITION 7: *In an FPE equilibrium, an income transfer between countries (i) raises pollution in the donor country, (ii) lowers pollution in the recipient country, (iii) has no effect on global pollution, and (iv) has no effect on welfare in either country.*

These surprising results are more general than might be thought, and they are quite closely related to Warr's theorem in public finance (Warr, 1983). Warr shows that the private provision of a public good, and the welfare levels of the providers, are unaffected by income transfers among agents. Warr's result and ours do not rely on identical or even homothetic tastes;[23] they do, however, require that each individual face the same trade-off between private-goods consumption and public-goods consumption at the margin. In our context, agents (nations) contribute to a pure public bad (pollution), and each makes a positive contribution; but more importantly, each faces the same trade-off between private goods and pollution abatement in any FPE equilibrium. Consequently, income transfers have no effect on any agent's welfare.

H. TERMS-OF-TRADE EFFECTS

Up until now, we have assumed that no individual country has an incentive to use its pollution policy to improve its terms of trade. In this section, we relax this assumption and allow each country to account explicitly for terms-of-trade effects when choosing its pollution quota.

---

[22] In contrast, when factor prices are not equalized by trade, one can show that a transfer from Northern to Southern countries lowers world pollution.

[23] The following argument demonstrates that we do not require homotheticity or identical preferences. See also the proof of Warr's result in Theodore C. Bergstrom et al. (1986). With FPE and constant returns to scale, consumers can be thought of as purchasing embodied factor services (see Footnote 11), and therefore the utility of a donor can be written as $V(\rho, I_o + \rho E^w, E^w)$, where $I_o = L - T - \rho E_{-i}$, E_{-i} denotes foreign pollution, and where we can think of the government as choosing E^w, subject to $E^w \geq E_{-i}$. The first-order condition is $\rho = -V_E/V_I \equiv \phi(\rho, I_o, E^w)$. If the South adjusts its pollution to fully offset the transfer (i.e., if $\Delta E^* = -\Delta T/\rho$), then $\Delta I_o = 0$. Moreover, $\phi(\rho, I_o, E^w)$ is unchanged. Consequently, North's optimal choice of E^w (and hence E_i) is unchanged. Thus, if all other countries adjust their pollution to fully offset the transfer, the remaining country's optimal response is to do the same. Finally, since all real incomes and trade-offs are unaffected, the aggregate demand for pollution services is unaffected, and hence the equilibrium price ρ is also the same as before.

We consider only the FPE case. Each country i chooses its own pollution E_i to maximize (Equation 8.25), treating foreign pollution ($E_{-i}^w \equiv E^w - E_i$) as given, subject to the market-clearing condition (Equation 8.15), which can be rewritten as

$$\rho \equiv \frac{\bar{\theta} L^w}{\left(1 - \bar{\theta}\right)\left(E_{-i}^w + E_i\right)}. \tag{8.29}$$

The first-order condition for country i's optimal choice is

$$\rho + V_E^i / V_I^i - M_i\left(\partial\rho/\partial E_i\right) = 0 \tag{8.30}$$

where M_i is net imports of embodied pollution services. Using (Equation 8.29) to solve for $\partial\rho/\partial E_i$, and rearranging, we can write (Equation 8.30) as

$$\rho - \beta\left(E^w\right)^{\gamma-1} I_i - \frac{\theta_i - \bar{\theta}}{E^w} I_i = 0 \tag{8.31}$$

where $I_i \equiv L_i + \rho E_i$ is national income of country i, and $\theta_i \equiv \rho E_i / I_i$ is the share of pollution-permit income in country i's national income (recall that $\bar{\theta}$ is the share of pollution charges in world income). Equation (8.31) differs from (8.6) only by the presence of the final term, which represents the terms-of-trade effect. Taking account of terms-of-trade effects tends to increase the marginal benefit of polluting for the North (a net importer of pollution services with $\theta < \bar{\theta}$) and tends to reduce it for the South (a net exporter with $\theta^* > \bar{\theta}$).

To find the new global pollution level, sum (Equation 8.31) over all countries, rearrange, and obtain (Equation 8.17): the level of world pollution is unaffected by the recognition of terms-of-trade effects. Moreover, using (Equation 8.15), we conclude that the equilibrium level of ρ is also unaffected. These two results might appear to be artifacts of our Cobb–Douglas specification, but they are not. In fact, they require only that aggregate demand and supply for pollution services be independent of the world distribution of income.

PROPOSITION 8: *Assume FPE, constant-returns-to-scale technologies which are identical across countries, and identical preferences which are homothetic over goods for any given level of world pollution. Consider two regimes: one where countries ignore terms-of-trade effects when choosing pollution levels, and the other where countries do take into account terms-of-trade effects. Then the equilibrium level of world pollution is the same in the two regimes.*
(See Appendix A for the proof.)

The intuition for this result is as follows. Notice from (Equation 8.30) that the introduction of terms-of-trade effects means that the net marginal benefit of polluting shifts up by $M\partial\rho/\partial E$ for a pollution importer, and down by $M^*\partial\rho/\partial E^*$ for a pollution exporter. But since these terms represent pure transfers of income, they must always sum to zero. Hence, on impact, the aggregate net marginal benefit of polluting (i.e., the aggregate "supply" of pollution) is unaffected by the introduction of terms-of-trade effects. If preferences are identical and homothetic over goods for given levels of global pollution, the ensuing income redistribution will also have no effect on the aggregate demand for pollution. Consequently, with both demand and supply unaffected, so too is world pollution.

This is quite a striking result because it tells us that global pollution levels will be unaffected if nations manipulate pollution policy to gain a terms-of-trade advantage. While this is quite surprising, it does rely heavily on our assumption that global pollution affects all countries equally.

If instead we were to assume that creating emissions within a county also leads to additional local effects, then the result would not hold. That is, like Warr's theorem, Proposition 8 relies on the purity of the public bad or good.

Even though terms-of-trade effects have no aggregate effect on pollution, country-specific levels are affected. Using $I_i = L_i + \rho E_i$ in (Equation 8.31), we find

$$E_i = \frac{\rho E^w - L_i \left[\beta \left(E^w \right)^\gamma - \bar{\theta} \right]}{\rho \left[1 + \beta \left(E^w \right)^\gamma - \bar{\theta} \right]}. \tag{8.32}$$

Using (Equation 8.32) and (Equation 8.17), the difference between Northern and Southern pollution is

$$E^* - E = \frac{L - L^*}{\rho} \left[\frac{\left(n + n^* - 1 \right) \bar{\theta}}{1 + \left(n - n^* - 1 \right) \bar{\theta}} \right] > 0. \tag{8.33}$$

As before, Southern countries produce more pollution than Northern countries in free trade. Moreover, Proposition 2 and 3 continue to hold: the South must always gain from trade, and the North must always lose (see the Appendix). Despite these similarities, it is straightforward to show that the gap between Northern and Southern pollution is smaller in (Equation 8.33) than in (Equation 8.14). Consequently, the North does not cut its pollution with the opening of trade by as much as before, and the South increases its pollution by less. As a result, the North's free-trade income is higher, and the South's is lower, when terms-of-trade effects are not ignored.

We previously noted that South's low income gives it a strategic advantage in the pollution game since it can credibly commit to polluting more with the opening of trade. Our analysis here reveals that being a net importer of pollution services gives the North a strategic advantage in the interaction over terms-of-trade effects: the North's incentive to improve its terms of trade allows it to commit credibly to pollute more, and this makes the South less aggressive in increasing its pollution. In terms of the graph shown in Figure 8.2, when terms-of-trade effects were ignored, the free-trade equilibrium was at point T; once countries actively manipulate the terms of trade, the new equilibrium point must be to the right of point T.

It is apparent that the relative strengths of these two strategic effects determine the division of the gains from trade. As mentioned earlier, the strategic advantage of a low income always dominates in our formulation, to leave the North worse off in trade. Since this result is likely to be model-specific, it is useful to examine how the strengths of the two opposing effects are determined. Since market size is an obvious determinant of the strength of terms-of-trade effects, let us consider the effects of increasing the number of countries, holding the ratio of Northern to Southern countries (n/n^*) fixed:

PROPOSITION 9: *Let $n^* = kn$, where k is a constant. Let V^T and V^A denote Northern utility in free trade and autarky. Then $d(V^T - V^A)/dn < 0$, and $d(V^{*T} - V^{*A})/dn > 0$, when terms-of-trade effects are not ignored by countries when choosing pollution policy; when terms-of-trade effects are ignored, $d(V^T - V^A)/dn = d(V^{*T} - V^{*A})/dn = 0$.*

(See Appendix A for the proof.) Proposition 9 follows because the North, as a pollution importer, gains a strategic benefit as terms-of-trade effects become stronger. Terms-of-trade effects strengthen as the number of countries shrinks, and consequently, Northern losses from trade decrease. Conversely, Southern gains from trade rise with the number of countries because terms-of-trade motivations diminish in importance, and hence the strategic advantage of low income is enhanced.

A further implication of this analysis is the following:

PROPOSITION 10: *The North prefers a regime that allows pollution to be used as an instrument of trade policy, whereas the South prefers that such actions be banned.*

This proposition suggests that GATT Article XX outlawing environmental policy as disguised trade policy works in favor of lower-income nations. A regime that removes the ability of net importers of pollution services to manipulate their terms of trade via pollution policy puts them at a strategic disadvantage relative to net exporters. Such a rule strengthens the South's commitment to pollute more in free trade, and this shifts the ownership of the world's pollution services to the advantage of the South. Again, what is at issue here is the division of the right to pollute across countries, since global pollution is unaffected.

A final interesting change created by terms-of-trade effects is that the normative part of the neutrality result on transfers (Proposition 7) now fails.[24] Consider the effect of a transfer T from a Northern country to a Southern country. It is straightforward to show that, as before, the transfer does not affect the level of world pollution. The welfare effects, however, are different. In the large-country case, we had $dE/dT = 1/\rho$, and hence the North's real income was unaffected by the transfer (since $dI/dT = \rho dE/dT - 1 = 0$). In the present case, we have (differentiating (Equation 8.31), and noting that ρ and E^w stay constant):

$$\frac{dE}{dT} = \frac{1}{\rho} - \left[\frac{I}{\rho(n + n^* - 1)\overline{\theta} + \theta} \right] \frac{d(\theta - \overline{\theta})}{dT} < \frac{1}{\rho}.$$

Once again, North's pollution response is dampened by its recognition of terms-of-trade effects. Moreover, since $dE < dT/\rho$, the North must lose from a transfer to the South. The intuition for this result is evident once we recall that the incentive to exploit terms-of-trade effects increases in the relative difference between countries. Since income is the only fundamental difference between countries, an income-disparity-reducing transfer from the North to the South tends to reduce the motive to use pollution to manipulate the terms of trade. (In terms of Equation 8.31, the absolute value of $\theta - \overline{\theta}$ falls with the transfer.) This reduces the North's incentive to raise its pollution level to compensate for the income-reducing effects of the transfer, and similarly dampens the South's pollution reduction.

In summary, the analysis of strategic trade-policy motives for pollution policy has allowed us to generate several new and interesting results. In particular it has highlighted the strategic advantage gained by a net importer of pollution services and has shown that the key qualitative implications of our model are not sensitive to the recognition of terms-of-trade effects.[25] Moreover, substituting (Equation 8.17) into (Equation 8.31), we see that, as $n + n^*$ gets large, the terms-of-trade effect, $[\theta_i - \overline{\theta}]/E^w$, approaches zero, while the marginal disutility of increased pollution, $\beta(E^w)^{\gamma-1}$, gets large.[26] Thus, as $n + n^*$ gets large, the model of this section converges to our earlier specification.

---

[24] Many other authors have noted that Warr's result relies on each agent valuing his or her contribution to the public good only so far as it contributes to the aggregate quantity of the public good. Once terms-of-trade effects are allowed, this requirement is no longer met.

[25] Moreover, as before, the North would prefer to commit to autarky pollution levels prior to trade, whereas the South would not (Proposition 4). The effects of pollution reform in Section F are only slightly different since each country has internalized its own terms-of-trade effect. If $n = n^* = 1$, no country has an incentive to make a unilateral reduction, since countries are initially at a Nash equilibrium. However, when there is more than one country of each type, the effects of pollution reduction by a coalition of countries are essentially the same: there is no first-order benefit or cost from the terms-of-trade effects of one's own reduction, but as before, each country in the coalition is affected by the terms-of-trade consequences of pollution reduction by all other members of the coalition, and also by the environmental benefits.

[26] The terms-of-trade effect $[\theta_i - \overline{\theta}]/E^w$ approaches zero as n approaches infinity because E^w is increasing in $n + n^*$, and $0 \le \theta_i \le 1$. Using (Equation 8.17) in (Equation 8.31), the marginal disutility of increased pollution is then $\beta[(n + n^*)\overline{\theta}/\beta]^{(\gamma-1)/\gamma}$. This increases with $n + n^*$ since $\gamma \ge 1$.

I. CONCLUSION

This chapter has examined the effects of trade and environmental policy on trade flows, pollution levels, and welfare. To focus on income effects, we have eliminated all motives for trade except those arising from income-induced differences in attitudes toward the environment. In addition we have modeled global pollution as a pure public bad where a country's physical size, population density, or weather pattern has no bearing on its exposure to global pollution. A resolution of the debate over the effects of trade on environment must of course examine the interaction among all of the many motives for trade and must consider how local and global pollution interact with each country's physical environment to generate the true impact of pollution. Nevertheless, while our model is very simple and stylized, we think it raises several interesting issues worthy of future examination.

First, we find that the pre-trade world distribution of income determines how trade will affect the environment. If the world distribution of income is highly skewed, then free trade harms the global environment; but if countries have relatively similar incomes, then free trade has no adverse effect on the environment. Second, we find that, because lower-income countries have a strategic advantage in setting pollution levels in a free-trade regime, they have an incentive to delay international pollution negotiations until after multilateral trade liberalization has been achieved. Third, we find that reductions in pollution by a coalition of counties may be Pareto improving, and that income transfers tied directly to pollution reduction can be welfare-enhancing. Untied transfers, however, may have no effect on global pollution levels, on prices, and most surprisingly, on either country's welfare. Finally, we find that many of these results continue to hold when countries use pollution policy to manipulate their terms of trade. However, terms-of-trade motivations for pollution policy do lessen the strategic advantage of lower-income countries.

Overall, our results underline the importance of endogenizing pollution policy within a general-equilibrium framework. While a complete resolution of the debate over the effects of trade on the environment must await further study and more general models, we have shown that income effects created by income transfers or by trade in goods or pollution permits have important and often surprising effects on pollution, trade flows, and welfare levels. Competitive trade theory is replete with examples of surprising results created by general-equilibrium income effects. Our contribution has been to show that these same income effects may also play a large role in determining how international trade affects the global environment.

APPENDIX A: PROOFS OF PROPOSITIONS

PROOF OF PROPOSITION 3:
Denote free trade variables with T and autarky variables with A. Then using (Equation 8.4), and the fact that E^w is not affected by trade, we have

$$V^T - V^A = \int_0^1 b(z) \ln\left[\frac{I^T/p^T(z)}{I^A/p^A(z)}\right] dz = \ln\left(\frac{L + \rho^T E^T}{L + \rho^A E^A}\right)\left(\frac{\rho^A}{\rho^T}\right)^{\bar{\theta}}. \qquad (8.34)$$

Derivation of (Equation 8.34) also uses (Equation 8.7) and the fact that $p(z) = c(z) = k(z)\rho^{\alpha(z)}w$, where $k(z)$ is an industry-specific constant. Let $s \equiv \bar{\theta}/(1 - \bar{\theta})$. Then from (Equation 8.10) and (Equation 8.16), we have

$$\rho^A E^A = sL$$
$$\rho^T E^W = SL^w \qquad (8.35)$$

Dividing these expressions and noting that $E^w = (n + n^*)E^A$ (from Equation 8.17 and Equation 8.12), we obtain

$$\frac{\rho^A}{\rho^T} = \frac{(n + n^*)L}{L^w} .$$ (8.36)

Inverting (Equation 8.8), evaluating at free trade, and rearranging yields

$$E^T = \frac{E^w}{\beta(E^w)^\gamma} - \frac{L}{\rho^T} .$$ (8.37)

Hence using (Equation 8.17), (Equation 8.35), and (Equation 8.36), we have

$$\rho^T E^T + L = sL^w \big/ (n + n^*)\overline{\theta} .$$ (8.38)

Now substitute (Equation 8.38), (Equation 8.35), and (Equation 8.36) into (Equation 8.34), and noting that $\overline{\theta}(1 + s) = s$, simplify to obtain

$$V^T - V^A = \ln\left(\frac{L^w}{(n + n^*)L}\right)^{1-\overline{\theta}} < 0$$ (8.39)

since $L^w = n^*L^* + nL < (n + n^*)L$. Similarly, $V^{T^*} - V^{A^*} > 0$ since $(n + n^*)L^* < L^w$.

PROOF OF PROPOSITION 5:
   Use (Equation 8.6) to eliminate permit prices from (Equation 8.23) and (Equation 8.24), to obtain:

$$E^N\left(E^N + E^S\right)^{\gamma-1} = \frac{n\theta(\overline{z})}{\beta\varphi(\overline{z})}$$

$$E^S\left(E^N + E^S\right)^{\gamma-1} = \frac{n^*\theta^*(\overline{z})}{\beta\varphi^*(\overline{z})}$$ (8.40)

Dividing yields $E^* = E^S/n^* > E^N/n = E$ in trade, that is,[27]

$$\frac{E^*}{E} = \frac{\theta^*(\overline{z})/\varphi^*(\overline{z})}{\theta(\overline{z})/\varphi(\overline{z})} > 1 .$$ (8.41)

---

[27] To confirm the inequality, note that because α is increasing in z, we have

$$\theta^*(\overline{z}) > \int_{\overline{z}}^1 \alpha(\overline{z})b(z)dz = \alpha(\overline{z})\varphi^*(\overline{z})$$

$$\theta(\overline{z}) < \int_0^{\overline{z}} \alpha(\overline{z})b(z)dz = \alpha(\overline{z})\varphi(\overline{z}).$$

Using these inequalities in (Equation 8.41) yields the result.

World pollution is determined by added the expressions in (Equation 8.40) and solving

$$E^N + E^S = \left(\frac{n^* \theta^*(\bar{z})/\varphi^*(\bar{z}) + n\theta(\bar{z})/\varphi(\bar{z})}{\beta} \right)^{1/\gamma}. \tag{8.42}$$

Subtracting (Equation 8.13) from (Equation 8.42) to obtain

$$E^N + E^S - E^{aw} = \left(\frac{n^* \theta^*(\bar{z})/\varphi^*(\bar{z}) + n\theta(\bar{z})/\varphi(\bar{z})}{\beta} \right)^{1/\gamma} - \left(\frac{(n + n^*)\bar{\theta}}{\beta} \right) > 0.$$

The result follows since the first term is increasing in z and equals the second term for $\varphi = n/(n + n^*)$; but $\varphi > n/(n + n^*)$ since $I > I^*$ in the non-FPE equilibrium.

PROOF OF PROPOSITION 8:

By constant returns to scale and FPE, the utility function can be written as $V(\rho, I_i, E^w)$, using an argument similar to that in Footnote 8.13. By homotheticity, $V^i = \psi(\rho, E^w)I_i$, for some function ψ. Using Roy's identity, the aggregate demand for pollution services is

$$E^w = -\sum_i \left(\psi_\rho / \psi \right) I^i = -I^w \left(\psi_\rho / \psi \right) = -\left(L^w + \rho E^w \right) \left(\psi_\rho / \psi \right). \tag{8.43}$$

The first-order condition for country i's supply of pollution is

$$-M^i \left(\partial \rho / \partial E^i \right) + \rho + \left(\psi_E / \psi \right) I^i = 0 \tag{8.44}$$

where M^i is i's net imports of pollution services. Summing over i (noting that $\Sigma_i M^i = 0$ and that, with FPE, $\partial \rho / \partial E^i = \partial \rho / \partial E^j \ \forall \ i,j$) yields:

$$\left(n + n^* \right) \rho + \left(\psi_E / \psi \right) \left(L^w + \rho E^w \right) = 0. \tag{8.45}$$

The level of ρ and E^w are determined by (Equation 8.43) and (Equation 8.45). If terms-of-trade effects are ignored, the term $-M^i(\partial \rho / \partial E^i)$ drops out of (Equation 8.44), but this has no effect on (Equation 8.45) or (Equation 8.43). Hence the same equations (Equation 8.43) and (Equation 8.45) determine E^w in each regime.

PROOF OF PROPOSITION 9:

From the proof of Proposition 3, we have $V^T - V^A = \ln(H)$, where

$$H \equiv \left(\frac{L + \rho^T E^T}{L + \rho^A E^A} \right) \left(\frac{\rho^A}{\rho^T} \right)^{\bar{\theta}} = \left[\frac{L + s(nL + n^* L^*)}{L + s(n + n^*)L} \right] \left[\frac{(n + n^*)L}{nL + n^* L^*} \right]^{s/(1+s)}$$

using an argument similar to that in the proof of Proposition 3, but using (Equation 8.32) instead of (Equation 8.37). Hence, letting $n^* = kn$, we now have $d(V^T - V^A)/dn = d(\ln(H))/dn = k_1(L^* - L) < 0$, where $k_1 > 0$. For the South, the roles of L^* and L are reversed, and we have $d(V^{*T} - V^{*A})/dn > 0$. For the case in which terms-of-trade effects are ignored, the result follows from (Equation 8.39), which is independent of n, once n/n^* is held constant.

PROPOSITION A1: *The South gains from trade, and the North loses from trade, under the assumption of Section H (i.e., FPE and that countries take into account terms-of-trade effects when setting pollution quotas).*

PROOF:

For the South, the argument behind Proposition 2 still works. For the North, referring to the proof of Proposition 9, note that $H = 1$ for $L = L^*$, and H is increasing in L^*. Hence $H < 1$ for $L^* < L$, and therefore $V^T - V^A = \ln(H) < 0$ for $L^* < L$.

APPENDIX B: CONDITIONS FOR FACTOR-PRICE EQUALIZATION

To determine the boundary between the FPE and non-FPE cases, start in an equilibrium where factor prices differ, and consider the effects of increasing L^*/L. This increases $B(z)$ for all z, but does not affect $T(z)$. As L^*/L rises, there is a point at which $B(z)$ and $T(z)$ intersect at $z = \hat{z}$. At this point, $\omega = 1$, and factor prices just equalize. Define

$$\delta \equiv \frac{\int_0^{\hat{z}} b(z)[1 - \alpha(z)]dz}{\int_{\hat{z}}^1 b(z)[1 - \alpha(z)]dz} .$$

Using (Equation 8.22) and (Equation 8.20), we have $B(\hat{z}) = T(\hat{z})$ when $n^*L^*\delta/nL = 1$, since by definition, $n^*\varphi(\hat{z}) = n\varphi^*(\hat{z})$. Thus we have a non-FPE equilibrium for $nL/n^*L^* > \delta$. By symmetry, if we reverse the roles of North and South, we will also have a non-FPE equilibrium if $n^*L^*/nL > \delta$, or if $nL/n^*L^* > 1/\delta$. For intermediate values of nL/n^*L^*, factor prices are equalized.

PROPOSITION B1: *Factor prices are equalized if and only if $1/\delta \leq nL/n^*L^* \leq \delta$. If $nL/n^*L^* > \delta$, then $\tau > \tau^*$, and the North specializes in relatively clean goods, while the South specializes in pollution-intensive goods. If $nL/n^*L^* < 1/\delta$, then $\tau < \tau^*$, and the North specializes in pollution-intensive goods, while the South specializes in clean goods.*

REFERENCES

Bergstrom, T. C., Blume, L., and Varian, H. R., On the private provision of public goods, *J. Public Econ.*, 29 (1), 25, 1986.

Copeland, B. R. and Taylor, M. S., North–South trade and the environment, *Q. J. Econ.*, 109 (3), 755, 1994.

Dean, J. M., Trade and the environment: a survey of the literature, in *International Trade and the Environment*, Low, P., Ed., World Bank discussion papers, World Bank, Washington, DC, 1992, 15.

Dixit, A. and Norman, V., *Theory of International Trade*, Cambridge University Press, Cambridge, 1980.

Dornbusch, R., Fischer, S., and Samuelson, P. A., Heckscher–Ohlin trade theory with a continuum of goods, *Q. J. Econ.*, 95, 203, 1980.

Grossman, G. M. and Krueger, A. B., Environmental impacts of a North American Free Trade Agreement, in *The Mexico–U.S. Free Trade Agreement*, Garber, P., Ed., MIT Press, Cambridge, 1993, 13.

Ludema, R. D. and Wooton, I., Cross-border externalities and trade liberalization: the strategic control of pollution, *Can. J. Econ.*, 27, 950, 1994.

Markusen, J. R., International externalities and optimal tax structures, *J. Int. Econ.*, 5, 15, 1975a.

Markusen, J. R., Cooperative control of international pollution and common property resources, *Q. J. Econ.*, 89, 618, 1975b.

Rauscher, M., National environmental policies and the effects of economic integration, *Eur. J. Political Econ.*, 7, 313, 1991.

Warr, P. G., The private provision of a public good is independent of the distribution of income, *Econ. Lett.*, 13, 207, 1983.

9 Environmental Policy and International Trade When Governments and Producers Act Strategically[1]

Alistair Ulph

Some environmentalists express concern that trade liberalization may damage the environment by giving governments incentives to relax environmental policies to give domestic producers a competitive advantage. Support for such concern may be given by models of imperfectly competitive trade where there may be "rent-shifting" incentives for governments to relax environmental policies. But there are also incentives for producers to act strategically, e.g., through their investment in R & D, and in this chapter I extend the literature on strategic environmental policy by allowing for strategic behavior by producers as well as governments. I show that (i) allowing for producers to act strategically on balance reduces the incentive for governments to act strategically; (ii) allowing governments to act strategically increases the incentive for producers to act strategically; (iii) welfare is lower when both parties act strategically; and (iv) strategic behavior by producers and governments is greater when governments use emission taxes than when they use emission standards.

© 1996 Academic Press, Inc.

A. INTRODUCTION

In the recent debates over the Uruguay Round, the Single European Market, and NAFTA, some environmentalists raised concern that trade liberalization might damage the environment. One concern is that the consequent expansion of consumption, production, and trade will lead to increased pollution and use of scarce natural resources unless corrective policies are taken. A second concern, which will be the focus of this chapter, is that in the absence of trade policy, governments may relax their environmental policies to give their domestic producers an advantage in the more competitive international markets (so-called "eco-dumping"). This leads to policy recommendations to harmonize environmental regulations across countries, or, if that is not achieved, that countries who impose tighter environmental regulations than their rivals should be able to impose counter-vailing tariffs on imports from countries with laxer environmental regulations. Not surprisingly, such policies frequently find favor with industries in the traded sector.

Economists have tended to dismiss such concerns and reject the policy conclusions as being covert protectionism.[2] With competitive markets, a small country whose production-related pollution caused only local, not transboundary, damages would have no incentive to distort its environmental policies.[3] If there were competitive markets, but a country had market power, then in the

---

[1] I am very grateful for financial support from the ESRC (Y320-25-3038) and the EC (EV5VCT920184); computing assistance from Laura Valentini; and helpful comments from an anonymous reviewer and participants in workshops in London and Oslo.

[2] See Low (1992) for a useful collection of papers.

[3] See, for example, Long and Siebert (1991).

---

absence of trade instruments governments will have incentives to distort their environmental policies: a country which is an exporter of a pollution intensive good will want to impose environmental regulations which are tougher than the first-best rule (set emissions so that marginal abatement cost equals marginal damage cost), while an importing country will set laxer environmental policies.[4] So there is no presumption that all countries will engage in eco-dumping.

To make sense of the concern about eco-dumping it is natural to turn to models of imperfectly competitive trade. There are now a number of studies[5] which have developed variants of the Brander and Spencer (1985) model of Cournot oligopolistic international markets to take account of environmental pollution. These studies show that indeed there can be "rent-shifting" incentives for governments of *all* producing countries to relax their environmental policies. While these studies appear to rationalize environmentalists' concern about eco-dumping, there are a number of reasons why the incentives for setting too lax environmental policies may be small or even reversed: (i) there is a welfare cost in terms of a dirtier environment to offset any strategic trade gain (Barrett, 1994); (ii) there may be general equilibrium effects in factor markets (Rauscher, 1994); and (iii) if producers compete using prices rather than quantities, then governments will want to set excessively tough environmental targets (Barrett, 1994).

All the papers referred to above concentrate on strategic behavior by governments, but producers will also have incentives to act strategically to try to shift rents in their favor, for example, through their investment in capacity or R & D. Ulph (1992, 1993a) showed that the incentives for producers to engage in such strategic behavior could be significantly affected by the government's choice of environmental policy instrument: emission standards reduce the incentive for strategic behavior relative to emission taxes. In these models governments were assumed to have exogenously determined targets for emissions so the scope for strategic behavior by governments was confined to the choice of policy instruments.[6]

In this chapter a simple model is set out which allows both governments and producers to act strategically (for convenience, the strategic variable for producers are referred to as R & D). Four results are shown:

1. allowing producers to act strategically reduces, but does not eliminate, the incentives for governments to relax environmental policy; I show that this is the net result of two offsetting effects:
   a. in the *output game* the fact that producers are acting strategically to lower costs means there is *less* need for governments to act strategically to lower costs;
   b. there is now an *R & D game* between producers, and governments have incentives to relax environmental policy to encourage domestic producers to do more R & D;
2. allowing governments to act strategically increases the incentive for producers to act strategically;
3. welfare is lower when both governments and producers act strategically than when only one party acts strategically, because there are now two sets of distortions in the economy;
4. strategic behavior by both governments and producers is greater when governments use emission taxes than when they use emission standards.

The structure of the chapter is as follows. In the next section I set out the structure of the model, which is a three-stage game. In Section C, I analyze three equilibria: the benchmark equilibrium

---

[4] See Rauscher (1994) among others.

[5] See Barrett (1994), Conrad (1993), and Kennedy (1994). Like the model I shall develop in this chapter, these papers all assume that the location of firms is fixed. Similar incentives for eco-dumping can arise if firms can vary locations (see Markusen et al. 1992, 1993; Ulph, 1994; and Rauscher, 1993).

[6] Copeland (1991) also noted the difference between emission taxes and emission standards when producers act strategically, but in the context of a closed economy.

where there is no strategic behavior by any party, the equilibrium where only governments act strategically, and the equilibrium where only producers act strategically. In Section D, the equilibrium where both producers and governments act strategically is analyzed and compared with the three equilibria in Section C. Section E offers some conclusions.

B. THE MODEL

A partial equilibrium model of a single industry is used in which there are only two producers of a homogeneous good, each located in a different country. Since only symmetric equilibria shall be considered, the two producers and countries are identical. These producers sell this good on a world market, which does not contain any consumers located in the two producing countries, and the total revenue function for a producer is denoted by $R(x,y) \equiv x(A - x - y)$, where x is own output and y is output of the rival.

Costs of producing output x are given by the *restricted* total cost function $C(x,\phi) \equiv \phi x^2/4$, where ϕ is a parameter representing the strategic variable chosen by firms such that reductions in ϕ shift down the firms total, average, and marginal restricted costs curves. The cost of ϕ is given by $1/\phi$. Think of ϕ as the level of cost corresponding to a particular level of technology, and $1/\phi$ as the cost of R & D required to develop that technology. If the producer is *not* acting strategically and is just choosing ϕ to minimize $C(x,\phi) + 1/\phi$ for any given level of output, the producer would set $\phi = 2/x$, and this would give the firm the *unrestricted* total cost function $K(x) \equiv x$. This will be referred to as the *efficient* choice of R & D.

Associated with the production of x is a pollutant. However, the producer does have technology available for abating this pollutant. Units are chosen such that if the producer chooses output level x and abatement level a, then emissions of the pollutant are $e \equiv x - a$. Total costs of abatement are $a^2/2$. The pollutant causes damage only to the local economy and total damage costs are $de^2/2$. The government can use one of two instruments for controlling pollution, an *emission standard*, where the government announces an upper limit on emissions, e, or an emission tax, where emissions are taxed at a rate t. We shall not analyze the game of choice of instrument by government but simply assume that either both governments set taxes or both governments set standards. The government's welfare function will be total revenue minus total costs of production minus total costs of abatement minus total costs of pollution damage.

To complete notation, while symmetric equilibria shall be studied, it will be necessary to refer to some of the decisions taken by the rival producer or government. We denote by y, ψ, ε, and τ the levels of output, R & D, emission standard, and emission tax set in the other country.

Finally, the move structures will be described. In the general case where both governments and both producers act strategically, there is a three-stage game, in which in stage 1 the two governments set their emission standards or their emission taxes, taking as given the level of policy instrument chosen by their rival; in stage 2 the two producers chooses their levels of R & D, again taking as given the level of R & D selected by their rival; in stage 3 the two producers choose their output levels non-cooperatively. We seek a subgame perfect Nash equilibrium of this three-stage game. When only the governments act strategically, we can collapse the second- and third-state games into a single stage in which producers have to choose simultaneously their level of output and R & D, and so R & D will be chosen efficiently. When only producers act strategically, the governments in stage 1 will be assumed to ignore the impact of their environmental policy instrument on the output or R & D levels of the rival producer. When neither producers nor governments act strategically we have what we shall define as a *first-best* equilibrium, which is first-best taking as given that the two producers and governments act noncooperatively.[7]

---

[7] Within the model set out here governments can obviously do better if they cooperate. Even with non-cooperative behavior the equilibrium described as first-best would no longer be first-best if there were consumers located in the countries where the firms are located.

C. THE FIRST THREE EQUILIBRIA

1. FIRST-BEST EQUILIBRIUM

Since the producers are not acting strategically, we have a two-stage game in which in the second stage producers just choose outputs using their efficient choices of R & D, and hence, their unrestricted total cost functions. There are then two cases, depending on whether the governments set standards or taxes.

Standards

At the second stage the producer takes as given its emission standard, e and the output of its rival, y, and chooses x to maximize

$$(A - x - y)x - x - 0.5(x - e)^2,$$

which leads to the first-order condition and reaction function

$$x = (A - 1 + e - y)/3. \tag{9.1}$$

Letting ε be the emission standard chosen by the second government, then we can solve (Equation 9.1) and its analogue for the rival producer to get the equilibrium output levels at the second stage

$$\left. \begin{array}{l} x = (2A - 2 + 3e - \varepsilon)/8 \\ y = (2A - 2 + 3\varepsilon - e)/8, \end{array} \right\} \tag{9.2}$$

from which we readily see that $\partial x/\partial e = 3/8 > 0$, $\partial x/\partial \varepsilon = -1/8 < 0$. In other words, an increase (slackening) of emission standard will raise the equilibrium output of the domestic producer and cut the output of the rival producer, other things being equal.

In the first-stage game, the government takes as given the emission standard ε set by the other government, and *for the case of non-strategic behavior, the output of the other producer*, and chooses e to maximize

$$W(e) \equiv (A - x - y)x - x - 0.5(x - e)^2 - 0.5de^2, \tag{9.3}$$

taking account of the dependence of x, but not y, on e. The first-order condition is

$$\{A - 1 - y + e - 3x\}\frac{\partial x}{\partial e} + x - e - de = 0,$$

and using (Equation 9.1) this yields the well-known *first-best rule*

$$x - e = de \tag{9.4a}$$

or

$$e = x/(1 + d) \tag{9.4b}$$

Equation (9.4a) is just the condition that marginal abatement cost equals marginal damage cost. This shows that although the government takes account of the fact that its emission standard can affect its domestic firm's output and hence profits, if the firm is already maximizing profits, that consideration is irrelevant in the final rule for setting emission standards.

To conclude, in a symmetric equilibrium we will have $e = \varepsilon$, so combining (Equation 9.2) and (Equation 9.4b) we get the first-best levels of output and emissions:

$$\left.\begin{aligned} x^* &= (A-1)(1+d)/(3+4d)\\ e^* &= (A-1)/(3+4d), \end{aligned}\right\} \tag{9.5}$$

Taxes

In the second-stage game the producer takes as given the level of emission tax set by the government, t, and the output of its rival, y, and chooses its own output, x, and abatement level, a, to maximize

$$(A - x - y)x - x - t(x - a) - 0.5a^2$$

which yields the first-order conditions and reaction function

$$x = (A - 1 - y - t)/2$$

$$a = t, \tag{9.6}$$

where the second condition is just the usual condition that the firm abates pollution to the point where marginal abatement cost equals the tax.

If we compare the reaction function when the government uses taxes to that when the government uses standards we see immediately that with standards $\partial x/\partial y = -1/3$, while with taxes $\partial x/\partial y = -1/2$; in other words, for a unit reduction in output by its rival the producer will raise output more if the government uses taxes than if it uses standards. The reason is that when the rival producer cuts output, that will raise the producer's marginal revenue, and so the producer will expand output until marginal cost again equals the new level of marginal revenue. With taxes, marginal cost is marginal production cost plus a tax which is independent of output, while with standards, marginal cost is marginal production cost plus marginal abatement cost, which rises with output. Thus, marginal cost is more steeply sloped with standards than with taxes, and so the producer will not expand output as much for a given increase in marginal revenue.

Solving the reaction functions of the two producers yields the equilibrium output levels

$$\left.\begin{aligned} x &= (A-1-2t+\tau)/3\\ y &= (A-1-2\tau+t)/3, \end{aligned}\right\} \tag{9.7}$$

where τ is the emission tax set by the other government. Again it is straightforward to see that $\partial x/\partial t = -2/3 < 0$, $\partial x/\partial \tau = 1/3 > 0$, so that an increase in emission tax will reduce the market share of the domestic producer and expand the market share of its rival.

Turning to the first-stage game, the government will take as given the tax rate of the rival government and the output of the rival producer, and will choose its emission tax to maximize

$$W(t) \equiv (A - x - y)x - x - 0.5t^2 - 0.5d(x - t)^2, \tag{9.8}$$

where the choice of abatement level by the producer ($a = t$) has already been included. The first-order condition is

$$\{A - y - 2x - 1 - d(x - t)\}\frac{\partial x}{\partial t} - t + d(x - t) = 0 ,$$

which, using (Equation 9.6), becomes

$$\left(1 - \frac{\partial x}{\partial t}\right)\{d(x - t) - t\} = 0$$

or

$$t = d(x - t) \qquad\qquad (9.9a)$$

or

$$t = dx/(1 + d) . \qquad\qquad (9.9b)$$

Equation (9.9a) is the usual *first-best* tax rule that says that the emission tax should be set equal to the marginal damage cost.

In a symmetric equilibrium, $t = \tau$, so solving (Equation 9.7) and (Equation 9.9b), and recalling that $a = t$, we have the first-best equilibrium

$$\left.\begin{array}{l} x^* = (A - 1)(1 + d)/(3 + 4d) \\ t^* = (A - 1)d/(3 + 4d) \\ e^* = (A - 1)/(3 + 4d). \end{array}\right\} \qquad\qquad (9.10)$$

Comparing (Equation 9.10) with the first-best using standards, (Equation 9.5) shows that in the first-best equilibrium taxes and standards are equivalent policy instruments.

2. ONLY GOVERNMENTS ACT STRATEGICALLY

Since producers act non-strategically, the second-stage games set out above continue to apply, but in the first-stage game the governments recognize that the output of the rival firm depends on the policy instrument it sets.

Standards

The government takes ε as given and chooses e to maximize $W(e)$ as defined in (Equation 9.3), recognizing that both x and y depend on e, so that we have the first-order condition

$$\{A - 1 - y + e - 3x\}\frac{\partial x}{\partial e} - x\frac{\partial y}{\partial e} + x - e - de = 0 .$$

Again using (Equation 9.1) this becomes

$$e(1 + d) = x - x\frac{\partial y}{\partial e} ,$$

where the second term on the right-hand side is the strategic element in the rule for setting the emission standard. Since $\partial y/\partial e = -1/8 < 0$, this strategic element leads the government to set too high a target for emissions compared to the first-best rule. To be precise, we now have the rule for setting emission standards

$$e = 9x/8(1+d) \tag{9.11}$$

compared to the first-best rule

$$e = x/(1+d).$$

Solving (Equation 9.11) and (Equation 9.2) for the symmetric equilibrium where $e = \varepsilon$, we get the equilibrium output and emission levels when only the government acts strategically:

$$\left.\begin{array}{l} \hat{x} = (A-1)(1+d)/\{3+4d-1/8\} \\ \hat{e} = 9(A-1)/\{8(3+4d)-1\}. \end{array}\right\} \tag{9.12}$$

Comparison with the first-best shows that we have higher output and higher emissions when governments act strategically.

Taxes

The government takes τ as given and chooses t to maximize $W(t)$ as defined by (Equation 9.8), recognizing that both x and y depend on t. This leads to the first-order condition

$$\{A - y - 2x - 1 - d(x-t)\}\frac{\partial x}{\partial t} - x\frac{\partial y}{\partial t} - t + d(x-t) = 0,$$

which, on using (Equation 9.6), becomes

$$\{t(1+d) - dx\} = x\frac{\partial y}{\partial t}\bigg/\left\{\frac{\partial x}{\partial t} - 1\right\},$$

where the term on the right-hand side represents the strategic distortion from the first-best rule. Since $\partial y/\partial t = 1/3 > 0$, $\partial x/\partial t = -2/3 < 0$, the term on the right-hand side is $-x/5 < 0$, so that the distortion is to set too low environmental taxes compared to the first-best rule. The rule now is

$$t = (d - 0.2)x/(1+d) \tag{9.13}$$

compared to the first-best rule

$$t = dx/(1+d).$$

Solving (Equation 9.13) and (Equation 9.7) for the symmetric case where $t = \tau$ yields the equilibrium values where only the governments act strategically using taxes:

$$\left.\begin{array}{l} \tilde{x} = (A-1)(1+d)/(3+4d-0.2) \\ \tilde{t} = (A-1)(d-0.2)/(3+4d-0.2) \\ \tilde{e} = (A-1)(1+0.2)/(3+4d-0.2), \end{array}\right\} \tag{9.14}$$

Comparing (Equation 9.14) with (Equation 9.10) we see that when only governments act strategically using taxes then we get higher output and emissions and lower emission taxes than those of the first-best equilibrium. More interestingly, comparing (Equation 9.14) with (Equation 9.12) it is readily seen that we get higher output and emissions when the government acts strategically using taxes than when it acts strategically using standards. The rationale is that what motivates governments to act strategically is that producers take as given the output of their rival, while governments recognize that if the domestic firm expands output the rival will reduce its output. Because of the different slopes of the reaction functions under emission taxes and emission standards, it is straightforward to see that a given outward shift in the domestic producer's reaction function will generate a greater reduction in rival output with emission taxes than with emission standards.[8]

3. ONLY PRODUCERS ACT STRATEGICALLY

This requires analysis of the full three-stage game where in the second stage producers choose their R & D and in the third stage they choose outputs.

Standards

At the third stage the producer knows the standard, e, set by the government in the first stage and the level of R & D, ϕ, it has chosen in the second period, takes as given the level of output, y, set by its rival, and chooses its own output, x, to maximize

$$\left(A - x - y\right)x - \phi x^2/4 - 0.5\left(x - e\right)^2,$$

where the second term is the producer's restricted total cost function. This leads to the first-order condition and reaction function

$$x = 2\left(A - y + e\right)/\left(6 + \phi\right). \tag{9.15}$$

Solving (Equation 9.15) and its analogue for the rival producer yields the equilibrium output levels

$$\left.\begin{array}{l} x = \lambda\left\{\left(8 + 2\psi\right)A + 2\left(6 + \psi\right)e - 4\varepsilon\right\} \\ y = \lambda\left\{\left(8 + 2\phi\right)A + 2\left(6 + \phi\right)\varepsilon - 4e\right\} \end{array}\right\}, \tag{9.16}$$

where

$$\lambda = 1/\left(32 + 6\phi + 6\psi + \phi\psi\right).$$

At the second stage the producer now chooses R & D level ϕ, taking as given its rival's choice ψ, but recognizing that both outputs in the third-stage game depend on its own choice of R & D. Thus it chooses ϕ to maximize

$$\left(A - x - y\right)x - \phi x^2/4 - 1/\phi - 0.5\left(x - e\right)^2$$

---

[8] It should be noted that this result does depend on the particular functional forms used here; Ulph (1992, 1993a) shows that different rankings of taxes and standards can arise depending on the technology of abatement.

with first-order condition, or R & D reaction function

$$\{A - y + e - 0.5(6 + \phi)x\}\frac{\partial x}{\partial \phi} - x\frac{\partial y}{\partial \phi} - x^2/4 + 1/\phi^2 = 0.$$

Using (Equation 9.15) this becomes

$$1/\phi^2 = x^2/4 + x\frac{\partial y}{\partial \phi}.$$

From (Equation 9.16), $\partial y/\partial \phi = 2\lambda x$, so the rule for setting the R & D level becomes

$$1/\phi^2 = \left(x^2/4\right)(1 + 8\lambda). \tag{9.17}$$

Recalling that the efficient choice of R & D is given by the rule

$$1/\phi = x/2,$$

we can see that strategic behavior by producers leads to excessive expenditure on R & D, because they aim to drive down their operating costs to gain a competitive advantage in the third-stage game.

The outcome of the second stage of the game is summarized by the following pair of R & D reaction functions:

$$\left.\begin{array}{l}1/\phi = 0.5x(1 + 8\lambda)^{1/2} \\ 1/\psi = 0.5y(1 + 8\lambda)^{1/2},\end{array}\right\} \tag{9.18}$$

where x and y are in turn defined by (Equation 9.16). ϕ and ψ depend on e and ε through the dependence of x and y on e and ε. It is not possible to solve these explicitly to obtain the equilibrium levels of R & D. Total differentiation of (Equation 9.18) and (Equation 9.16) with some elaborate calculation establishes that $\partial\phi/\partial e < 0$, $\partial\phi/\partial\varepsilon > 0$; i.e., relaxing environmental policy will increase the R & D done by the domestic producer and reduce the R & D done by the rival producer. The rationale is that by reducing domestic producer's costs this increases the incentive to do R & D, and the subsequent expansion of domestic producer's output reduces the profitability of R & D by the rival producer.

Finally, we turn to the first-stage game, where we shall return to the assumption that the government acts non-strategically. The government's welfare function is

$$W(e) \equiv (A - x - y)x - \phi x^2/4 - 1/\phi - 0.5(x - e)^2 - de^2. \tag{9.19}$$

By the government acting non-strategically, we mean that the government recognizes that x and ϕ depend on e (in the first case x depends on e both directly and through its dependence on ϕ), and that y depends on ϕ, but neglects the fact that y depends directly on e and that ψ depends on e. In short, the government does not believe it can directly affect either the R & D or the output level in the rival country. Maximizing $W(e)$ with respect to e, taking ε as given, yields the first-order condition

$$\left[(A - y + e) - 0.5x(6 + \phi)\right]\left(\frac{\partial x}{\partial e} + \frac{\partial x}{\partial \phi}\frac{\partial \phi}{\partial e}\right) + \left\{-x\frac{\partial y}{\partial \phi} - x^2/4 + 1/\phi^2\right\}\frac{\partial \phi}{\partial e} + x - e - de = 0.$$

By the first-order conditions for the choice of x and ϕ the first two terms are zero, so we are left with the simple first-best rule

$$e = x/(1+d) \tag{9.20}$$

as we would expect.

Taking the symmetric equilibrium where $e = \varepsilon$, we can now solve (Equation 9.16), (Equation 9.18), and (Equation 9.20) to get the solution when only the producers act strategically and the government uses standards

$$x' = 2A/(2D + \phi')$$

$$e' = x'/(1+d) \tag{9.21}$$

where $D \equiv (3 + 4d)/(1 + d)$, and ϕ' solves the quadratic equation

$$8/\left[(4+\phi')(8+\phi')\right] = \left[4D^2 + 4D\phi' - (A^2 - 1)\phi'^2\right]/A^2\phi'^2 \,,$$

for which it can be checked that there is a unique real positive root.

It is straightforward to show that both output and emissions will be higher than they are in the first-best equilibrium, but it is not possible to say anything about output or emissions compared to the cases where only the government acts strategically. Note, however, that we would expect emissions to be lower than in the case where only the government acts strategically, because the only reason why emissions are higher than in the first-best equilibrium is because output is higher than in the first-best equilibrium, and the first-best rule relates emissions to output.

It may be asked why, if it was optimal for the government acting nonstrategically to set emissions standards at the first-best *level* when producers acted nonstrategically, it is not optimal to set the standards at the same *level* when producers act strategically, since neither damage costs nor abatement costs have changed. The reason is that there are two ways of reducing pollution: incurring abatement costs and cutting output, or the firm will trade these off against each other so that marginal abatement cost equals marginal loss of profit on a unit of output. The government recognizes that because producers are acting strategically, and hence reducing their operating costs, the marginal loss of profits to the economy from having to reduce output is now higher than it was when producers acted nonstrategically. So when equilibrium occurs with marginal damage cost equal to marginal abatement cost equal to marginal profit this will be at a higher level of output and emissions when producers act strategically than when they act non-strategically.

Taxes

In the third-stage game the producer knows the emission tax, t, set by the government at stage 1 and the level of R & D set in stage 2, ϕ, takes as given the output of its rival, y, and chooses its output, x, and abatement, a, to maximize

$$(A - x - y)x - \phi x^2/4 - t(x - a)^2 - a^2/2 \,,$$

yielding first-order conditions and reaction function

$$a = t$$

$$x = 2(A - y - t)/(4 + \phi) \tag{9.22}$$

Comparing (Equation 9.22) and (Equation 9.15) we see that, again, the reaction functions have different slopes, so that in response to a cut in its rival's output, the producer will expand output more when the government sets an emission tax than when it sets an emission standard. The reason is exactly the same as provided in the discussion of the case where the producers act nonstrategically. Solving (Equation 9.22) and its rival's analogue yields the equilibrium outputs in stage three

$$\left.\begin{array}{l} x = \mu\big[(4+2\psi)A - 2(4+\psi)t + 4\tau\big] \\ y = \mu\big[(4+2\phi)A - 2(4+\phi)\tau + 4t\big], \end{array}\right\} \tag{9.23}$$

where

$$\mu = 1/(12 + 4\phi + 4\psi + \phi\psi).$$

Turning now to the second-stage game, the producer recognizes that x and y depend on ϕ, and taking ψ as given chooses ϕ to maximize

$$(A - x - y)x - \phi x^2/4 - 1/\phi - t(x - t) - 0.5t^2$$

with first-order condition

$$\big[A - y - t - 0.5x(4+\phi)\big]\frac{\partial x}{\partial \phi} - x\frac{\partial y}{\partial \phi} - x^2/4 + 1/\phi^2 = 0.$$

Using (Equation 9.22) this becomes

$$1/\phi^2 = x^2/4 + x\frac{\partial y}{\partial \phi}.$$

From (Equation 9.23), $\partial y/\partial \phi = 2\mu x$, so that the rule for choice of R & D becomes

$$1/\phi = 0.5x(1 + 8\mu)^{1/2} \tag{9.24}$$

compared to the rule for efficient investment

$$1/\phi = 0.5x.$$

Comparing (Equation 9.24) and (Equation 9.17), and noting that, for a given ϕ and ψ, $\mu > \lambda$, there will be more strategic overinvestment in R & D when the government uses an emission tax than when it uses an emission standard. The rationale is the same as was given for government strategic behavior.

The outcome of the second-stage game of choice of R & D is summarized by the pair of R & D reaction functions

$$\left.\begin{array}{l} 1/\phi = 0.5x(1 + 8\mu)^{1/2} \\ 1/\psi = 0.5y(1 + 8\mu)^{1/2}. \end{array}\right\} \tag{9.25}$$

Again it is not possible to solve explicitly for equilibrium R & D levels but total differentiation of (Equation 9.25) and (Equation 9.23) with some elaborate calculation establishes that $\partial \phi/\partial t > 0$,

$\partial \phi / \partial \tau < 0$; i.e., raising emission taxes reduces the domestic producer's R & D and increases the rival producer's R & D.

Finally, in the first-stage game, the government takes as given the tax rate, τ, set by the other government and chooses t to maximize

$$W(t) \equiv (A - x - y)x - \phi x^2/4 - 1/\phi - 0.5t^2 - 0.5d(x - t)^2 , \tag{9.26}$$

where we have substituted $a = t$, and where the government recognizes that x depends on t and ϕ, that ϕ depends on t, that y depends on ϕ (and hence, indirectly, on t), but does not recognize that t can affect directly the levels of y or ψ. The first-order condition is

$$\{A - y - 0.5x(4 + \phi) - d(x - t)\}\left(\frac{\partial x}{\partial t} + \frac{\partial x}{\partial \phi}\frac{\partial \phi}{\partial t}\right) + \frac{\partial \phi}{\partial t}\left\{1/\phi^2 - x^2/4 - x\frac{\partial y}{\partial \phi}\right\} - t + d(x - t) = 0 .$$

Using the first-order conditions for the choices of x and ϕ this becomes

$$\{t - d(x - t)\}\left\{\frac{\partial x}{\partial t} + \frac{\partial x}{\partial \phi}\frac{\partial \phi}{\partial t} - 1\right\} = 0 ,$$

or the first-best rule

$$t = dx/(1 + d) . \tag{9.27}$$

For the symmetric case where $t = \tau$, we can then solve (Equation 9.23), (Equation 9.25), and (Equation 9.27) to yield the equilibrium values for the case where only the producers act strategically and the government chooses taxes:

$$\left.\begin{array}{l} x'' = 2A/(2D + \phi'') \\ e'' = x''/(1 + d) \\ t'' = dx''/(1 + d), \end{array}\right\} \tag{9.28}$$

where ϕ'' solves the quadratic equation

$$8/\{(2 + \phi'')(6 + \phi'')\} = \{4D^2 + 4D\phi'' - (A^2 - 1)\phi''^2\}/(A^2\phi''^2)$$

which, again, can be shown to have a unique real positive root.

Comparison of (Equation 9.28) with the first-best shows that we get higher output, and emissions, and lower taxes than those in the first-best, while comparison with (Equation 9.21) shows that output, R & D, and emissions will be higher when producers act strategically and the government sets taxes rather than standards. The rationale is the same as for the case when the government acts strategically.

D. GOVERNMENTS AND PRODUCERS ACT STRATEGICALLY

In this section, the modes of the previous section are put together to study the case where both producers and governments act strategically. The stage-three and stage-two games of Section C3 remain unchanged, but the behavior of the government in the first-stage game needs to be modified.

Standards

The levels of outputs and R & D expenditures determined in the third- and second-stage games are given in (Equation 9.16) and (Equation 9.18), respectively, while the welfare function for the government is given in (Equation 9.19). In maximizing $W(e)$, taking ε as given, the government recognizes that x, y, ϕ, and ψ depend on e. This yields the first-order condition

$$0 = \left\{ \left[A - y + e - 0.5x(6+\phi) \right] \left[\frac{\partial x}{\partial e} + \frac{\partial x}{\partial \phi}\frac{\partial \phi}{\partial e} + \frac{\partial x}{\partial \psi}\frac{\partial \psi}{\partial e} \right] \right\} - x\frac{\partial y}{\partial e}$$

$$+ \left\{ 1/\phi^2 - x\frac{\partial y}{\partial \phi} - x^2/4 \right\} \frac{\partial \phi}{\partial e} - x\frac{\partial y}{\partial \psi}\frac{\partial \psi}{\partial e} + (x-e) - de.$$

The first term is zero by the first-order condition for the choice of x, while the third term is zero by the first-order condition for the choice of ϕ, so we are left with the rule for the choice of e:

$$e(1+d) = x\left\{ 1 - \frac{\partial y}{\partial e} - \frac{\partial y}{\partial \psi}\frac{\partial \psi}{\partial e} \right\}. \tag{9.29}$$

So the incentive to deviate from the first-best rule for setting emission standards derives from the strategic incentive to try to influence the output of the rival producer, both *directly* (the second term in parentheses in (Equation 9.29)), and *indirectly* (the third term in Equation 9.29), by influencing the rival's investment in R & D.

   From (Equation 9.16) $\partial y/\partial e = -4\lambda$, $\partial y/\partial \psi = -\lambda y(6+\phi)$, while we have already established that $\partial \psi/\partial e > 0$, so that slackening emission standards will directly reduce the output of the rival but will also reduce its R & D level, which will indirectly reduce its output in the third-stage game. So both the *direct* and *indirect* incentives are for the government to slacken its environmental standards relative to the first-best. However, if we compare (Equation 9.29) with (Equation 9.11) it is clear that the direct effect when producers act strategically is *smaller* than when producers do not act strategically. Because the expression for $\partial \psi/\partial e$ is extremely complex it is difficult to tell analytically whether the reduction in the direct effect outweighs the indirect effect. Numerical results will be discussed shortly.

Taxes

The levels of output and R & D expenditures derived in the third- and second-stage games are given by (Equation 9.23) and (Equation 9.25), respectively, and the welfare function, $W(t)$, for the government in the first-stage game is given by (Equation 9.26). Now when the government maximizes $W(t)$, taking τ as given, it recognizes that x, y, ϕ, and ψ all depend on t. The first-order condition for the government's problem is

$$0 = \left\{ \left[A - y - 0.5x(4+\phi) - d(x-t) \right] \left[\frac{\partial x}{\partial t} + \frac{\partial x}{\partial \phi}\frac{\partial \phi}{\partial t} + \frac{\partial x}{\partial \psi}\frac{\partial \psi}{\partial t} \right] \right\} - x\frac{\partial y}{\partial t}$$

$$+ \left\{ 1/\phi^2 - x^2/4 - x\frac{\partial y}{\partial \phi} \right\} \frac{\partial \phi}{\partial t} - x\frac{\partial y}{\partial \psi}\frac{\partial \psi}{\partial t} - t + d(x-t).$$

Using (Equation 9.22) and (Equation 9.24) this can be simplified to

$$t(1+d) = dx - x\left[\frac{\partial y}{\partial t} + \frac{\partial y}{\partial \psi}\frac{\partial \psi}{\partial t}\right]\Big/\rho, \tag{9.30}$$

where

$$\rho = \left\{1 - \frac{\partial x}{\partial t} - \frac{\partial x}{\partial \phi}\frac{\partial \phi}{\partial t} - \frac{\partial x}{\partial \psi}\frac{\partial \psi}{\partial t}\right\}. \tag{9.31}$$

The second term on the right-hand side of (Equation 9.30) gives the strategic incentive for the government to manipulate its environmental policy, and, as in the case of standards, it comprises two terms: the *direct* effect of the emissions tax on the output of the rival producer, and the *indirect* effect of changing the R & D level of the rival producer and hence the output of the rival.

From (Equation 9.23) it is readily calculated that $\partial x/\partial t = -2\mu(4 + \phi)$, $\partial y/\partial t = 4\mu$, $\partial x/\partial \psi = \partial y/\partial \phi = 2\mu x$, $\partial x/\partial \phi = \partial y/\partial \psi = -\mu x(4 + \phi)$; in Section C3, I showed that $\partial \phi/\partial t > 0$, $\partial \psi/\partial t < 0$. Thus, raising the emission tax will directly lower domestic output and raise rival's output, while it will indirectly reduce domestic R & D and increase rival's R & D, which will in turn lower domestic output and raise rival's output. So in (Equation 9.31), $\rho > 1$, while in (Equation 9.30) the factor in square brackets is positive. Thus because both the direct and indirect effects of raising an emission tax are to lower domestic output and raise rival's output, the strategic incentive for governments is again to lower the emission tax compared to the first-best rule.

Comparing (Equation 9.30) with (Equation 9.13) also shows that the direct effect of the emissions tax is *smaller* when the producers are acting strategically than when the producers are not acting strategically, but of course there is the additional indirect effect. As with standards, the complex expression for the indirect effect makes it difficult to calculate analytically whether the overall reduction in the emissions tax when both governments and producers act strategically is greater than when only the government acts strategically.

It has not been possible to answer analytically the four questions posed in the introduction: (i) if producers act strategically will this increase or decrease the incentives for governments to relax their environmental policies; (ii) if governments act strategically will this increase or decrease the incentives for producers to overinvest in R & D; (iii) are distortions greater when governments use emission taxes or emission standards; and (iv) what are the implications of the answer to (i)–(iii) for welfare? To answer these questions numerical solutions were computed for a large number of values of the two parameters in this model: A, the level of demand, and d, the level of pollution damage costs. The results [9] can be summarized as follows:

1. If producers act strategically, this always *reduces*, but does not reverse, the incentive for governments to relax their environmental policies. In terms of the earlier discussion, this means that the reduction in the direct effect outweighs the indirect effect. For orders of magnitude, instead of emissions being about 25% higher than they would be under first-best when only governments act strategically, they would be about 15% above first-best when both act strategically.
2. If governments act strategically, this always increases the incentives for producers to overinvest in R & D; this really follows from (1) since on balance governments still relax their environmental policies and this induces domestic producers to do more R & D. However this effect is small, consistent with the finding in (1) that the indirect effect is small.

---

[9] For details see Ulph (1993b).

3. When both governments and producers act strategically, distortions to both environmental policy and R & D are larger when governments use emission taxes than when they use emission standards.

4. Welfare is lower when both governments and producers act strategically than when only one party act strategically, even though environmental policy is less distorted; this is because there are now two sources of distortion in the economy. Not surprisingly, welfare is also lower when governments use emission taxes than when they use emission standards.[10]

E. CONCLUSIONS

We have shown that if producers act strategically, this reduces but does not eliminate the incentive for governments to relax their environmental policies. The model used is very simple, both in using special functional forms and in omitting a number of other considerations. Almost all the arguments go through for more general functional forms. The exception is the comparison between the size of distortions under emission taxes and emission standards which is sensitive to the modeling of abatement technology. The omission of other considerations was dictated not just by reasons of tractability. Thus, transboundary pollution was ignored because we wanted the only source of strategic incentive to slacken environmental policy to come from trade considerations; we assumed a single producer in each country to focus the strategic considerations on competition with other countries rather than trying to induce cooperative behavior among domestic producers; we ignored domestic consumers to ensure that, in the absence of environmental considerations, governments share the concern of producers to maximize profits; we assumed non-cooperative behavior by governments because we did not want to give them scope for cooperative behavior which was not available to producers directly. Dropping any of these assumptions, or the assumption of Cournot behavior, is likely to change the results significantly, but many of these changes can be predicted from analyses that are already available in the literature, although it would be useful to check that out. It should also be noted that it has been assumed that R & D reduces production costs, rather than emissions. Ulph (Equation 9.17) explored the latter case and showed that this could cause the indirect effect to go in the opposite direction of the direct effect. We hope to report shortly the results of a model with both types of R & D.

A final topic for further research is the design of institutional structure to reduce the incentives for strategic distortion of environmental policy.[11] It must be emphasized that this chapter provides no support for the kind of policy recommendations made by environmentalists cited in the Introduction.

REFERENCES

Barrett, S., Strategic environmental policy and international trade, *J. Public Econ.*, 54 (3), 325, 1994.

Brander, J. and Spencer, B., Export subsidies and international market share rivalry, *J. Int. Econ.*, 18, 83, 1985.

Conrad, K., Taxes and subsidies for pollution-intensive industries as trade policy, *J. Environ. Econ. Manage.*, 25, 121, 1993.

Copeland, B., Taxes versus standards to control pollution in imperfectly competitive markets, mimeo, University of British Columbia, 1991.

Kennedy, P., Equilibrium pollution taxes in open economies with imperfect competition, *J. Environ. Econ. Manage.*, 27, 49, 1994.

---

[10] Note that this does not imply that governments will choose to use emission standards. That depends on an analysis of a prior stage game involving choice of policy instruments, which I have not conducted. For this particular model I conjecture that the results of Ulph (1993a) will hold, and that the outcome would be a Prisoner's Dilemma in which governments choose taxes but would be better off with standards.

[11] See Ulph (1996) for a preliminary discussion.

Long, N. and Siebert, H., Institutional competition versus ex-ante harmonisation: the case of environmental policy, *J. Inst. Theor. Econ.*, 147, 296, 1991.

Low, P., Ed., *International Trade and the Environment*, World Bank, Washington, DC, 1992.

Markusen, J., Morey, E., and Olewiler, N., Noncooperative equilibria in regional environmental policies when plant locations are endogenous, NBER working paper 4051, 1992.

Markusen, J., Morey, E., and Olewiler, N., Environmental policy when market structure and plant locations are endogenous, *J. Environ. Econ. Manage.*, 24, 69, 1993.

Rauscher, M., Environmental regulation and international capital allocation, *"Nota di Lavoro 79.93,"* FEEM, Milan, 1993.

Rauscher, M., On ecological dumping, *Oxford Econ. Pap.*, 46, 822, 1994.

Ulph, A., The choice of environmental policy instruments and strategic international trade, in *Conflicts and Cooperation in Managing Environmental Resources*, Pethig, R., Ed., Springer–Verlag, Berlin, 1992.

Ulph, A., Environmental policy and strategic international trade, discussion paper in *Economics and Econometrics 9304*, University of Southampton, 1993a.

Ulph, A., Environmental policy and international trade when governments and producers act strategically, discussion paper in *Economics and Econometrics 9318*, University of Southampton, 1993b.

Ulph, A., Environmental policy, plant location and government protection, in *Trade, Innovation, Environment*, Carraro, C., Ed., Kluwer, Dordrecht, 1994a, 123.

Ulph, D., Strategic innovation and strategic environmental policy, in *Trade, Innovation, Environment*, Carraro, C., Ed., Kluwer, Dordrecht, 1994b, 205.

Ulph, A., Strategic environmental policy, international trade and the single European market, in *Environmental Policy with Economic and Political Integration: The European Community and the United States*, Braden, J., Folmer, H., and Ulen, T., Eds., Edward Elgar, Cheltenham, 1996, 235.

10 Wildlife, Biodiversity, and Trade[1]

Edward B. Barbier and Carl–Erik Schulz

This chapter develops a model of wild-resource exploitation that includes both the standard bioeconomic properties of growth and harvesting and a species-area relationship linked to habitat conversion. The model is developed for both a closed and an open economy. In the closed-economy model the characteristics of the long-run equilibrium are analyzed in three versions of the model: the basic bioeconomic model, the inclusion of habitat conversion, and the addition of biodiversity value. In the open-economy model, we also examine the potential effects of trade interventions on the optimal exploitation of species and habitat conversion in the long run. The results confirm that the inclusion of habitat conversion as well as biodiversity value in a model of wild-resource exploitation yields significantly different outcomes than a basic bioeconomic model.

A. INTRODUCTION

The world's remaining natural areas and habitats for wild species are believed to be located mainly in the developing regions of the world. However, the demand for resources and land to meet the needs of growing populations and developing economies has resulted in many of these wild resources and wildlands coming under threat. At the same time, it is economically feasible and worthwhile to maintain some natural areas broadly in their original state and to conserve wildlife species while allowing important human uses of these resources (Swanson and Barbier, 1992). Sustainable management regimes can be extremely diverse, ranging from commercial wildlife ranching and harvesting to non-timber forest product extraction to community-based wildlife and wildlands development to tourism and recreation, and so on. They may take the form of traditional uses of wild resources by local communities (IIED 1994). Alternatively, they may be deliberate attempts to reconcile conservation and development aspirations through integrated community-based development projects (Wells and Brandon, 1992) or through major international aid programs such as biodiversity reserves funded by the Global Environmental Facility (Munasinghe, 1993).

Economics clearly has an important role to play in the assessment of the "sustainable" exploitation and management of wildlife and wildlands. However, it is important to recognize two facets to the economic analysis of optimal exploitation of wildlife species. First, there is the familiar bioeconomic problem of determining the long-run optimal harvest or use rate, given certain characteristics of the population dynamics of wild species such as their growth or regeneration rates. This approach does not include any cost for land (or sea) used as habitat, and is particularly appropriate in the analysis of intertemporal allocation of marine resources for which it was originally developed (Clark, 1976). It is less applicable to allocation decisions involving terrestrial wildlife resources, which require determination of not only the optimal stock and harvesting rate of species biomass but also the optimal size or area of the natural habitat that maintains the species (Swanson, 1994).

Thus, a second important facet to the economics of wildlife use or exploitation is the optimal allocation of natural habitat. To determine the optimal size or area of habitat, it is necessary to

---

[1] Paper prepared for the European Association for Environmental and Resource Economics Sixth Annual Conference, Umeå, Sweden, 17–20 June 1995. We are grateful for comments by Rögnvaldur Hannesson, Charles Perrings, and three anonymous referees; however, the usual disclaimer applies.

compare the benefits of maintaining habitat with its opportunity costs, which are none other than the foregone economic activities dependent on the conversion of natural habitat to other uses (Swanson and Barbier, 1992; Barrett, 1993). However, knowledge of the optimal area to be preserved as natural habitat alone still gives us only half the story; it tells us nothing about the optimal rate of exploitation of those species found within the preserves. Thus, if we are interested in the possibility of sustainable utilization of species beyond either their complete preservation in a biosphere reserve or only their nonconsumptive uses such as for tourist viewing, then we must consider the problems of optimal habitat provision and species exploitation together.

Two additional dimensions to the problem are worth considering: the role of trade interventions to promote sustainable management, and the wider social values of biodiversity as a public good for humankind. Although many wild resources are used locally, some products of natural areas are important traded goods, e.g., wildlife products (skins, horns, ivory, and even the animals themselves), timber logs and products (tropical hardwoods), and non-timber products (rattan, honey, resins, etc.). In recent years, trade interventions imposed either unilaterally or through multilateral agreements have been proposed to control overexploitation of the traded products derived from wild resources and thus encourage their sustainable management within developing countries (Barbier et al., 1990; Burgess, 1994). Environmental pressure groups in major importing countries are also advocating sanctions and interventions in the tropical timber trade as a means of coercing producer countries into reducing tropical deforestation and the consequent loss of environmental values (Barbier et al., 1994).

However, in addition to their direct use or exploitation, the wild species and resources contained in the natural areas of developing countries are also considered to have wider social values that are usually linked to stock size, not the harvesting size. These values are diverse, and could include the non-consumptive uses of species (e.g., for tourism, recreation, and scientific study); their ecological functional role that may indirectly support or protect economic activity (control of pests, genetic diversity, improving productivity); their potential future use values (option and quasi-option values); and their non-use or pure "existence" value. Various studies indicate that non-consumptive use and ecological values of wild species and resources can be significant and, if captured by a developing country, could make an important contribution to overall welfare (see Table 10.1). Some of these values are attributed to a single species, but in many cases it is the entire stock of species in a wild area and its inherent diversity, or biodiversity, that are the source of value. Moreover, many important social values of biodiversity may be underestimated at present, as there is a lot we still do not know about the potential ecological role of species and biodiversity in supporting ecosystem functioning and "resilience" (Holling et al., 1995).

In this chapter, we attempt to integrate some of these concerns by developing a model of wild or natural species exploitation that extends the traditional bioeconomic model for renewable resources to include conversion of habitat and the wider social value of biodiversity. The model is developed for both a closed and an open economy. In the case of the closed economy we demonstrate how extending the bioeconomic model of species exploitation to account explicitly for the opportunity costs of habitat conservation plus any non-consumptive biodiversity values produces a different optimal extraction for wild species than if only the harvesting problem were considered. The open-economy model is an extension of this model, and is used to examine the potential effects of trade interventions on the optimal exploitation of species and habitat conversion in the long run, as a means of determining whether such interventions can encourage less exploitation of wild resources.

B. BASIC MODEL OF SPECIES EXPLOITATION AND HABIBAT CONVERSION

In the following model, we assume that the key allocation problem is to determine the optimal stock and rate of exploitation of wild species or resources, given that this total resource stock is a function of a natural rate of population growth or regeneration, harvesting and the total area of

TABLE 10.1
Various Valuation Estimates for Non-Consumptive Uses of Wild Resources in Developing Countries (U.S. $)

Activity	Estimated Net Present Value
Viewing value of elephants, Kenya (Brown and Henry, 1993)	$25,000,000/year
Ecotourism, Costa Rica (Tobias and Mendelsohn, 1990)	$1,250/ha
Tourism, Thailand	$385,000–860,000/year
Research/education, Thailand (Dixon and Sherman, 1990)	$38,000–77,000/year
Tourism, Cameroon	$19/ha
Genetic value, Cameroon (Ruitenbeek, 1989)	$7/ha
Eco-tourism, Madagascar (Kramer, 1993)	$800,000–2,160,000
Net private return to pharmaceutical prospecting, Costa Rica (Barbier and Aylward, 1996)	$4,810,000/product
Impact on crop productivity of pest control by Anolis lizard, Greater and Lesser Antilles (Narain and Fisher, 1994)	$670,000/year in lost output per 1% decline in population

natural habitat or "wildlands." To do this, we combine a standard bioeconomic model of species exploitation with an approach developed by Barrett (1993), which was used principally to determine the optimal allocation of habitat between preservation and conversion to other economic uses.[2] Our approach requires setting up the problem so that both the biological growth of populations and changes in habitat area affect the aggregate stock of species.

For example, let $S(t)$ be the total stock of species (wild resources) comprising the various populations or species found in a given area, aggregated in some appropriate way.[3] This aggregate stock will consist of many animal and plant populations with the capacity for reproduction and growth. As these populations regenerate and grow, so must the aggregate stock of these populations, $S(t)$. However, the size of $S(t)$ will also be affected by the total number of species comprising this aggregate stock, and as research from island biogeography indicates, the total number of species should be related to the total area of habitat available to these species. For our model, this implies that changes in habitat area will affect the total number of species, and hence determine changes in the total species stock, $S(t)$. We therefore introduce an important modification to the standard bioeconomic model: changes in the total species stock are not only determined by the aggregate biological growth rate across these species, but are also affected by the expansion in the number of species as the size of the natural area or habitat changes. Finally, we assume that changes in the total species stock, $S(t)$, must be net of any such harvesting offtake.

Thus, defining $A(t)$ as the total area of natural habitat or wildlands, and $y(t)$ as the aggregate rate of harvest of species or "wild" resources, then the change in total species stock, $dS(t)/dt$, can be depicted as

$$\frac{dS(t)}{dt} = \dot{S} = G\big(S(t), A(t)\big) - y(t) = G_1\big(S(t)\big) + G_2\big(A(t)\big) - y(t). \tag{10.1}$$

---

[2] Barrett (1993) does not include the bioeconomic growth of the species stock or harvesting from this stock in his habitat allocation model.

[3] For example, the stock of species could be aggregated by the number of individuals in each distinct population or species. Alternatively, the total "biomass" of each population or species could be aggregated across all species. Some form of "weighted average" combination of the two approaches could also be employed.

That is, change in the total species stock is determined by a growth function, G, which is in turn a function of the size of the stock, $S(t)$, and the total habitat area, $A(t)$, as well as the total offtake, $y(t)$. We assume that the growth function G is additively separable into subcomponents G_1 and G_2. This separability condition facilitates our ability to depict our results graphically in a two-dimensional (S,y) space. The assumption of separability is not essential to our main results, however. In Appendix B we illustrate this with an alternative formulation of the allocation problem that includes a non-separable version of the growth function, G.

In Equation (10.1), $G_1(S(t))$ can be defined as the aggregate growth or regeneration function of the total species stock.[4] This function is assumed to display the standard bioeconomic properties: G_1 is defined over the interval $S(t) \geq 0$, and there exist values $\underline{S} < S_{max}$ such that $G_1(\underline{S}) = G_1(S_{max}) = 0$ and $d^2G_1(S(t))/dS(t)^2 < 0$.

However, $G_2(A(t))$ is an additional component to the standard bioeconomic properties of a renewable-resource growth function. G_2 represents the growth in total species stock that is attributable to an expansion in the number of species based on a species-area relationship. The basic premise of this relationship is that the total number of species found in a given habitat will increase with the size or area of the habitat. Originally developed through the study of island biogeography, species-area relationships were initially employed to estimate the number of species in biologically diverse habitats such as tropical forests. Now, however, such curves are being used to forecast the extent of species loss associated with current trends in tropical deforestation. As noted by Barrett (1993), a typical species-area relationship employed by ecologists is

$$S(t) = \alpha A(t)^{\beta} \quad \text{or} \quad A(t) = \alpha^{-1/\beta} S(t)^{1/\beta}, \qquad 0 < \beta < 1. \tag{10.2}$$

The parameter α may depend on the type of species, its population density and the biogeography region. The parameter β determines the shape of the species-area curve, and for most regions of the world has been estimated to range between 0.16 and 0.39 (Lugo et al., 1993). Although β has been observed to increase as habitat area becomes smaller, in the following model we assume it to be fixed.[5] As shown in Equation (10.2), by inverting the species-area relationship one can also determine the habitat area required to maintain a given number of species.

Ignoring G_1 in Equation (10.1) for the moment, following Barrett's approach we can determine from (Equation 10.2) the growth in the total species stock that is attributable to the effect of a change in habitat area on the species-area relationship.

$$G_2(A(t)) = \beta \alpha A(t)^{\beta-1} r(t), \qquad \text{where } r(t) = dA(t)/dt = \dot{A}$$

$$= \beta \alpha^{1/\beta} S(t)^{(\beta-1)/\beta} r(t) \tag{10.3}$$

$$= aS(t)^{-b} r(t), \qquad \text{where } a = \beta \alpha^{1/\beta} \text{ and } b = (1-\beta)/\beta > 0$$

Thus Equation (10.3) suggests that the growth in total species stock attributed to the species-area relationship can be expressed in terms of the rate of change in habitat area, $r(t)$, and the size of the total stock, $S(t)$. It therefore follows that (Equation 10.1) can be rewritten as

---

[4] It is possible that G_1 aggregated from individual species' growth rates in some meaningful way, such as taking the weighted average of growth rates across all the species comprising the total species stock at time t. Alternatively, G_1 could be simply the observed aggregate growth rate of the total species stock at time t.

[5] Precise estimation of the species-area curve has become an important and controversial issue in ecology, given the tendency to use such curves to predict species extinction associated with tropical deforestation. For example, if β is assumed to be low (< 20%), then the relationship predicts that more than 50% of the land area can be deforested before additional deforestation results in rising rates of extinction. Conversely, at high values for β (> 60%), extinction rates are almost proportional to deforestation rates. For further discussion, see Lugo et al. (1993).

$$\dot{S} = G\big(S(t), A(t)\big) - y(t) = G_1\big(S(t)\big) + aS(t)^{-b} r(t) - y(t) \,. \tag{10.1'}$$

Equation (10.1') is the basic equation of motion of our model, which links changes in the total stock of species, or wild resources, to biological growth, habitat area, and harvesting offtake.

As noted in the previous section, in developing countries the main threat to natural habitat is conversion to other economic activities. Again, following Barrett (1993), this can be incorporated in our model. Let $D(t)$ be defined as the total developed area arising from conversion of natural habitat, then

$$D(t) = D_0 + \int_0^t r(\tau)d\tau = D_0 + A_0 - A(t) \,, \tag{10.4}$$

if the total land area is fixed and $r(t) \leq 0$. That is, natural habitat is converted into land used for other activities at the rate $r(t)$ over time. This implies that $r(t)$ is non-positive in Equation (10.1'). Conversion of habitat occurs because the resulting economic activities produce goods and services for consumption, which in turn contributes to economic welfare.[6] Let $F(D(t))$ be the aggregate production function for economic activities dependent on converted land, which is assumed to have the normal properties $dF(D(t))/dD(t) > 0$ and $d^2F/dD(t)^2 < 0$. It therefore follows that from (Equation 10.2) and (Equation 10.4)

$$F\big(D(t)\big) = F\Big(D_0 + A_0 - \alpha^{-1/\beta} S(t)^{1/\beta}\Big) = f\big(S(t)\big) \,, \tag{10.5}$$

with $df/dS(t) < 0$ and $d^2f/dS(t)^2 < 0$ for $0 < \beta < 1$. That is, an increase in the total species stock due to an expansion in the number of species has an increasingly negative impact on aggregate production of output from converted habitat land. This is not surprising, given that the species-area relationship suggests that the number of species can increase only if there is more habitat area, which of course implies less converted land and a corresponding fall in aggregate output from this land.

Finally, if we assume that all output is consumed and that total consumption, $c(t)$, can be defined as $c(t) = F(D(t)) = f(S(t))$, then we can take into account explicitly in the social utility function of the country the trade-off between, on the one hand, the harvesting or utilization of wild species from their conserved natural habitat and, on the other, the conversion of habitat land to produce other commodities. This trade-off can be further elaborated if we assume also that any positive stock externalities, or biodiveristy values, associated with $S(t)$ enter directly into our utility function. Defining the latter as U, then

$$U = U\big(y(t), c(t), S(t)\big) \,, \tag{10.6}$$

which is assumed to be additively separable and to have the standard properties with respect to its partial derivatives, $\partial U/\partial i > 0$, $\partial^2 U/\partial i^2 < 0$ and $\lim_{i \to 0} \partial U/\partial i \to \infty$, $i = y(t), c(t), S(t)$. It is clear from all the above relationships (Equation 10.1)–(Equation 10.6) that an increase in consumption and its consequent social benefits must occur at the expense of the gains from additional harvesting of wild species and biodiversity values.

Equations (Equation 10.1)–(Equation 10.6) represent the basic relationships of the model. In the next two sections we analyze formally the intertemporal allocation problem assuming closed and open economy conditions respectively. In the following optimal control problems, notation is

---

[6] We ignore any windfall gain from the conversion.

simplified by omitting the argument of time-dependent variables, by representing a derivative of a function by its prime, by employing numbered subscripts to indicate partial derivatives of a function, and by denoting the time derivatives of a variable by a dot.

C. CLOSED-ECONOMY MODEL

We initially assume a closed-economy model. This may be a realistic representation of the problem if all consumption from converted land and harvesting of wild species are non-tradable commodities. We relax this assumption in the next section. Our principal aim in developing the closed-economy model is to demonstrate explicitly how the inclusion of the opportunity costs of habitat conversion plus any non-consumptive value of biodiversity in a bioeconomic model of species exploitation affects the equilibrium stock and thus the optimal harvesting path.

The developing country is assumed to maximize the present value of future welfare, W,

$$Max_y W = \int_0^\infty \left[U(y,c,S) - w(S)y \right] e^{-\delta t} dt , \tag{10.7}$$

subject to (Equation 10.1′) and (Equation 10.5). Let $w(S)$ be defined as the unit social cost of harvesting, with $w' \le 0$ and $w'' \ge 0$, and $\delta > 0$ be the social discount rate. For simplicity, we assume that the rate of land conversion, $r(t) \le 0$, is a constant which we define as $-r$. Finally, the initial and terminal boundaries of the state variable, S, are respectively $S(0) = S_0$ and $\lim_{t \to \infty} S(t) \ge 0$. We exclude the situation with total conversion of the habitat.[7]

The current-value Hamiltonian of the above problem is

$$H = U\left(y, f(S), S\right) - w(S)y + \lambda \left[G_1(S) - aS^{-b}r - y \right], \tag{10.8}$$

where the co-state variable λ can be interpreted as the shadow price of the total species stock, S. Assuming an interior solution, the first-order conditions of the control problem are

$$dH/dy = U_y - w(S) - \lambda = 0 \qquad \text{or} \qquad U_y = w(S) + \lambda \tag{10.9}$$

$$-dH/dS = \dot{\lambda} - \delta\lambda \qquad \text{or} \qquad \dot{\lambda} = \left(\delta - G_1' - baS^{-(b+1)}r\right)\lambda + yw' - U_c f' - U_s \tag{10.10}$$

$$dH/dy = \dot{S} = G_1(S) - aS^{-b}r - y . \tag{10.11}$$

Condition (Equation 10.9) indicates that, along the optimal trajectory, the marginal benefit of harvesting or utilizing wild species for domestic consumption, U_y, must equal its marginal cost, which consists of marginal harvesting costs, $w(S)$, as well as the marginal costs of stock depletion, λ. Condition (Equation 10.11) simply recovers Equation (10.1′) for determining the change in the total species stock, which we have discussed above. Finally, condition (Equation 10.10) resembles the standard renewable resource dynamic condition for denoting the change in the value of the stock of wild resources, S, but it includes not only the standard (net) marginal bioeconomic costs for retaining species, $(\delta - G_1')\lambda + yw'$, but also the net marginal benefits of "holding on" to land as habitat, $-baS^{-(b+1)}r\lambda - U_c f'$, and the marginal social value of an additional unit of the total species

---

[7] Equation (10.7) includes the opportunity cost of land in the developed sector, and additionally the assumption that the unit cost function, $w(S)$, represents the contribution of harvest to costs.

stock, U_s. As this condition is important for the analytical results of our model, we state its interpretation formally as

PROPOSITION 1. *Although the optimal path the rate of change in the shadow price of the total species stock, $d\lambda/dt$, equals the difference between the marginal "bio-economic" costs of holding on to an additional unit of the total species stock, $(\delta - G_1')\lambda + yw'$, less both the marginal net benefits of holding on to natural habitat for that additional unit of total stock, $-baS^{-(b+1)}r\lambda - U_cf'$, and the marginal social value of that additional unit, U_s.*

The long-run equilibrium for the closed economy can be determined by assuming that in the steady state the shadow price of the stock of wild resources, the rates of harvesting and land conversion, and the total species stock are constant: $d\lambda/dt = dy/dt = r = dS/dt = 0$. From the first-order conditions, the long-run equilibrium can be characterized as

$$\delta = G_1' + \frac{\left[-yw' + U_cf' + U_s\right]}{\left[U_y - w(S)\right]} \quad \text{for} \quad \dot{\lambda} = \dot{y} = r = \dot{S} = 0 \tag{10.12}$$

$$y = G_1(S) \quad \text{for} \quad \dot{S} = r = 0. \tag{10.13}$$

Equation (10.12) is the standard long-run equilibrium for holding on to renewable resources, but in this case with some important modifications. This is best seen by interpreting the different components on the right-hand side of (Equation 10.12). Let

$$R_1(S, y) \equiv G_1' + \frac{-yw'}{\left[U_y - w(S)\right]} > 0 \tag{10.14}$$

$$R_2(S, y) \equiv \frac{U_cf'}{\left[U_y - w(S)\right]} < 0 \tag{10.15}$$

$$R_3(S, y) \equiv \frac{U_s}{\left[U_y - w(S)\right]} > 0. \tag{10.16}$$

Then it follows that R_1 is the standard marginal net "bioeconomic" return for holding on to species, R_2 is the (relative) marginal opportunity cost of holding on to land as natural habitat for wild species, and R_3 is the (relative) marginal non-consumptive value of biodiversity. If a solution exists for the long-run equilibrium, then it suggests

PROPOSITION 2. *In a closed economy, the equilibrium total species stock, S^*, is determined by equating the social discount rate, δ, and biodeconomic returns net of the opportunity cost of natural habitat plus the non-consumptive value of biodiveristy, $R_1 + R_2 + R_3$.*

To verify Proposition 2 it is necessary to determine whether an optimal saddle path and solution exist for the long-run equilibrium characterized by Equations (10.12) and (10.13). For our purposes, it is useful to conduct this analysis of the long-run equilibrium in stages, using three versions of the model: the basic bioeconomic model, the inclusion of habitat conversion, and the addition of biodiversity value. As will become evident, each version has different implications for the optimal long-run stock of wildlife species and harvesting rates in the closed economy. The derivation of the results is shown in Appendix 1, and the results are summarized below.

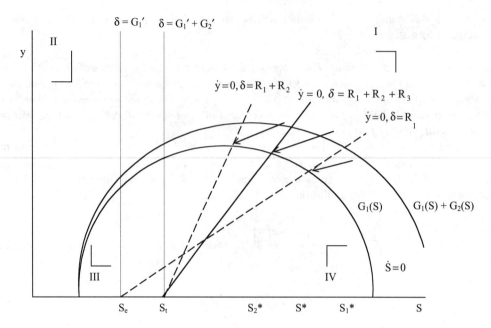

FIGURE 10.1 The long-run equilibrium of the closed economy.

Case 1. Bioeconomic model ($\delta = R_1$)

We begin with the basic bioeconomic model with no costs of holding onto land as habitat and no positive stock externalities associated with biodiversity, i.e., $R_2 = R_3 = 0$. As indicated in Appendix A, the result is a standard modification of the conventional bioeconomic harvesting model, where the optimal steady-state species stock is determined by equating the social discount rate and bioeconomic returns net of harvesting costs that vary with stock size.

The locally stable equilibrium (S_1^*, y_1^*) for case 1 is depicted in Figure 10.1. An optimal trajectory is represented graphically, assuming $G_1' < 0$ and $S_0 > S_1^*$. Note that for any $y_1^* > 0$ it follows that $S_1^* > S_e$, i.e., the equilibrium level of wild-species stock in case 1 must lie to the right of the vertical $\delta = G_1'$ curve in Figure 10.1. This result is caused by the decreasing unit costs as the stock increases. We therefore note that if in our closed-economy model there is no opportunity cost of holding on to land as natural habitat and if there is non-consumptive value of biodiversity, then we obtain

PROPOSITION 2.1. *In a closed economy if $R_2 = R_3 = 0$, there exists an equilibrium harvesting rate $y_1^* > 0$ such that equilibrium total species biomass stock occurs to the right of $\delta = G_1' > 0$, i.e., $S_1^* > S_e$.*

Case 2. Habitat conversion model ($\delta = R_1 + R_2$)

In this version the basic bioeconomic model is extended to include the costs of habitat conversion, i.e., R_2 of Equation (10.15). As shown in Figure 1 and demonstrated in Appendix A, S_2^* and y_2^* are the new equilibrium levels of total species biomass and harvesting respectively. Corresponding to these values must be an equilibrium level of natural habitat area, A_2^*. Appendix A indicates that the equilibrium of case 2 is a saddle point for an upward-sloping dy/dt curve. In Figure 10.1, the optimal trajectory assuming $G_1' < 0$ and $S_0 > S_2^*$ is shown.

In Appendix A, the equilibria of cases 1 and 2 are compared, and the result suggests that $S_1^* > S_2^* > S_f > S_e$, i.e., the equilibrium level of wild-species stock in case 2 must be less than that derived in case 1. Intuitively, the reasoning is fairly straightforward: now that there is an opportunity cost for holding on to land as natural habitat, in the long run a smaller total stock of species will be maintained.

This result is an important modification of the standard bioeconomic model of renewable resource utilization. We have suggested that the growth in total species stock is dependent not only on the natural biological regeneration rate of species, but also on the total number of species in a given habitat area. However, holding land as a natural habitat also implies an opportunity cost, which in our case 2 version of the model is represented by forgone consumption from development activities that use converted habitat land. Because conventional bioeconomic models do not take explicitly into account the opportunity of maintaining "wild" lands as habitat for species, the equilibrium level of species stock — and thus implicitly habitat area — is higher than if the opportunity cost of maintaining habitat were included. Consequently, this outcome can be stated more formally as

PROPOSITION 2.2. *In a closed economy if $R_3 = 0$, there exists an equilibrium harvesting rate $y_2^* > 0$ such that equilibrium total species stock occurs to the right of $\delta = G_1' + G_2' > 0$ but is less than the equilibrium stock of the bioeconomic model, i.e. $S_1^* > S_2^* > S_f > S_e$.*

Case 3. Inclusion of biodiversity value ($\delta = R_1 + R_2 + R_3$)

Finally, we extend the model to include $S(t)$ in the social utility function to represent the stock externalities associated with non-consumptive biodiversity values, thus arriving at the original optimal problem and corresponding first-order conditions (Equation 10.9), (Equation 10.10) and (Equation 10.11). As shown in Figure 10.1 and demonstrated in Appendix A, S^* and y^* are the new equilibrium levels of total species stock and harvesting respectively, which in turn imply a new equilibrium level of natural habitat area, A^*. Appendix A confirms that the equilibrium of the full version of the model (case 3) is a saddle point for an upward-sloping $dy/dt = 0$ curve, thus verifying Proposition 2. The optimal trajectory is depicted in Figure 10.1 assuming $G_1' < 0$, $S_0 > S^*$ and $S_1^* > S^* > S_2^*$.

As indicated in Appendix A, comparison of the results of case 3 with those of cases 1 and 2 yields two interesting conclusions. First, comparing cases 2 and 3, the inclusion of non-consumptive biodiversity values in the model will tend to offset the opportunity costs of holding on to habitat, resulting in a higher long-run equilibrium level of species stock and habitat area. That is, in the full version of the model with biodiversity values included (case 3), the equilibrium wildlife stock and habitat area will always exceed that of case 2 which excludes these values. Second, comparing cases 1 and 3, it is not clear whether the equilibrium level of species stock (and habitat area) in the full model will be less or greater than the equilibrium stock determined by the simple bioeconomic model. The outcome will depend on whether in the long run the marginal value of biodiversity, U_s, compensates sufficiently for the marginal opportunity cost of maintaining habitat, $-U_c f'$. Only under special conditions will the bioeconomic model (case 1) actually be able to predict the long-run equilibrium of the full model (case 3), i.e., when $U_s = -U_c f'$.

It is worth summarizing these results with a further addition to Proposition 2:

Proposition 2.3. *In a closed economy if the $\dot{y} = 0$ curve is defined by $\delta = R_1 + R_2 + R_3$, there exists an equilibrium harvesting rate $y^* > 0$ such that equilibrium total species biomass stock occurs to the right of $\delta = G_1' = G_2' > 0$ and exceeds the equilibrium stock of the habitat conversion model, i.e., $S^* > S_2^* > S_f > S_e$; however, whether S^* is less than, equals or exceeds the equilibrium species stock of the bioeconomic model, S_1^*, will depend on whether U_s is less than, equals or exceeds $-U_c f'$.*

D. OPEN-ECONOMY MODEL

In this section we develop the model to allow for trade. The basic assumption is that some wildlife products are exported, which in turn allows the importation of consumption goods in addition to domestic production from converted habitat land. As noted in the introduction, if wildlife or natural products enter international markets then the possibility exists that trade interventions may be used to influence the optional level of harvesting and species biomass stocks. The likely scenario is that importing nations may use such interventions to control exploitation of wildlife stocks and conversion of habitat land. Following the approach of Barbier and Rauscher (1994), we extend the full version of our model to examine the likely implications of such trade interventions, and compare them to the impacts of an alternative policy, such as international transfer.

We assume the following additional relationships for the open-economy model: $x(t)$ is exports of wildlife or natural resource products (in terms of biomass); $m(t)$ is consumption of imported goods; $y(t) - x(t)$ is the domestic consumption of wildlife or natural resource products; and p is the terms of trade, p_x/p_m.

As before, two outputs are produced domestically, $f(S(t))$ from converted habitat land and $y(t)$ from harvesting wildlife resources. By assumption, habitat land is a net exporter of resource products: $x(t) > 0$. All domestic products and imports are assumed to be homogeneous. Consequently, the open-economy model includes a trade balance equation and a new definition of total domestic consumption, c:

$$px(t) = m(t) \tag{10.17}$$

$$c(t) = f(S(t)) + m(t). \tag{10.18}$$

From substituting (Equation 10.17) and (Equation 10.18) into (Equation 10.7), the open-economy welfare-maximizing problem for the developing country is now

$$Max_{y,x} W = \int_0^\infty \left[U(y - x, f(S) + px, S) - w(S)y \right] e^{-\delta t} dt \tag{10.19}$$

subject to (Equation 10.1′) and (Equation 10.5) and the initial and terminal conditions as defined before. Thus the open-economy problem contains an additional dimension beyond the social decision to trade off wildlife utilization with habitat conversion: on the one hand, holding on to natural habitat implies an opportunity cost of forgone consumption from less habitat conversion; on the other hand, trade in wildlife and natural resource products allows consumption of imported goods which can substitute for commodities produced from converted habitat.

The current-value Hamiltonian of the problem is

$$H = U\left[y - x, f(S) + px, S\right] - w(S)y + \lambda\left[G_1(S) - aS^{-b}r - y\right], \tag{10.20}$$

which yields the following first-order conditions, assuming an interior solution:

$$dH/dy = U_1 - w(S) - \lambda = 0 \quad \text{or} \quad U_1 = w(S) + \lambda \tag{10.21}$$

$$dH/dx = -U_1 + pU_2 = 0 \quad \text{or} \quad U_1 = pU_2 \tag{10.22}$$

$$-dH/dS = \dot{\lambda} - \delta\lambda \quad \text{or} \quad \dot{\lambda} = \left(\delta - G_1' - baS^{-(b+1)}r\right)\lambda + yw' - U_2 f' - U_3 \tag{10.23}$$

$$dH/dy = \dot{S} = G_1(S) - aS^{-b}r - y \tag{10.11}$$

where $U_1 = \partial U/\partial(y - x) > 0$, $U_2 = \partial U/\partial(c + m) > 0$ and $U_3 = \partial U/\partial S > 0$. We assume as before that these marginal utilities are subject to diminishing returns and will tend to infinity as their arguments approach zero.

Condition (Equation 10.21) indicates that, along the optimal trajectory, the marginal benefit of harvesting or utilizing wild species for domestic consumption, U_1, must equal its marginal cost, which consists of marginal harvesting costs, $w(S)$, as well as the marginal costs of stock depletion, λ. However, the only difference with condition (Equation 10.9) of the closed economy is that domestic consumption is now defined as harvested output net of exports, $y - x$. Condition (Equation 10.22) is entirely new, and it indicates that, if the developing country is a price-taker in international trade, then the relative marginal value of domestic consumption of wildlife and natural resource products to imported-good consumption must be equated with the terms of trade, p. Finally, condition (Equation 10.23) is almost identical with the renewable-resource dynamic condition (Equation 10.10) for denoting the change in the value of the total species stock in the closed economy, and it includes not only the standard (net) marginal bioeconomic costs for holding on to species, $(\delta - G_1')\lambda + yw'$, but also the net marginal benefits of "holding on" to land as habitat, $-baS^{-(b+1)}r\lambda - U_2 f'$, and the marginal social value of an additional unit of biodiversity, U_3.

By combining (Equation 10.21) and (Equation 10.23) we obtain

$$\dot{y} - \dot{x} = \frac{\left[\left(\delta - G_1' + baS^{-(b+1)}r\right)\left(U_1 - w(S)\right) + w'\left(G_1(S) - aS^{-b}r\right) - U_2 f' - U_3\right]}{U_{11}} \tag{10.24}$$

Utilizing (Equation 10.24) and the first-order conditions (Equation 10.21), (Equation 10.22), (Equation 10.23) and (Equation 10.11), one can solve for the following system of equations describing an equilibrium where $dy/dt = dx/dt = dS/dt = r = 0$

$$U_1\left(y^+, c^+\right) = pU_2\left(S^+, x^+\right) \tag{10.25}$$

$$\delta - G_1'\left(S^+\right) = \frac{-y^+ w'\left(S^+\right) + U_2\left(S^+, x^+\right)f'\left(S^+\right) + U_3\left(S^+\right)}{\left(U_1\left(y^+\right) - w\left(S^+\right)\right)} > 0, \qquad \dot{y} = \dot{x} = \dot{S} = r = 0$$

$$\text{or} \qquad \delta = V_1\left(S^+, y^+, x^+\right) + V_2\left(S^+, y^+, x^+\right) + V_3\left(S^+, y^+, x^+\right) \tag{10.26}$$

or

$$G_1\left(S^+\right) = y^+, \qquad \dot{S} = r = 0, \tag{10.27}$$

where the equilibrium values of y, x and S are denoted by obelisks (+), and $V_1 = G_1'(S^+) - y^+ w'(S^+)/[U_1(y^+, x^+) - w(S^+)]$, $V_2 = U_2(S^+, x^+)f'(S^+)/[U_1(y^+, x^+) - w(S^+)]$ and $V_3 = (S^+)/[U_1(y^+, x^+) - w(S^+)]$.

Total differentiation of the system of Equations (10.25)–(10.27) can lead to characterization of the equilibrium state and trajectories leading to the equilibrium. As shown in Appendix A, the equilibrium is a saddle point, and the saddle path is positively sloped. Using these results, the optimal solution is represented graphically in Figure 10.2 for $S_0 > S^+$ and $G_1' < 0$. This solution suggests

PROPOSITION 3. *In an open economy if the $dy/dt = dx/dt = 0$ curve is defined by $\delta = V_1 + V_2 + V_3$ there exists an equilibrium harvesting rate $y^+ > 0$ such that equilibrium total species stock, S^+, occurs to the right of $\delta = G_1' + G_2' > 0$.*

It is useful to compare the equilibrium (S^*, y^*) of the full (case 3) version of the closed-economy model with the new equilibrium (S^+, y^+) obtained in the open economy. If we assume that the marginal utilities of consumption and biodiversity values are the same in the two models, then $x = 0$ implies that $U_2(f(S^+)) = U_c(f(S^*))$, $U_1(y^+) = U_y(y^*)$ and $U_3(S^+) = U_s(S^*)$. Consequently,

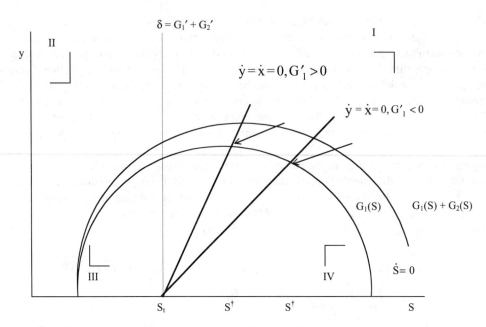

FIGURE 10.2 The long-run equilibrium of the open economy.

equilibrium in the closed and open economy will be the same: $S^+ = S^*$ and $y^+ = y^*$. However, for any $x > 0$ then $U_2(f(S^*) + x) < U_2(f(S^*))$ and $U_1(y^* - x) > U_1(y^*)$. From inspection of Equation (10.26), it is unclear in which direction S must change in order to maintain the equilibrium condition. This is because, on the one hand, exports of wildlife products reduce the need for domestic consumption from habitat conversion activities, but on the other hand, increased wildlife exports mean less domestic consumption of these products. If the marginal utility of the latter effect outweighs that of the former, then the right-hand side of (Equation 10.26) may increase, and S must therefore also increase to keep the equilibrium condition. However, if the increase in U_1 is the more powerful effect, then the right-hand side of (Equation 10.26) could decrease and S would have to rise correspondingly. Thus it is unclear whether the open-economy equilibrium for $x^+ > 0$ will yield a long-run equilibrium species stock, S^+, that will be less than, equal to or greater than the equilibrium, S^*, under the closed economy.

There is also an intuitive explanation of this ambiguous result. For the economy to begin trading in wildlife products, p must rise sufficiently high to make exports worthwhile. Starting from autarky this has two effects. First, the development of an external market means that it is more profitable to harvest from wildlife resources, as the external terms of trade with respect to consumption goods are now higher than the domestic terms of trade. This may increase exploitation of the resource base to produce wildlife exports. On the other hand, as noted above, there is less pressure for conversion of land to other uses, as it is now less expensive to obtain consumption goods from world markets. We shall evaluate these effects in the next section.

E. GLOBAL BIODIVERSITY VALUES: TRADE INTERVENTIONS VS. TRANSFERS

Comparative static analysis of the long-run equilibrium can be employed to determine the impacts of trade interventions in influencing the developing country's decision on how much wildlife species to maintain. Our model already includes positive *domestic* biodiversity values, such as the non-consumptive values that the developing country may capture from maintaining species and natural habitat. However, it is unlikely that the developing country will take into account any positive *global* biodiversity values associated with wildlife and habitat conservation, for example the values that the rest of the world attaches to "conserving" genetic material for future unknown uses; the global educational and scientific research value of biodiversity; and the "existence" value of living creatures and natural systems.[8] If the global values of biodiversity are being ignored by domestic policy-makers, then it is likely that the international community would want the developing country to conserve more species and habitat in the long run than the country would do on its own.

There are essentially two broad approaches that the international community could take to achieve this objective. As described in the introduction, it could through trade interventions such as import bans, tariffs or other controls on wildlife and natural resource products trade try to influence the developing country's decision to "hold on" to species and habitat. The alternate policy would be to encourage the developing country to "internalize" any global biodiversity values through an international transfer of funds to reduce the need to exploit and convert wild lands. The effects of both approaches can be characterized through comparative static analysis of our open-economy model.

Following Barbier and Rauscher (1994), we assume that import bans, controls, and restrictions have the effect of reducing the terms of trade for the wildlife trade of the developing country. In our open-economy model, a change in the terms of trade, p, has the following impacts on the long-run equilibrium of total species biomass stock:

$$\frac{dS^+}{dp} = -\frac{\left[\Psi(1+\eta_2)U_2 - f'U_{22}x(U_{11} + p^2U_{22})\right]}{\Delta} \tag{10.28}$$

where $\eta_2 < 0$ is the elasticity of marginal utility, U_2, with respect to imported consumption goods, m and $\psi = (\delta - G_1')U_{11} + f'U_{22}p$. If the absolute value of η_2 is large, i.e., $|\eta_2| > 1$, then marginal welfare in the developing country is highly responsive to a change in imported commodities. The situation with $|\eta_2| > 1$ occurs only when the exports supply curve of the developing country is backward-bending. Hence $|\eta_2| < 1$ is the normal case in the trade sector of the model.

From (Equation 10.28) it follows that

$$\frac{dS^+}{dp} < 0 \quad \text{when } |\eta_2| < 1 \text{ and } \psi < \frac{f'U_{22}x(U_{11} + p^2U_{22})}{(1+\eta_2)U_2} < 0, \quad \text{or}$$

$$\frac{dS^+}{dp} < 0 \quad \text{when } |\eta_2| > 1 \text{ and } \psi > \frac{f'U_{22}x(U_{11} + p^2U_{22})}{(1+\eta_2)U_2} > 0.$$

These conditions are not necessarily fulfilled. For example, if $\psi(1 + \eta)U_2 > f'U_{22}x(U_{11} + p^2U_{22})$, we get $dS/dp > 0$, quite contrary to the results of standard bioeconomic models (Clark, 1976). A situation with $dS/dp > 0$ will always occur for a positive ψ when $|\eta_2| < 1$, and for a negative ψ when $|\eta_2| > 1$. Under these conditions, trade interventions are counterproductive to conserving species and habitat, and therefore are the wrong strategy for persuading the developing country to internalize global biodiversity values. Thus, without further information on the value of η_2 or the sign of ψ it is impossible to determine whether trade interventions that reduce the terms of trade, p, will increase or decrease the equilibrium value of species stock, S^+. The ambiguous stock effects make trade interventions a poor policy instrument for securing biodiversity conservation.

A large transfer of international funds to the developing country, perhaps through the provisions of the international Convention of Biodiversity or the Global Environmental Facility (GEF), could assist the country in wildlife and habitat conservation. For example, the GEF has a mandate to fund biodiversity conservation projects in developing countries that yield global values but not sufficient domestic net benefits to be funded by these countries. Such international transfers have the effect

---

[8] See WCMC (1992) for further discussion and summary of such global values.

of freeing up domestic financial resources for other purposes. The developing country is essentially receiving a subsidy from the rest of the world to 'internalize' global biodiversity values. In our model, these transfers can be represented by an increase in foreign exchange available for consumption of imported goods. This has a double benefit in terms of species and habitat conservation: it not only supplements wildlife export earnings, but also reduces consumption from habitat conversion activities. Thus from (Equation 10.17) and (Equation 10.18) consumption now becomes

$$c = f(S) + px + i \,, \tag{10.29}$$

where i is the amount of international transfer. An increase in the subsidy has the following effect on the long-run equilibrium of total species biomass stock:

$$\frac{dS^+}{di} = -\frac{\left[\Psi p U_{22} - f' U_{22} \left(U_{22} + p^2 U_{22} \right) \right]}{\Delta} \,. \tag{10.30}$$

A transfer of funds will increase the long-run stock, $dS^+/dt > 0$, if $(\delta - G_1') > f'/p$. This is always the case if the equilibrium is to the right of $\delta = G_1'$ as shown in Figure 10.2, and assuming no impact of the transfer on p. The effect of the international transfer is therefore to make the receiving country better off, with increased income to import from abroad. The subsidy not only replaces wildlife exports as a means of importing consumption goods but also puts less pressure on the species stock through increased harvesting for domestic consumption. Consequently, an increased international transfer leads to a greater long-run equilibrium total species stock.[9]

From the discussion of the effects of trade interventions and a direct international transfer it is worth stating that:

PROPOSITION 4. *In an open economy, if $(\delta - G_1') > f'/p$ then a direct international transfer, i, will increase the long-run equilibrium total species stock, $S^\dagger$, unambiguously; however, the effect of a trade intervention is ambiguous.*

F. CONCLUSION

We have attempted to explore two issues in this chapter. First, by developing several versions of a closed-economy model of optimal species utilization, we have examined the extent to which the long-run equilibrium of the basic bioeconomic model is affected by the inclusion of habitat conversion and then additionally biodiversity value as a positive externality. Second, by developing an open-economy version of the full model we were able to compare the likely impacts of trade interventions as opposed to international transfers on the developing country's long-run decision to conserve wildlife species and habitat. Several important results emerged from this analysis.

The inclusion of habitat conversion proves to be an important modification of the standard bioeconomic model of renewable resource utilization. We have suggested that the growth in total species stock is dependent not only on the natural biological regeneration rate of species but also on the total number of species contained in a given habitat area. However, holding land as natural habitat also implies an opportunity cost, which is represented by foregone consumption from development activities that use converted habitat land. Because conventional bioeconomic models

---

[9] Note that if $\psi < 0$ and $|\eta_2| < 1$ then (Equation 10.30) and (Equation 10.32) suggest that an international subsidy will still be preferred to a trade intervention so long as $pU_{22} > -(1 + \eta_2)U_2$. However, it is not necessary for ψ to be negative in order for a subsidy to be preferred to a trade intervention. For example, letting $\omega = f'U_{22}(-U_{11} - p^2U_{22})$, if $\psi > 0$ and $|\eta_2| > 1$ then a subsidy is still preferred if $\omega \geq \psi(1 + \eta_2)U_2$ and $\omega > \psi p U_{22}$. Similarly, if $\psi > 0$ and $|\eta_2| < 1$ then a subsidy is preferred if $\omega > \psi p U_{22}$.

do not take explicitly into account the opportunity cost of maintaining wild lands as habitat for species, the equilibrium level of species stock, and thus implicitly habitat area, is higher than if the opportunity cost of maintaining habitat were included. Consequently, when applied to wildlife utilization problems, conventional bioeconomic models may over-estimate the optimal long-run levels of wildlife species and habitat.

Our analysis shows that the additional inclusion of biodiversity values as a positive stock externality will counteract, to some extent, the effects of the opportunity cost of maintaining habitat for species. If the marginal value of biodiversity is sufficiently high, then despite the marginal opportunity cost of maintaining habitat, the optimal equilibrium levels of species and habitat can equal or even exceed the levels determined by the conventional bioeconomic model.

These results for the closed-economy model highlight the importance of extending the conventional bioeconomic model for wild species and resource utilization to include both the opportunity cost of maintaining species habitat and positive biodiversity values if these costs and values are considered significant. If they are instead excluded, the conventional bioeconomic model will predict the correct long-run equilibrium level of species stock (and habitat area) only under very special conditions.

The open-economy model provides an additional dimension to the social decision to trade off wildlife utilization with development benefits from converted habitat land: on the one hand, holding on to natural habitat implies an opportunity cost of foregone consumption from less habitat conversion; on the other hand, trade in wildlife and natural resource products allows consumption of imported goods which can substitute for commodities produced from converted habitat. However, it is unclear, when compared to the closed-economy equilibrium, whether the open economy leads to a higher long-run optimal level of species and habitat conservation. Exports of wildlife products appear to reduce the need for domestic consumption from habitat conversion activities, but exports also mean less domestic consumption of these products.

Comparative static analysis of the long-run equilibrium of the open-economy model suggest caution in the use of trade interventions as an international policy for forcing a developing country to internalize externality values of biodiversity. Under certain conditions, trade interventions may be unambiguously counterproductive, whereas the alternative policy of an international transfer of funds would lead to greater long-run conservation of species and habitat. Although in our model it is possible for trade interventions (trade bans, tariffs, or trade subsidies) to have the desired effect of increasing long-run equilibrium species stock and habitat area, we need to know the specific conditions underlying the bioeconomic, habitat conversion, and trade sectors of the model in order to determine the precise influence of trade policy interventions. Hence, we conclude that trade intervention is a second-best approach for global biodiversity conservation.

Finally, it is worth commenting that the results of the open-economy model might be affected if it were extended to reflect other aspects of trade-environment dimensions to resource exploitation problems in developing countries. For example, Chichilnisky (1994) examines the interactions of ill-defined property rights regimes and international trade in resource products from developing countries; Barbier and Rauscher (1994) expand their basic model of trade and tropical deforestation to allow for market power and foreign asset accumulation (or debt); and Schulz (1994) examines the effects of trade on wildlife utilization in developing countries by explicitly accounting for species interactions, such as predator-prey relationships. Extensions to our model along these lines may further increase its applicability to resource utilization problems in developing countries.

APPENDIX A

SOLUTIONS OF THE CLOSED- AND OPEN-ECONOMY MODELS

This appendix derives the long-run equilibrium solutions and stability conditions for the various versions of the model developed in this chapter.

Case 1. Biodiversity model ($\delta = R_1$)
We begin with the basic bioeconomic model of the closed economy with no costs of holding on to land as habitat and no positive stock externalities associated with biodiversity: $R_2 = R_3 = 0$. Thus the following equations of motion can be derived from the first-order conditions (Equation 10.9), (Equation 10.10) and (Equation 10.11):

$$\dot{y} = \frac{\left(\delta - G_1'\right)\left(U_y - w(S)\right) + yw' + w'\dot{S}}{U_{yy}}$$

$$= \frac{\left(\delta - G_1'\right)\left(U_y - w(S)\right) + w'G_1(S)}{U_{yy}}$$

$$\dot{S} = G_1(S) - y,$$

which yield the following long-run equilibrium condition for $dy/dt = dS/dt = 0$:

$$\delta - G_1'\left(S_1^*\right) = \frac{-y_1^* w'\left(S_1^*\right)}{\left(U_y\left(y_1^*\right) - w\left(S_1^*\right)\right)} > 0, \qquad \dot{y} = \dot{S} = 0$$

$$G_1\left(S_1^*\right) = y_1^*, \qquad \dot{S} = 0,$$

where S_1^* and y_1^* are the equilibrium levels of the total species stock and harvesting respectively. This equilibrium for case 1 is depicted in Figure 10.1. Note that from the above equations $y_1^* = 0$ results in $\delta = G_1'$, which means that the origin of the $dy/dt = 0$ curve for case 1 in Figure 10.1 is $(S_e, 0)$. Therefore, for any $y_1^* > 0$ it follows that $S_1^* > S_e$, i.e., the equilibrium level of wild-species stock in case 1 must lie to the right of the vertical $\delta = G_1'$ curve in Figure 10.1. This result is caused by the decreasing unit costs as the stock increases.

Through linearizing the above two equations of motion and evaluating the resulting matrix of coefficients at the equilibrium (S_1^*, y_1^*), it can be shown that the equilibrium is a saddle point if the determinant of the matrix is negative, and the curve $dy/dt = 0$ is positively sloped and cuts the curve $dS/dt = 0$ from below. Through linearizing we obtain

$$\dot{S} = G_1'\left(S - S_1^*\right) - \left(y - y_1^*\right)$$

$$\dot{y} = \frac{X\left(S - S_1^*\right) + Z\left(y - y_1^*\right)}{U_{yy}},$$

where $X = -(\delta - G_1')w' - (Uy - w(S))G'' + G_1'w'' > 0$, $Z = (\delta - G_1')U_{yy} + w' < 0$ and U_{yy} is assumed to be constant.[10] The determinant Δ of the matrix of coefficients of the above equations is

$$\Delta = \frac{G_1'Z = X}{U_{yy}}.$$

---

[10] We take advantage of the fact that at the equilibrium $G_1'dS_1^* = dy_1^*$.

It follows that $\Delta < 0$ if $G_1' < 0$ or if $G_1' > 0$ and $-X/U_{yy} > Z/U_{yy}$. However, if these conditions hold, then comparing the $dy/dt = 0$ and $dS/dt = 0$ curves shows that

$$\left.\frac{dy}{dS}\right|_{\dot{y}=0} = -\frac{X}{Z} > G_1' = \left.\frac{dy}{dS}\right|_{\dot{S}=0}.$$

Consequently, the requirement that $dy/dt = 0$ is upward-sloping and cuts the $dS/dt = 0$ curve from below is a necessary condition for $\Delta < 0$. The equilibrium (S_1^*, y_1^*) is therefore locally stable (a saddle point). Thus, as shown in Figure 10.1, for case 1 isosectors I and II are convergent and contain optimal trajectories to the locally stable equilibrium (S_1^*, y_1^*). An optimal trajectory is represented graphically, assuming $G_1' < 0$ and $S_0 > S_1^*$.

Case 2. Habitat conversion model $(\delta = R_1 + R_2)$
In this version the basic bioeconomic model is extended to include the costs of habitat conversion, i.e., R_2 of Equation (10.15). The equations of motion for the closed economy are now modified to

$$\dot{y} = \frac{(\delta - G_1' - G_2')(U_y - w(S)) + yw' + w'\dot{S} - U_c f'}{U_{yy}}$$

$$= \frac{(\delta - G_1' + baS^{-(b+1)})(U_y - w(S)) + w'(G_1(S) - aS^{-b}r) - U_c f'}{U_{yy}}$$

$$\dot{S} = G_1(S) + G_2(S) = G_1(S) - aS^{-b}r - y.$$

This leads to the following long-run equilibrium conditions for $dy/dt = dS/dt = r = 0$

$$\delta - G_1'(S_2^*) = \frac{-y_2^* w'(S_2^*) + U_c f'(S_2^*)}{(U_y(y_2^*) - w(S_2^*))} > 0, \qquad \dot{y} = \dot{S} = r = 0$$

$$G_1(S_2^*) = y_2^*, \qquad \dot{S} = r = 0,$$

where S_2^* and y_2^* are the new equilibrium levels of the total species stock and harvesting respectively, as depicted for case 2 in Figure 10.2. From Equation (10.2) in the text it is clear that corresponding to these values must be an equilibrium level of natural habitat area, A_2^*. Since we have added $G_2(S)$ to the growth function, we must change the origin $(y_2^* = 0)$ of the $dy/dt = 0$ curve to S_f, such that $\delta = G_1'(S_f) + G_2'(S_f)$ and $S_f > S_e$. Therefore, as shown for case 2 in Figure 10.1, if the $dy/dt = 0$ curve is upward-sloping, then for any $y_2^* > 0$ if follows that $S_2^* > S_f$; i.e., the equilibrium level of wild-species stock will be to the right of the point S_f.

By linearizing the above equations of motion and evaluating them at the equilibrium point (S_2^*, y_2^*) it is possible to prove that the above condition holds for case 2. Given that at this equilibrium $r = 0$, we obtain through linearizing

$$\dot{S} = G_1'(S - S_2') - (y - y_2^*)$$

$$\dot{y} = \frac{[X - U_{cc}(f')^2 - U_c f^*](S - S_2^*) + Z(y - y_2^*)}{U_{yy}}$$

where $X > 0$ and $Z < 0$ as before and $-U_{cc}(f')^2 - U_c f'' > 0$. The determinant Δ of the matrix of coefficients of the above linearized equations is

$$\Delta = \frac{G_1' Z + \left(X - U_{cc}(f')^2 - U_c f''\right)}{U_{yy}}.$$

It follows that $\Delta < 0$ for $G_1' < 0$. If $G_1' > 0$, then $\Delta < 0$ requires $-(X - U_{cc}(f')^2 - U_c f'')/Z > G_1'$. The result of such conditions is

$$\left.\frac{dy}{dS}\right|_{\dot{y}=0} = -\frac{\left(Z - U_{cc}(f')^2 - U_c f''\right)}{Z} > G_1' = \left.\frac{dy}{dS}\right|_{\dot{S}=0}.$$

Thus for case 2 the requirement that $dy/dt = 0$ is upward-sloping and cuts the $dS/dt = 0$ curve from below is a necessary condition for $\Delta < 0$. The equilibrium (S_2^*, y_2^*) is therefore locally stable (a saddle point). In Figure 10.1, the optimal trajectory assuming $G_1' < 0$ and $S_0 > S_2^*$ is shown.

In comparing the equilibria of cases 1 and 2, note that in case 2 if in equilibrium $-w'G_1(S) > -U_c f'$, then $(\delta - G_1') > 0$ and the $dy/dt = 0$ curve is upward-sloping. This result suggests that $S_1^* > S_2^* > S_f > S_e$, i.e., the equilibrium level of wild-species stock in case 2 must be less than the equilibrium level derived in case 1.[11] That is, because in case 2 there is an opportunity cost for holding on to land as natural habitat, in the long run a smaller total stock of species and habitat will be maintained.

Case 3. Inclusion of biodiversity value ($\delta = R_1 + R_2 + R_3$)
Incorporating $S(t)$ in the social utility function to represent the stock externalities associated with non-consumptive biodiversity values yields the original optimal control problem and corresponding first-order conditions (Equation 10.9), (Equation 10.10) and (Equation 10.11). The equation of motion for optimal harvesting derived in case 2 now becomes

$$\dot{y} = \frac{(\delta - G_1' - G_2')\left(U_y - w(S)\right) + yw' + w'\dot{S} - U_c f' - U_S}{U_{yy}}$$

and the long-run equilibrium conditions for $dy/dt = dS/dt = r = 0$ are the same as (Equation 10.12) and (Equation 10.13) in the text

$$\delta - G_1'(S^*) = \frac{-y^* w'(S^*) + U_c f'(S^*) + U_S(S^*)}{\left(U_y(y^*) - w(S^*)\right)} > 0, \qquad \dot{y} = \dot{S} = r = 0$$

$$\text{or} \qquad \delta = R_1(S^*, y^*) + R_2(S^*, y^*) + R_3(S^*, y^*)$$

$$G_1(S^*) = y^*, \qquad \dot{S} = r = 0,$$

---

[11] It is fairly easy to show why this equilibrium condition for case 2 is necessary to yield an upward-sloping $dy/dt = 0$ curve. In case 2, if $-w'G_1(S) = -U_c f'$, then $\delta = G_1'$. It follows that $y_2^* = y_f = G_1(S_f)$. Alternatively, if $-w'G_1(S) < -U_c f'$, then $\delta < G_1'$ and $S_2^* < S_f$ and $y_2^* < y_f$. However, the result would be $-U_c f' \to \infty$, $S_2^* \to \underline{S}$ and $G_1(\underline{S}) = y = 0$. That is, there is complete conversion of habitat in the initial time period $t = 0$. Thus a necessary condition for an upward-sloping $dy/dt = 0$ curve is $-w'G_1(S) > -U_c f'$.

where S^* and y^* are the new equilibrium levels of the total species stock and harvesting respectively, which in turn imply an equilibrium level of natural habitat area, A^*. Note that once again $y^* = 0$ results in $\delta = G_1'(S_f) + G_2'(S_f)$; i.e., as shown in Figure 10.1 for case 3 the $dy/dt = 0$ curve originates at S_f, and the equilibrium level of wild-species stock will lie to the right of this point.

The stability condition for this case is similar to that of case 2. That is, the inclusion of biodiversity value means that the determinant Δ of the matrix of coefficients of the linearized equations of motion evaluated at the equilibrium (S^*, y^*) is

$$\Delta = \frac{G_1'Z + \left(X - U_{cc}(f')^2 - U_c f'' - U_{SS}\right)}{U_{yy}}.$$

It follows that $\Delta < 0$ for $G_1' < 0$. If $G_1' > 0$, then $\Delta < 0$ requires $-(X - U_{cc}(f')^2 - U_c f'' - U_{SS})/Z > G_1'$. The result of such conditions is

$$\left.\frac{dy}{dS}\right|_{\dot{y}=0} = -\frac{\left(X - U_{cc}(f')^2 - U_c f'' - U_{SS}\right)}{Z} > G_1' = \left.\frac{dy}{dS}\right|_{\dot{S}=0}.$$

Thus for case 3 the requirement that $dy/dt = 0$ is upward-sloping and cuts the $dS/dt = 0$ curve from below is a necessary condition for $\Delta < 0$. The equilibrium (S^*, y^*) is therefore locally stable (a saddle point). In Figure 10.1, the optimal trajectory assuming $G_1' < 0$ and $S_0 > S^*$ is shown.

It is useful to compare the equilibrium species stock derived in cases 3 with that resulting from cases 1 and 2. In case 3, if $-w'G_1(S) + U_S > -U_c f'$, then $(\delta - G_1') > 0$ and the $dy/dt = 0$ curve is upward-sloping.[12] An immediate result of this condition must be that $G_1'(S_2^*) > G_1'(S^*)$, which in turn implies $S^* > S_2^*$. That is, in the full version of the model with biodiversity values included (case 3), the equilibrium wildlife stock exceeds that of case 2 which excludes biodiversity values. However, the relationship between S^* and S_1^* is more ambiguous and is governed by the following condition:

$$\text{If} \quad U_S \lessgtr -U_c f' \quad \text{then} \quad S^* \lessgtr S_1^*$$

That is, if the marginal value of biodiversity is sufficiently high relative to the marginal opportunity cost of maintaining natural habitat, then the optimal equilibrium level of species, S^*, can equal or even exceed the equilibrium level of the bioeconomic model. In Figure 10.1, the optimal trajectory assumes that $S^* < S_1^*$ which implies $U_S < U_c f'$.

The Open-Economy Model

The stability conditions for the open-economy model can be derived in the following manner. Total differentiation of the simultaneous equation system Equations (10.25)–(10.27) characterizing the equilibrium (S^+, y^+, x^+) yields

---

[12] As in the previous note, for case 3 if $-w'G_1(S) + U_S = -U_c f'$, then $\delta = G_1'$ and $y^* = y_f = G_1(S_f)$. Similarly, if $-w'G_1(S) + U_S < -U_c f'$, then $\delta < G_1'$, $S^* < S_f$ and $y^* < y_f$. However, as $-U_c f' \to \infty$, $S^* \to \underline{S}$ and $G_1(S) = y = 0$, which again suggests complete conversion in the initial time period $t = 0$. Thus for case 3, a necessary condition for an upward-sloping $dy/dt = 0$ curve is $-w'G_1(S) + U_S > -U_c f'$.

$$U_{11}dy - \left(U_{11} + p^2 U_{22}\right)dx - \left(pU_{22}f'\right)dS - \left(U_2 + pU_{22}x\right)dp = 0$$

$$\left[(\delta - G_1')U_{11} + w'\right]dy - \psi dx + \phi dS - f'U_{22}x dp = 0$$

$$-dy + G_1'dS = 0,$$

where $\psi = (\delta - G_1')U_{11} + f'U_{22}p$ and $\phi = -(U_1 - w(S))G_1'' - (\delta - G_1')w' + yw'' - U_{22}(f')^2 - U_2 f''$ $- U_{33} > 0$. The determinant of the Hessian matrix of coefficients for dy, dx and dS in the above system of equations is

$$\Delta = G_1'\left[\left((\delta - G_1)U_{11} + w'\right)\left(U_{11} + p^2 U_{22}\right) - \psi U_{11}\right] + \psi p U_{22}f' + \phi\left(U_{11} + p^2 U_{22}\right).$$

Signing Δ again requires comparing the slopes of the $dy/dt = dx/dt = 0$ and $dS/dt = 0$ curves. Substituting for dx in the above system of equations yields

$$\left.\frac{dy}{dS}\right|_{\dot{S}=0} = G_1'$$

$$\left.\frac{dy}{dS}\right|_{\dot{y}=\dot{x}=0} = -\frac{\psi p U_{22}f' + \phi\left(U_{11} + pU_{22}\right)}{\left((\delta - G_1')U_{11} + w'\right)\left(U_{11} + p^2 U_{22}\right) - \psi U_{11}}.$$

It follows that if $G_1' < 0$ and the $dy/dt = dx/dt = 0$ curve is upward-sloping, then $\Delta < 0$. However, if $G_1' > 0$ then not only must the $dy/dt = dx/dt = 0$ curve be upward-sloping, but also the slope of this curve must exceed the slope of the $dS/dt = 0$ curve.[13] If these conditions are met, then the equilibrium (S^+, y^+, x^+) is locally stable (a saddle point). Finally, from Equation (10.26) in the text it is clear that $y^+ = 0$ implies that the $dy/dt = dx/dt = 0$ curve originates at S_f, which was derived for case 2 of the closed-economy model. Using all these results, the optimal solution is depicted graphically in Figure 10.2 for $S_0 > S^+$ and $G_1' < 0$.

APPENDIX B

AN ALTERNATIVE FORMULATION OF THE ALLOCATION PROBLEM

Equation (10.1) in the text specifies a growth function for the total species stock, S, which is additively separable into subcomponents, $G_1(S)$ and $G_2(A)$, that indicate the influence respectively of stock size and total habitat area, A, on growth. This appendix demonstrates that it is possible to specify a non-separable version of the growth function, $G(S,A)$, and an alternative formulation of the biodiversity and habitat allocation problem without affecting significantly the main results.[14]

For example, following more conventional bioeconomic models, if total habitat area, A, is effectively interpreted as the maximum carrying capacity of the total stock of species, S, then (1) could reasonably be specified as a non-separable logistic function, i.e.,

---

[13] However, an additional restriction in both cases is that the numerator of the slope of the $dy/dt = dx/dt = 0$ curve must be negative and the denominator positive.

[14] We are grateful to an anonymous referee for suggesting the following logistic function as an alternative non-separable version of the growth function $G(S,A)$. This referee raised doubts about the empirical relevance of the original specification of Equation (10.1) in the text as a separable function, but also noted that this assumption does not materially affect the subsequent results. We hope that this appendix goes some way towards illustrating this latter point.

$$\frac{dS(t)}{dt} = \dot{S} = G\big(S(t), A(t)\big) - y(t) = \alpha S(t)\big(A(t) - S(t)\big) - y(t).$$

Similarly, if it is assumed that land is still allocated through conversion, r, from habitat area and developed land, D, used to produce consumption goods, c, then the optimal allocation problem for the closed economy can be re-specified as

$$Max_{y,r} W = \int_0^\infty \big[U(y,c,S) - w(S)y\big]e^{-\delta t} dt,$$

subject to $\dot{S} = \alpha S(t)\big(A(t) - S(t)\big) - y(t)$, $S(0) = S_0$, $\lim_{t \to \infty} S(t) \geq 0$

$$\dot{A} = -r, \qquad A(0) = A_0$$

$$\dot{D} = r, \qquad D(0) = D_0$$

$$c = F(D), \qquad F' > 0, \qquad F'' < 0,$$

where r is now a control variable in the problem. Specifying μ and ξ as the co-state variables for habitat area and developed land respectively, the following current-value Hamiltonian and first-order conditions result:

$$H = U\big(y, F(D), S\big) - w(S)y + \lambda\big[\alpha S(A - S) - y\big] - \mu r + \xi r$$

$$dH/dy = U_y - w(S) - \lambda = 0 \qquad \text{or} \qquad U_y = w(S) + \lambda$$

$$-dH/dS = \dot{\lambda} - \delta\lambda \qquad \text{or} \qquad \dot{\lambda} = (\delta + 2\alpha S)\lambda - U_S + w'y$$

$$dH/dr = \xi - \mu = 0 \qquad \text{or} \qquad \xi = \mu$$

$$-dH/dA = \dot{\mu} - \delta\mu \qquad \text{or} \qquad \dot{\mu} = \delta\mu - \lambda\alpha S$$

$$-dH/dD = \dot{\xi} - \delta\xi \qquad \text{or} \qquad \dot{\xi} = \delta\xi - U_c F'.$$

The last three conditions can be combined with the first to derive $\lambda = U_y - w(S) = U_c F'/\alpha S$, i.e., along the optimal trajectory the marginal net benefits of harvesting an additional unit of the species stock must equal the net marginal costs of maintaining that unit. Utilizing this relationship and the remaining first-order conditions, the equations of motion for the system are

$$\dot{y} = \frac{\delta\big(U_y - w(S)\big) + 2U_c F' + w'\big(\alpha S(A - S)\big) - U_S}{U_{yy}}$$

$$\dot{S} = \alpha S(A - S) - y$$

$$\dot{A} = -\dot{D} = -r.$$

Assuming as before that in equilibrium $dy/dt = dS/dt = r = 0$, the long-run steady-state conditions are

$$\delta = \frac{-y^* w'\left(S^*\right) - 2U_c F'\left(D^*\right) + U_S\left(S^*\right)}{U_y\left(y^*\right) - w\left(S^*\right)} \quad \text{for} \quad \dot{y} = \dot{S} = r = 0$$

$$G\left(S^*, A^*\right) = \alpha S^*\left(A^* - S^*\right) = y^* \quad \text{for} \quad \dot{S} = r = 0.$$

Comparison with the equilibrium conditions (Equation 10.12) and (Equation 10.13) derived for the version of the model presented in the text indicates the similarity between the two sets of conditions. That is, Proposition 2 also holds for the above steady state; i.e., in the closed economy the equilibrium total species stock, S^*, is determined by equating the social discount rate, δ, and bioeconomic returns net of the opportunity cost of habitat plus the non-consumptive value of biodiversity. Although not shown here, it is possible to demonstrate that the above equilibrium is locally stable in (S,y) space, and that a necessary requirement for stability is that the $dy/dt = 0$ curve is upward-sloping and cuts the $dS/dt = 0$ from below. Similarly, it is also fairly straightforward to extend the above version of the model to verify Proposition 3 and 4 derived for the open-economy model, although this is not attempted here.[15]

REFERENCES

Barbier, E. B. and Aylward, B. A., Capturing the pharmaceutical value of bio-diversity in a developing country, *Environ. and Resour. Econ.*, 8, 157, 1996.

Barbier, E. B. and Rauscher, M., Trade tropical deforestation and policy interventions, *Environ. and Resour. Econ.*, 4, 75, 1994.

Barbier, E. B., Burgess, J. C., Swanson, T. M., and Pearce, D. W., *Elephants, Economics and Ivory*, Earthscan, London, 1990.

Barbier, E. B., Burgess, J. C., Bishop, J. T., and Aylward, B. A., *The Economics of the Tropical Timber Trade*, Earthscan, London, 1994.

Barrett, S., Optimal economic growth and the conservation of biological diversity, in *Economics and Ecology: New Frontiers and Sustainable Development*, Barbier, E. B., Ed., Chapman & Hall, London, 1993.

Brown, G. and Henry, W., The viewing value of elephants, in *Economics and Ecology: New Frontiers and Sustainable Development*, Barbier, E. B., Ed., Chapman & Hall, London, 1993.

Burgess, J. C., The environmental effects of trade in endangered species, in *The Environmental Effects of Trade*, OECD, Paris, 1994.

Chichilnisky, G., North–South trade and the global environment, *Am. Econ. Rev.*, 84, 851, 1994.

Clark, C., *Mathematical Bioeconomics: The Optimal Management of Renewable Resources*, John Wiley & Sons, New York, 1976.

Dixon, J. A. and Sherman, P., *Economics of Protected Areas: A New Look at Benefits and Costs*, Earthscan, London, 1990.

Holling, C. S., Schindler, D. W., Walker, B. W., and Roughgarden, J., Biodiversity in the functioning of ecosystems: An ecological primer and synthesis, in *Biodiversity Loss: Ecological and Economic Issues*, Perrings, C., Mäler, K. -G., Folke, C., Holling C. S., and Jansson B. -O., Eds., Cambridge University Press, 1995.

International Institute for Environment and Development (IIED), *Whose Eden? An Overview of Community Approaches to Wildlife Management: A Report to the UK Overseas Development Administration*, IIED, London, 1994.

---

[15] For example, it is clear that the open-economy model would simply add another equilibrium condition, $U_1(y^*, x^*) = pU_2(D^*, x^*)$, to those already derived in this appendix. The similarity with equilibrium conditions (Equation 10.25)–(Equation 10.27) is readily apparent.

Kramer, R. A., Tropical forest protection in Madagascar, paper presented for Northeast Universities Development Consortium, Williams College, Williamstown, MA, October 1993.

Lugo, A. E., Parrotta, J. A., and Brown, S., Loss of species casued by tropical deforestation and their recovery through management, *AMBIO*, 22, 106, 1993.

Munasinghe, M., Environmental economics and biodiversity management in developing countries, *AMBIO*, 22(2–3), 126, 1993.

Narain, U. and Fisher, A. C., Modelling the value of biodiversity using a production function approach, in *Biodiversity Conservation: Policy Issues and Options*, Perrings, C., Mäler, K.- G., Folke, C., Jansson B. -O., and Holling, C. S., Eds., Kluwer Academic Publishers, Dordrecht, 1994.

Ruitenbeek, H. J., Social cost-benefit analysis of the Korup Project, Cameroon, report prepared for the World Wide Fund for Nature and the Republic of Cameroon, WWF, London, 1989.

Schulz, C. -E., Trade policy and ecology, mimeo, Department of Economics, NFH, University of Tromsø, 1994.

Swanson, T. M., The economics of extinction revisited and revised: A generalized framework for the analysis of the problems of endangered species and biodiversity losses, *Oxford Econ. Pap.*, 46, 800, 1994.

Swanson, T. M. and Barbier, E. B., Eds., *Economics for the Wilds: Wildlife, Wildlands, Diversity and Development*, Earthscan, London, 1992.

Tobias, D. and Mendelsohn, R., Valuing ecotourism in a tropical rainforest reserve, *AMBIO*, 20, 91, 1990.

Wells, M. P. and Brandon, K. E., *People and Parks: Linking Protected Area Management with Local Communities*, The World Bank, Washington, D.C., 1992.

World Conservation Monitoring Center (WCMC), *Global Biodiversity: Status of the Earth's Living Resources*, Chapman & Hall, London, 1992.

11 Games Governments Play: An Analysis of National Environmental Policy in an Open Economy[1]

Amitrajeet A. Batabyal

In this chapter we study some of the consequences of national environmental policy in a strategic international setting. Two broad questions are analyzed. First, we examine the circumstances under which the pursuit of environmental policy by a country in a Cournot game will make that country worse off when the incidence of pollution is domestic. Second, we study the effects of environmental regulation by means of alternate price control instruments in a Cournot game in which national governments care about international pollution, but polluting firms do not. It is shown that there are plausible theoretical circumstances in which the pursuit of environmental policy in a strategic setting is not necessarily a desirable objective. Further, it is shown that in choosing between alternate pollution control instruments, national governments typically face a tradeoff between instruments which correct more distortions but are costly to implement and instruments which correct fewer distortions and are less costly to implement. In particular, a dominance result for a tariff policy is obtained; this result favors protection, i.e., the use of tariffs, from an informational standpoint alone.

A. INTRODUCTION

Few issues have evoked as much debate in recent times as the conduct of national environmental policy in an open economy. The greening of public consciousness across the world has forced governments to act on hitherto relatively insignificant environmental issues. At the same time, there has been a concerted attempt by business and industry to scale down the implementation of regulatory action designed to mitigate the damage stemming from environmental externalities. The fact that the conduct of environmental policy might affect trade flows between nations has added additional spice to the ongoing debate.[2] Indeed, as Leidy and Hoekman (1994, p. 241) have observed, "... there has been a dramatic rise in instances of sector-specific administered protection from foreign competition."

Given this situation, in this chapter we study some aspects of the question of international environmental regulation from a game theoretic perspective. Two broad questions are addressed. We first examine the circumstances in which environmental policy, pursued strategically by a country in a Cournot game, will make that country worse off when the incidence of pollution is domestic. Second, we study the effects of regulating environmental pollution via alternate means in a Cournot game in which national governments are affected by international pollution, but

[1] I thank Larry Karp, John Kim, and anonymous referees for their input; I alone am responsible for the output. This research was supported by (i) the Giannini Foundation and by (ii) the Utah Agricultural Experiment Station, Utah State University, Logan, UT, 84322-4810, by way of project UTA 024. Approved as journal paper no. 4873.

[2] See Morkre and Kruth (1989) for some of the issues involved.

polluting firms within nations are not. Put differently, this latter issue involves the analysis of a kind of caring behavior by the respective national governments.[3]

The rest of this chapter is organized as follows. In Section B, the pertinent literature relating to the questions to be addressed in this chapter is reviewed. In Section C, we discuss the effects of strategic environmental policy when the incidence of pollution is domestic. In Section D, caring behavior by national governments is analyzed. In this setting, we shall be particularly interested in analyzing the efficacy of pollution control via alternate instruments. The three instruments that we shall consider are an import tariff (a trade policy instrument), a production tax (a domestic policy instrument), and a combination of these two instruments (the joint policy instrument). In Section E, the principal findings of this chapter are summarized.

B. INTERNATIONAL ENVIRONMENTAL REGULATION: A SYNOPSIS OF FINDINGS

Pethig (1976) and Asako (1979) have both shown that under certain conditions, when a nation's pollution intensive good is exported, increased trade can diminish that country's welfare. Siebert et al. (1980) have examined the relationship between environmental quality, environmental policy, and international trade in a two-country world in some detail. In a non-strategic context, these authors have identified conditions for (i) an increase in resource use in pollution abatement and (ii) a fall in national income in the pollution controlling country.

Batabyal (1991, 1993) has studied the conditions under which environmental policy pursued by a large country will make that country worse off. These two papers contain results regarding the effects of environmental policy on (i) the post-policy terms of trade, (ii) the post-policy producer[4] price ratio in the taxing country, and on (iii) the effect of the pollution tax on "environmentally adjusted" national income in the taxing country.

Collectively, the clear message of these papers is that plausible theoretical circumstances exist in which the unilateral conduct of environmental policy by a large country in a non-strategic context can make that country worse off.

A number of authors have analyzed the question of international pollution control. Markusen (1975) has derived optimal taxes for pollution control when the incidence of pollution is international. However, his analysis is conducted in a static, competitive framework, and hence this analysis does not address the important issues of imperfect competition and possible caring government behavior. Merrifield (1988) has studied transboundary pollution control in the context of the U.S.–Canada acid deposition issue. Merrifield's analysis is static and is conducted in a competitive framework. Consequently, his analysis does not take into account the strategic aspects of the U.S.–Canada acid deposition issue. Dockner and Long (1993) have studied the transboundary pollution control problem by formulating the problem as a differential game between two countries. Interestingly, they show that under certain circumstances cooperative pollution control policies can be supported without recourse to retaliation.

Finally, the choice of alternate price control instruments has been little studied. Recently, a small number of researchers have begun to address this question theoretically and empirically. Ulph (1992) has shown that when international trade is modeled as a Cournot game, the choice of standards dominates the choice of taxes. Conrad and Schröder (1991) have studied the resource costs of attaining a given level of environmental quality by means of emission standards, subsidies, and emission taxes in an applied general equilibrium model. They have shown that the use of an emission tax involves the lowest resource cost, followed by subsidies and then by emission standards.

---

[3] One can also think of this phenomenon as a kind of transboundary pollution.
[4] Here, producer refers to the producer of the polluting good.

None of these authors have modeled the fact that, in an international dispute, governments typically pay considerable attention to the actions of other governments.[5] In other words, governments recognize that the behavior of other governments matters, and hence they care about such behavior. While this fact is quite well understood, it is rarely modeled. Even less modeled is the question of the effects of alternate price control instruments when national governments display this kind of caring behavior. As a result, a key objective of this chapter is to study the pros and cons of alternate price control instruments in a Cournot game in which national governments display such caring behavior.[6] We now proceed to the main analysis of this chapter.

C. THE GAME MODEL WITH DOMESTIC POLLUTION

1. PRELIMINARIES

There are two countries, A and B. In each country there is a government which chooses a tax to control pollution, a monopolistic firm which produces a good for domestic and foreign consumption, and consumers who are affected by pollution differently and who buy on the domestic market from the A firm or the B firm. Further, there is a single good whose production causes pollution. This good will be denoted by q with the appropriate superscripts. The total quantity of the good in A is $Q^A = q^{AH} + q^{BX}$, where q^{AH} and q^{BX} denote the quantity produced by the A firm for home consumption and the quantity exported by the B firm. We shall denote the pollution taxes by t^A and t^B, respectively. Observe that the two countries are identical on the production side. The only difference between them arises from the unequal marginal social disutilities from pollution.

In this chapter we shall work with linear functional forms. As we shall see, even with the imposition of this additional structure, unambiguous results will in general not be forthcoming. Most of the results are in the form of inequalities. In certain cases, these inequalities can be easily understood; in other cases, the inequalities are harder to interpret.

Recall that the A government levies a pollution tax on the production of the polluting good q. The B government is allowed to retaliate. B retaliates due to two reasons. First, although there is pollution in B and B would like to control pollution, B is reluctant to take measures to do so unilaterally. A's actions give B a rationale for pollution prevention. Second, B retaliates because B fears that by allowing A's actions to go unchallenged, B will be worse off from the perceived shift in the terms of trade in A's favor subsequent to the imposition of a pollution tax by A.[7] Our goal is to characterize the optimal taxes and to explore the implications of using such taxes in the subgame perfect Nash Equilibrium (SPNE) of a two-stage Cournot game.

2. THE REGULATION OF DOMESTIC POLLUTION

The total quantity of the good in A is $Q^A = q^{AH} + q^{BX}$, where q^{AH} is the quantity produced by the monopolistic A firm for home consumption and q^{BX} is the quantity exported by the monopolistic B firm. In B the total quantity on the market is $Q^B = q^{BH} + q^{AX}$. The market clearing price in A and B is $P^A = a - Q^A$, and $P^B = a - Q^B$, respectively. The A and B firms have two kinds of costs: the cost of producing the good and the cost associated with tax payment. The first cost for the A firm is $C^A(\bullet) = c(q^{AH} + q^{AX})$, where c is the constant marginal cost. The second cost for the A firm is $t^A(q^{AH} + q^{AX})$. The government collects this second cost as tax revenue. This revenue is then transferred to consumers in A in a lump sum manner. The B firm has similar cost structure. The social welfare function in each country is $W^i(\bullet) = (1/2)(Q^i)^2 + \pi^i(\bullet) + t^i(q^{iH} + q^{iX}) - v^i(q^{iH} + q^{iX})$, $i = A, B$, where $\pi^A(\bullet) = \{q^{AH}(a - q^{AH} - q^{BX} - c - t^A)\} + \{q^{AX}(a - q^{AX} - q^{BH} - c - t^A)\}$ is the profit

---

[5] See Behr (1994) and Simone (1994) for more on this in the context of U.S.–Canada agricultural trade disputes.

[6] For an analysis of similar questions in the context of a Stackelberg game, see Batabyal (1996a).

[7] Batabyal (1991, 1993) has shown that, in some circumstances, these concerns are legitimate.

function of the A firm, $\pi^B(\bullet) = \{q^{BH}(a - q^{BH} - q^{AX} - c - t^B)\} + \{q^{BX}(a - q^{BX} - q^{AH} - c - t^B)\}$ is the profit function of the B firm and v^i is the national disutility from pollution parameter. In other words, social welfare is the sum of consumer surplus, firm profits, tax revenue less the disutility from pollution, all measured in dollar terms.

The timing of the Cournot game is as follows. First, governments simultaneously choose taxes t^A and t^B. Second, the two firms observe the taxes and then simultaneously choose the quantities to be produced for domestic and foreign consumption. That is, the A firm chooses (q^{AH}, q^{AX}), and the B firm chooses (q^{BH}, q^{BX}). Third, the players receive their payoffs which are profits (π^A, π^B) for the firms and social welfare (W^A, W^B) for the governments. Since this is a game of complete and perfect information, we can use the backward induction method of solving such games. We assume that there are no multiple equilibria in each stage game. Then this method is guaranteed to yield the SPNE (see Kreps, 1989, p. 421 for details).

Suppose that the governments have chosen t^A and t^B. If the 4 tuple $(q_*^{AH}, q_*^{AX}, q_*^{BH}, q_*^{BX})$ is a Nash equilibrium in the remaining game between the two firms, then it must be true that (q_*^{AH}, q_*^{AX}) solve $\max_{\{q^{AH}, q^{AX}\}} \pi^A(\bullet)$, and that (q_*^{BH}, q_*^{BX}) solve $\max_{\{q^{BH}, q^{BX}\}} \pi^B(\bullet)$. Now $\pi^A = \pi^{AH} + \pi^{AX}$. Thus, the A firm's original problem can be written as a pair of optimization problems. In the home market, let $q_*^{AH} = \arg\max_{q^{AH}}\{q^{AH}(a - q^{AH} - q_*^{BX} - c - t^A)\}$. Then

$$q_*^{AH} = (1/2)\left(a - c - q_*^{BX} - t^A\right). \tag{11.1}$$

In the export market, let $q_*^{AX} = \arg\max_{q^{AX}}\{q^{AX}(a - q^{AX} - q_*^{BH} - c - t^A)\}$. Then

$$q_*^{AX} = (1/2)\left(a - c - q_*^{BH} - t^A\right). \tag{11.2}$$

In a similar fashion, we can write $\pi^B = \pi^{BH} + \pi^{BX}$ for the B firm. Letting q_*^{BH} and q_*^{BX} be the solutions to the two different B firm optimization problems, we get

$$q_*^{BH} = (1/2)\left(a - c - q_*^{AX} - t^B\right) \tag{11.3}$$

and

$$q_*^{BX} = (1/2)\left(a - c - q_*^{AH} - t^B\right). \tag{11.4}$$

Equations (11.1)–(11.4) are the reaction functions of the two firms. Solving (Equation 11.1) and (Equation 11.4) simultaneously and (Equation 11.2) and (Equation 11.3) simultaneously, we get

$$q_*^{BH} = q_*^{BX} = (1/3)\left(a - c - 2t^B + t^A\right), \tag{11.5}$$

and

$$q_*^{AH} = q_*^{AX} = (1/3)\left(a - c - 2t^A + t^B\right). \tag{11.6}$$

Equations (11.5) and (11.6) represent the equilibrium production levels for the two firms. We now proceed to solve the first stage game between the two governments. The two governments will choose taxes to maximize social welfare in their respective countries. Thus, the ith government, $i = A, B$, solves

$$\max_{t^i}(1/2)\big(Q^i\big)^2 + \pi^i(\bullet) + t^i\big(q_*^{iH} + q_*^{iX}\big) - v^i\big(q_*^{iH} + q_*^{iX}\big). \tag{11.7}$$

The solutions to (Equation 11.7) are

$$t_*^A = (1/4)\big(2c - 2a + 7v^A - v^B\big), \tag{11.8}$$

and

$$t_*^B = (1/4)\big(2c - 2a + 7v^B - v^A\big). \tag{11.9}$$

Equations (11.8) and (11.9) give us the SPNE taxes for A and B, respectively. Since v^A is not necessarily equal to v^B, the two equilibrium taxes are not necessarily equal. The strategic aspect of this Cournot game is clearly captured by the dependence of t_*^A on v^B and the dependence of t_*^B on v^A. In the sense of Bulow et al. (1985), each firm treats the output of its competitor as a "strategic substitute." Substituting (Equation 11.8) and (Equation 11.9) into (Equation 11.5) and (Equation 11.6) gives us the SPNE outcome of this pollution tax game. This outcome is given by (Equation 11.8) and (Equation 11.9), $q_*^{AH} = q_*^{AX} = (1/4)(2a - 2c - 5v^A + 3v^B)$, and $q_*^{BH} = q_*^{BX} = (1/4)(2a - 2c - 5v^B + 3v^A)$.

Inspection of (Equation 11.8) and (Equation 11.9) is instructive. Let $a > c$, and consider (Equation 11.8). We see that when B's marginal social disutility from pollution is considerably greater than A's, i.e., when $v^B > 7v^A$, A will subsidize the production of q. On the other hand, if $a < c$, then whether or not A will subsidize production depends on the relative magnitudes of the cost and demand parameters (c, a) and the disutility parameters (v^A, v^B).

To determine the effect of the SPNE taxes on consumer surplus and firm profits in each country, we can substitute the equilibrium values of the four quantities and the two taxes into the corresponding defining functions. First, note that the pre-tax consumer surplus in B > post-tax consumer surplus.[8] The intuition for this result lies in the fact that, for the linear inverse demand function that we have been using, consumer surplus is monotonically increasing in total market output. Second, the post-tax profits for the B firm $\leq$ the pre-tax profits if and only if $\{52(2ac - a^2 - c^2)\}$ $\geq \{(12a - 12c)v^A + (276c - 276a)v^B + 25(v^A)^2 + 169(v^B)^2 + 130v^A v^B\}$. We see that the effect of the tax on profits is ambiguous in general.[9] However, the role of the various model parameters is evident. If the disutility parameters, i.e., v^A and v^B, are much larger than the demand and cost parameters, i.e., a and c, then the post-tax profit will exceed the pre-tax profit. On the other hand, if the cost and demand parameters are much larger than the disutility parameters, then the pre-tax profit will exceed the post-tax profit.

This analysis strengthens the conclusions of Batabyal (1991, 1993), and shows that when governments conduct environmental policy in a two stage Cournot game, there exist circumstances in which taxing countries can be worse off. Indeed, consumer surplus in both countries declines unambiguously. A key determinant of the question of being worse off would appear to be the different social disutilities from pollution in the two countries. For instance, since t^A is increasing in v^A and decreasing in v^B, and t^B is increasing in v^B and decreasing in v^A, significant differences in these parameters can be expected to lead to very different taxes and thence to very different levels of consumer surplus and firm profits. In concluding this section we note that even when one explicitly incorporates linear functional forms into the analysis, it is not, in general, possible to

---

[8] While this result is stated for B, it should be clear that an analogous result holds for A as well. In the rest of this paper, all results will be stated for B only.

[9] In the rest of this chapter, we shall not provide results for the status quo versus post-policy social welfare levels because these results are always ambiguous.

make unqualified statements about the effects of government pollution control policy. However, for both countries, the possibility of being worse off remains.

D. THE GAME MODEL WITH CARING GOVERNMENT BEHAVIOR

1. PRELIMINARIES

In this section we shall analyze the efficacy of alternate pollution control instruments when the government in one country cares about pollution in the second country. The governments in A and B are restricted to choosing between a domestic policy instrument (a production tax), a trade policy instrument (an import tariff), and a combination of these two instruments (the joint policy instrument). Since the underlying issues are now fairly involved, before studying the implications of these three policy instruments we shall first discuss the issues intuitively.

The first issue concerns world welfare. When there are a number of distortions present in the world economy (these are discussed below) and national governments attempt to correct for these distortions by means of the instruments mentioned above in a non-cooperative game, the ensuing equilibrium is typically optimal in a myopic sense. While individual country welfare is maximized by the respective governments, world welfare is not. In other words, the "correct" taxes and tariffs are those that arise from coordinated play by the respective governments. In practice, this involves coordination of environmental policy by all the players involved. This is something that is not only fairly well understood by analysts but also something that we typically do not observe for a variety of well-known reasons.[10] It is worth noting that all the subsequent results are myopic in the sense of this discussion.

In determining which policy instrument to adopt, the government in each country will attempt to weigh the effects of a particular policy on the three distortions that are present in the two-country world economy being studied. First, there is the pollution distortion. A production tax will reduce pollution by reducing the output of the good which causes pollution. However, this only reduces domestic pollution but does not reduce foreign pollution. An import tariff, on the other hand, reduces foreign pollution by making the post-tariff purchase of the foreign good undesirable and increasing the costs of foreign producers. However, since the amount of pollution that is reduced by means of a tariff is probably less than the amount of pollution reduced by means of a production tax,[11] as far as the pollution distortion is concerned, an optimally set production tax is likely to be the superior policy instrument of the two considered so far. The joint policy instrument would curb both domestic and foreign pollution. Therefore, as far as the pollution distortion alone is concerned, of the three instruments under consideration, *ex ante*, this instrument would appear to be the best means of pollution control.

The second distortion concerns the domestic share of total output. It is very likely that the monopolistic firm in each country does not sell the "correct" amount in the home market. In determining which policy instrument to adopt, the two governments would presumably like to increase the domestic market share of total good production and hence reduce imports. By discounting the purchase of the foreign good, an import tariff would certainly increase the domestic market share of the total output. A production tax, on the other hand, would not achieve this goal since a production tax would unambiguously curtail all domestic production. The joint policy instrument can be expected to result in total output which is bounded below by the tax output and above by the tariff output. Thus, as far as this distortion is concerned, *ex ante*, a tariff would appear to be the best policy instrument.

The third distortion concerns monopoly rents. The second stage game in the model is game between two monopolists earning excess economic rent. The government in each country would

---

[10] See Batabyal (1996b) for a discussion of this and related issues.
[11] The actual answer depends on the magnitudes of the relevant parameters.

presumably like to capture some of this rent. By collecting some of the revenue which otherwise would go to the foreign monopolist, each government can capture some of the monopoly rent by means of an import tariff. Likewise, a production tax can also transfer some of the surplus collected by the home monopolistic firm to the government setting the tax. The joint policy instrument can also be expected to have the same qualitative impact as a tariff or a tax except that its quantitative impact will certainly be different. Thus as far as this distortion is concerned, all three policy instruments can work in the right direction. The reader should note that, in general, the properties of the three instruments are likely to be quite different. We now formally analyze the related questions of the effects of environmental policy and the choice of policy instrument issue in a Cournot game in which governments care about international pollution.

2. INTERNATIONAL ENVIRONMENTAL POLICY WITH ALTERNATE CONTROL INSTRUMENTS

The notation used here is the same as in Section C except that the tariffs levied by the two countries are now denoted by r^A and r^B respectively. Let us begin with the general case. In this case, each government controls pollution by means of the joint policy instrument, i.e., an instrument which is a combination of an import tariff and a production tax. Recall that $\pi^A = \pi^{AH} + \pi^{AX}$ and $\pi^B = \pi^{BH} + \pi^{BX}$. On solving these optimization problems, we get four reaction functions.[12] These are

$$q_*^{AH} = (1/3)\left(a - c - 2t^A + r^A + t^B\right), \tag{11.10}$$

$$q_*^{AX} = (1/3)\left(a - c - 2r^B - 2t^A + t^B\right), \tag{11.11}$$

$$q_*^{BH} = (1/3)\left(a - c + r^B + t^A - 2t^B\right), \tag{11.12}$$

and

$$q_*^{BX} = (1/3)\left(a - c + t^A - 2r^A - 2t^B\right). \tag{11.13}$$

Equations (11.10)–(11.13) give us the equilibrium quantities which will be produced by the A firm and B firm in the domestic market and in the export market respectively.

Recall that the A and B governments both care both about the pollution in the other country. This is modeled by including the B government's disutility from pollution in the A government's objective and vice versa. With this in mind, the first stage of the two-stage game can be analyzed. In this stage, the A and B governments solve

$$\max_{r^A, t^A}(1/2)\left(Q^A\right)^2 + \pi^A(\bullet) + \left(t^A - v^A\right)\left(q_*^{AH} + q_*^{AX}\right) + \left(r^A - v^B\right)q_*^{BX} - v^B q_*^{BH} \tag{11.14}$$

and

$$\max_{r^B, t^B}(1/2)\left(Q^B\right)^2 + \pi^B(\bullet) + \left(t^B - v^B\right)\left(q_*^{BH} + q_*^{BX}\right) + \left(r^B - v^A\right)q_*^{AX} - v^A q_*^{AH} \tag{11.15}$$

---

[12] In this subsection, we shall solve for most of the equilibrium values of the variables of interest in terms of other endogenous variables and not in terms of the model parameters a, c, v^A and v^B exclusively. This will facilitate comparison between the three policy instruments under consideration.

respectively. The solutions to problems (Equation 11.14) and (Equation 11.15) are

$$r_*^A = (1/9)\left(3a - 3c - 3t_*^B + 7t_*^A - 3v^A + 6v^B\right),$$ (11.16)

$$t_*^A = (1/7)\left(4c - 4a + 3r_*^A - t_*^B + 2r_*^B + 12v^A - 6v^B\right),$$ (11.17)

$$r_*^B = (1/9)\left(3a - 3c - 3t_*^A + 7t_*^B - 3v^B + 6v^A\right),$$ (11.18)

and

$$t_*^B = (1/7)\left(4c - 4a + 3r_*^B - t_*^A + 2r_*^A + 12v^B - 6v^A\right),$$ (11.19)

Equations (11.16)–(11.19) give us the SPNE tariffs and taxes when the two governments care about pollution in each other's country and when they use the joint policy instrument to regulate pollution. Using (11.16)–(11.19), the four SPNE quantities can be determined. These are

$$q_*^{AH} = (1/567)\left(360a - 360c - 108r_*^A - 9t_*^B - 27r_*^B - 873v^A + 774v^B - 120t_*^A\right),$$ (11.20)

$$q_*^{AX} = (1/567)\left(171a - 171c - 108r_*^A - 267t_*^B - 27r_*^B - 1062v^A + 774v^B - 126t_*^A\right),$$ (11.21)

$$q_*^{BH} = (1/567)\left(360a - 360c - 27r_*^A + 120t_*^B - 108r_*^B + 774v^A - 873v^B - 9t_*^A\right),$$ (11.22)

and

$$q_*^{BX} = (1/567)\left(171a - 171c - 27r_*^A + 99t_*^B - 240t_*^A - 1062v^B + 774v^A - 108r_*^B\right).$$ (11.23)

Equations (11.16)–(11.23) give us the SPNE outcome of this joint policy instrument game. The reader will note an asymmetry in this SPNE outcome. Inspection of (11.16)–(11.19) tells us that while the optimal tariffs depend on the endogenous taxes, the optimal taxes depend on the endogenous taxes *and* the tariffs. This asymmetry will be highlighted even more starkly in the sequel.

We shall now compare the joint policy outcome with the status quo, i.e., the outcome with no government intervention of any kind. For country B consumer surplus with the joint policy $\leq$ consumer surplus with no intervention as $\{r_*^B + t_*^A + t_*^B\} \geq 0$. The B firm's profits with the joint policy instrument $\geq$ its profits with no policy as $\{2ar_*^B + 4at_*^A - 8at_*^B - 4ar_*^A - 2cr^B - 4ct_*^A + 8ct_*^B + 4cr_*^A - 4r_*^B t_*^B - 8t_*^A t_*^B + 8(t_*^B)^2 + (r_*^B)^2 + 8r_*4_*^B + 2r_*^B t_*^A - 4r_*^A t_*^A + 4(r_*^A)^2 + 2(t_*^A)^2 \geq 0$. While consumer surplus with the joint policy declines as long as the sum of r_*^B, t_*^A, and t_*^B is positive, the profits comparison is ambiguous. Because of this ambiguity and to facilitate a comparison of the implications of alternate forms of pollution control, we now consider two special cases of the above general case. In the first case, the two governments play a Cournot tariff game. In this scenario, the only pollution control instrument available to the two governments is an import tariff. In the second case, the two governments play a Cournot tax game. In this latter scenario, the sole pollution control instrument available to the A and B governments is a production tax.

a. The Tariff Only Game

The SPNE outcome of the Cournot tariff game can be determined by substituting $t^A = t^B = 0$ in (Equation 11.10)–(Equation 11.13), (Equation 11.16), and (Equation 11.18). This yields

$$r_*^A = (1/3)(a - c - v^A + 2v^B), \tag{11.24}$$

$$r_*^B = (1/3)(a - c - v^B + 2v^A), \tag{11.25}$$

$$q_*^{AH} = (1/9)(4a - 4c - v^A + 2v^B), \tag{11.26}$$

$$q_*^{AX} = (1/9)(a - c + 2v^B + 4v^A), \tag{11.27}$$

$$q_*^{BH} = (1/9)(4a - 4c - v^B + 2v^A), \tag{11.28}$$

and

$$q_*^{BX} = (1/9)(a - c + 2v^A - 4v^B). \tag{11.29}$$

Straightforward computation reveals that the status quo consumer surplus in B > the post-tariff consumer surplus. Further, the B firm's post-tariff profits $\leq$ the status quo level of profits if and only if $\{4ar_*^A + 2cr_*^B\} \geq \{4(r_*^A)^2 + (r_*^B)^2 + 4cr_*^A + 2ar_*^B\}$.

b. The Tax Only Game

In similar fashion, the SPNE outcome of the Cournot tax game can be determined by substituting $r^A = r^B = 0$ in (Equation 11.10)–(Equation 11.13), (Equation 11.17), and (Equation 11.19). This yields

$$t_*^A = (1/7)(4c - 4a - t_*^B + 12v^A - 6v^B), \tag{11.30}$$

$$t_*^B = (1/7)(4c - 4a - t_*^A + 12v^B - 6v^A), \tag{11.31}$$

$$q_*^{AH} = q_*^{AX} = (1/21)(11a - 11c - 30v^A + 24v^B - t_*^A + 2t_*^B), \tag{11.32}$$

and

$$q_*^{BH} = q_*^{BX} = (1/21)(11a - 11c - 30v^B + 24v^A - t_*^B + 2t_*^A). \tag{11.33}$$

Comparing the outcome of this tax game to the status quo outcome, it is straightforward to verify that the post-tax consumer surplus < the status quo consumer surplus. Further, the post-tax B firm's profits $\leq$ the status quo profits if and only if $\{4at_*^B + 2ct_*^A + 4t_*^A t_*^B\} \geq \{at_*^A + 4(t_*^B)^2 + 4ct_*^B + 2at_*^A\}$.

C. A COMPARATIVE ANALYSIS OF THE THREE POLICY INSTRUMENTS

Comparing the equilibrium tariff in the tariff only game with the equilibrium tax in the tax only game, we see that with respect to the status quo in B, the tariff (tax) increases consumer surplus provided that $r_*^B < (>) \{t_*^A + t_*^B\}$, and increases the firm's profits provided that $\{4(r_*^A)^2 + (r_*^B)^2 + 2(c-a)(2r_*^A - r_*^B)\} > (<) \{2(t_*^A)^2 + 8(t_*^B)^2 + 4(c-a)(2t_*^B - t_*^A) - 8t_*^A t_*^B\}$.

Similarly, comparing the outcome of the joint policy game with the tariff only game, we see that the consumer surplus in B with the joint policy instrument $\leq$ consumer surplus with the tariff as $\{r_*^B + t_*^A + t_*^B\}^{\text{Joint Policy}} \geq \{r_*^B\}^{\text{Tariff}}$. Further, the B firm's profits with the joint policy instrument $\geq$ profits with the tariff as $\{2ar_*^B + 4at_*^A - 8at_*^B - 4ar_*^A - 2cr_*^B - 4ct_*^A + 8ct_*^B + 4cr_*^A - 4r_*^B t_*^B - 8t_*^A t_*^B + 8(t_*^B)^2 + 8r_*^A t_*^B + 2r_*^B t_*^A - 4r_*^A t_*^A + 2(t_*^A)^2\}^{\text{Joint Policy}} \geq \{4(r_*^A)^2 + (r_*^B)^2 + 4cr_*^A - 4ar_*^A + 2ar_*^B - 2cr_*^B\}^{\text{Tariff}}$.

Finally, comparing the outcome in the tax only game to the outcome in the joint policy game, we see that consumer surplus in B with the joint policy $\leq$ consumer surplus with the tax as $\{r_*^B + t_*^A + t_*^B\}^{\text{Joint Policy}} \geq \{t_*^A + t_*^B\}^{\text{Tax}}$. The B firm's profits with the joint policy $\geq$ its profits with a tax as $\{2ar_*^B + 4at_*^A - 8at_*^B - 4ar_*^A - 2cr_*^B - 4ct_*^A + 8ct_*^B + 4cr_*^A - 4r_*^B t_*^B - 8t_*^A t_*^B + 8(t_*^B)^2 + 8r_*^A t_*^B + (r_*^B)^2 + 2r_*^B t_*^A - 4r_*^A t_*^A + 4(r_*^A)^2 + 2(t_*^A)^2\}^{\text{Joint Policy}} \geq \{2(t_*^A)^2 - 8at_*^B + 8(t_*^B)^2 + 8ct_*^B - 4ct_*^A + 4at_*^A - 8t_*^A t_*^B\}^{\text{Tax}}$.

An initial examination of the above results suggests that the choice of instrument issue is hopelessly tangled. However, a few insights can be gained by performing numerical analyses of this issue. Table 11.1 presents the results of two such case studies.[13] These results use the conditions identified in the previous three paragraphs. First, we see that the results are quite sensitive to the specific values assigned to the relevant parameters and the variables. Second, consumers will in general prefer the tariff or the tax to the joint policy instrument. This is because in neither of the two cases analyzed does the joint policy instrument have a desirable positive effect on consumer surplus. Third, consumers and the firm can be expected to have opposite preferences as far as the question of instrument choice is concerned. This stems from the fact that, in five of six pairwise comparisons, the instrument which has a greater positive impact on consumer surplus does not also have a similar impact on firm profits.

A comparison of (Equation 11.24) and (Equation 11.25) with (Equation 11.30) and (Equation 11.31) shows that while the equilibrium tariffs in the tariff only game constitute dominant strategies for the two governments, such is not the case for the equilibrium strategies in the tax only and joint policy games. In other words, in the special case in which governments do not levy production taxes, choosing tariffs given by (Equation 11.24) and (Equation 11.25) constitute dominant strategies

TABLE 11.1

A Numerical Analysis of the Effects of the Three Policy Instruments on Consumer Surplus and on Firm Profits in B

Variable of Interest	Tariff Versus Tax	Joint Policy Instrument Versus Tariff	Joint Policy Instrument Versus Tax
Case I: $a = c, r_*^A = 3, r_*^B = 4, t_*^A = 5, t_*^B = 6$			
Consumer surplus	Higher with tariff	Higher with tariff	Higher with tax
Firm profits	Higher with tax	Higher with joint policy instrument	Higher with joint policy instrument
Case II: $a = c, r_*^A = 7, r_*^B = 9, t_*^A = 3, t_*^B = 5$			
Consumer surplus	Higher with tax	Higher with tariff	Higher with tax
Firm profits	Higher with tariff	Higher with tariff	Higher with joint policy instrument

[13] The values assigned to the various parameters and variables in Table 11.1 are purely for illustrative purposes.

for the A and B governments. That is, optimal tariff setting by one government does not require knowledge of the tariff set by the second government. In general, one cannot expect countries to be able to choose which Cournot game they will play. However, this analysis suggests that the lower informational requirements of optimal tariff setting can provide a rationale for protection.

In every comparative exercise that we have undertaken, the intuition for the consumer surplus result lies in recognizing that consumer surplus is monotonically increasing in total output. Consequently, any government policy which reduces total output below the status quo level of output will necessarily reduce consumer surplus.

When the instrument choice question is restricted to one between the tariff and the tax, we see that when $r_*^B = t_*^A + t_*^B$, consumers in B are indifferent between a tariff and a tax because both instruments produce identical changes in their surplus. Now consider the reference point in which $r_*^B < t_*^A + t_*^B$. This is a situation where consumers are better off with a tariff. This tariff, say $\bar{r}_*^B$, is clearly less than $r_*^B = t_*^A + t_*^B$. Thus the tariff which makes consumers better off is lower than the tariff which makes consumers indifferent. The government's preference, as far as tariff or tax is concerned, will depend on which instrument leads to higher social welfare. Now view this question from the tax perspective. At the point of indifference, $t_*^B = r_*^B - t_*^A$. With the tax, consumers in B are better off when $r_*^B > t_*^A + t_*^B$, which implies that, like the tariff case, the tax which makes consumers in B better off is smaller than the tax which makes them indifferent.[14] In this last case, the B firm can be expected to be opposed to the tax and the A firm can be expected to be in favor of the tax.

Of the three instruments under consideration, the informational requirements of the joint policy instrument are the greatest. For instance, from (Equation 11.16) to (Equation 11.19), we see that in order to set r_*^B and t_*^B, the B government must know t_*^A and r_*^A and t_*^A respectively. While the joint policy instrument is probably the most desirable in terms of correcting the different distortions alluded to in Section D1, given its informational requirements, its implementation is likely to be rather costly. This completes our discussion of the instrument choice question.

E. CONCLUSIONS

In this chapter, a strategic analysis of two important issues concerning optimal environmental policy in an open economy was undertaken. We first explored the effects of conducting environmental policy in a two-stage Cournot game. It was shown that when governments interact strategically, there are theoretical circumstances in which the conduct of environmental policy can make a nation worse off. This finding is a possible explanation as to why governments are often loath to conduct environmental policy *even* when the incidence of pollution is domestic.

Next, we studied the issue of "prices versus prices," i.e., the implications of pollution regulation by means of alternate price control instruments when national governments care about government behavior and thence pollution in other countries. Our numerical analysis showed that it is not possible to resolve the instrument choice issue unambiguously. Nevertheless, we identified several inequalities which determine whether consumers and producers are likely to gain or lose from the pursuit of a specific policy. As we discussed, a basic issue faced by the two governments concerns the tradeoff between policy efficacy on the one hand, and the cost of policy implementation on the other.

The analysis of this chapter can be extended in a number of directions. With some adjustments, our model can be used to study the fractious agricultural/environmental policy formation process between trading entities such as the U.S. and the E.C. Second, this model can also help us understand why the policy response in large developing countries to environmental externalities has been slow if not altogether nonexistent. Finally, our analysis can be made richer by making the models truly dynamic and by explicitly incorporating uncertainty into the decision making process. This will enable one to analyze issues related to threats and punishments, governmental reputation, and long run policy formation.

---

[14] This discussion assumes that the B tax is positive.

REFERENCES

Asako, K., Environmental pollution in an open economy, *Econ. Rec.*, 55, 359, 1979.

Batabyal, A. A., Environmental pollution and unilateral control in an open economy, *Nat. Resour. Modeling*, 5, 445, 1991.

Batabyal, A. A., Should large developing countries pursue environmental policy unilaterally? *Indian Econ. Rev.*, 28, 191, 1993.

Batabyal, A. A., Game models of environmental policy in an open economy, *Ann. Reg. Sci.*, 30, 185, 1996a.

Batabyal, A. A., Development, trade, and the environment: which way now? *Ecological Econ.*, 13, 83, 1996b.

Behr, P., US, Canada settle fight over wheat, *Washington Post* (Washington, D.C.), August 2, p. D1, D4, 1994.

Bulow, J., Geanakoplos, J. D., and Klemperer, P. D., Multimarket oligopoly: strategic substitutes and complements, *J. Political Economy*, 93, 488, 1985.

Conrad, K. and Schröder, M., Controlling air pollution: the effects of alternative approaches, in *Environmental Scarcity: The International Dimension*, H. Siebert, Ed., Tubingen, Mohr, 1991.

Dockner, E. J. and Long, N. V., International pollution control: cooperative versus non-cooperative strategies, *J. Environ. Econ. Manage.*, 25, 13, 1993.

Kreps, D. M., *A Course in Microeconomic Theory*, Princeton University Press, Princeton, NJ, 1989.

Leidy, M. P. and Hoekman, B. M., "Cleaning up" while cleaning up? Pollution abatement, interest groups and contingent trade policies, *Public Choice*, 78, 241, 1994.

Markusen, J. R., International externalities and optimal tax structures, *J. Int. Econ.*, 5, 15, 1975.

Merrifield, J. D., The impact of selected abatement strategies on transnational pollution, the terms of trade, and factor rewards: a general equilibrium approach, *J. Environ. Econ. Manage.*, 15, 259, 1988.

Morkre, M. E. and Kruth, H. E., Determining whether dumped or subsidized imports injure domestic industries, *Contemporary Policy Issues*, 7, 78, 1989.

Pethig, R., Pollution, welfare, and environmental policy in the theory of comparative advantage, *J. Environ. Econ. Manage.*, 2, 160, 1976.

Siebert, H., Eichberger, J., Gronych, R., and Pethig, R., *Trade and Environment: A Theoretical Inquiry*, Elsevier, Amsterdam, 1980.

Simone, M., U.S. and Canada: The nature of ag trade disputes, *Agricultural Outlook AO–210*, August, 28, 1994.

Ulph, A. M., The choice of environmental policy instruments and strategic international trade, in *Conflicts and Cooperation in Managing Environmental Resources*, Pethig, R., Ed., Springer–Verlag, Berlin, Heidelberg, 1992.

12 Industrial Pollution Abatement: The Impact on Balance of Trade

H. David Robison

This chapter uses an ex-post partial-equilibrium framework to measure the impact of marginal changes in industrial pollution abatement costs on the U.S. balance of trade in general, and balance of trade with Canada in particular. The impacts are found to be negative for most industries, growing with trade volume, and small relative to domestic consumption. In addition, some evidence is found that pollution control programs have changed the U.S. comparative advantage such that more high-abatement-cost goods are imported and more low-abatement-cost goods are exported.

© Canadian Economics Association

A. INTRODUCTION

Although economists agree that by raising the prices of domestically produced products, pollution abatement expenditures alter a country's comparative advantage and trading pattern, no formal measurement of the impact has been made. In this chapter, we attempt to measure on an *ex post* basis the impact of marginal changes in industrial pollution abatement on U.S. balance of trade in general, and balance of trade with Canada as a special case. More specifically, Baumol and Oates's (1975) theoretical work on the trade impacts of industrial pollution abatement is extended from a two final-good model to a seventy-eight sector model with all interindustry effects considered. Using this extended model, measurements of the trade impacts of marginal increases in abatement expenditures are made for each sector. These measurements are an upper bound to potential impacts on the balance of trade because all mitigating variables such as improved terms of trade, offsetting governmental policies, and adjustments in exchange rates are ignored. The results for 1977 range from –$0.1 through –$80.9 to –$566.6 million for transportation services, electric utilities, and ferrous metals, respectively.

In addition, this chapter uses Walter's (1973) concept of the abatement content of trade to test the proposition, discussed by Baumol and Oates among others,[1] that undertaking pollution abatement will reduce the abating country's comparative advantage in producing high-abatement-cost goods and improve the comparative advantage in low-abatement-cost goods. Results indicate that the abatement content of imports has grown more rapidly than that of exports, thus supporting this proposition.

Baumol and Oates define and discuss the conditions by which increased industrial abatement costs can improve or hurt a country's balance of trade in a two-country, two-good world.[2] No intermediate product flows are considered in the Baumol–Oates world, which would be sufficient if most abatement costs were incurred in producing final goods. However, most industrial pollution abatement (IPA) costs occur at basic levels of production, requiring the incorporation of interindustry effects, as is done in this study.

Measuring the price changes caused by IPA is a straightforward application of input-output analysis as discussed in Leontieff (1970) and performed in Walter (1973), Pasurka (1984), and Robison (1985). The Robison method for measuring the price change is used in the current study.

---

[1] See D'Arge (1971), Mutti and Richardson (1977), and Magee and Ford (1972) for additional discussions.

[2] A less general approach to this issue can be found in Magee and Ford (1972).

A general-equilibrium framework, including an input-output table, would seem the appropriate approach for this type of study. D'Arge (1971), Mutti and Richardson (1977), Evans (1973), and two commercial economic consulting firms[3] provide general equilibrium-based forecasting studies designed to examine the impact of IPA on the macro economy and trade. There are, however, at least two problems with each of these studies. First, because the studies use *ex-ante* forecasting, they must assume or estimate abatement costs. Second, assumptions are made as to how the abatement costs are allocated between return to capital, labor, etc., in order to run the forecasting routines.[4] By choosing an *ex post* partial-equilibrium approach, these problems are avoided.

B. MODEL FRAMEWORK

The Baumol and Oates work derives the balance of trade conditions for a two-good two-country world. In doing so, one good has pollution associated with its production, while the other does not. In this simple world only direct abatement costs affect prices and trade. Here more general conditions are derived for the seventy-eight industries of the INFORUM interindustry macro model.

The balance of trade is obtained by

$$BT = \sum_{i=1}^{78} \left(P_i EX_i - P_i IM_i \right) \tag{12.1}$$

where BT = balance of trade, P_i = U.S. domestic price of good i, EX_i = U.S. exports for sector i, IM_i = U.S. imports for sector i, and i = 1, 2, 3, ..., 78. If environmental regulations affect the price of products produced by industry j, the change in the balance of trade is found by

$$\partial BT/\partial P_j = \sum_{i=1}^{78} \left[\left(\partial P_i/\partial P_j \right) EX_i + P_i \left(\partial EX_i/\partial P_j \right) - \left(\partial P_i/\partial P_j \right) IM_i - P_i \left(\partial IM_i/\partial P_j \right) \right] \tag{12.2}$$

Note:

$$\left(\partial EX_i/\partial P_j \right) = \left(\partial EX_i/\partial P_i \right) \left(\partial P_i/\partial P_j \right) \tag{12.3}$$

and

$$\left(\partial IM_i/\partial P_j \right) = \left(\partial IM_i/\partial P_i \right) \left(\partial P_i/\partial P_j \right) \tag{12.4}$$

substituting (Equation 12.3) and (Equation 12.4) into (Equation 12.2) and rewriting:

$$\partial BT/\partial P_j = \sum_{i=1}^{78} \left[\left(1 + \eta_i E \right) EX_i \left(\%\Delta P_i/\%\Delta P_j \right) \left(P_i/P_j \right) - \left(1 + \eta_i I \right) IM_i \left(\%\Delta P_i/\%\Delta P_j \right) \left(P_i/P_j \right) \right], \tag{12.5}$$

where $\eta_i E$ = export own price elasticity for industry i, and $\eta_i I$ = import own price elasticity for industry i.

In Equation (12.5) the ratios of the percentage change in price of good i to the percentage change in price of good j are the terms through which the interindustry impacts are felt. These

---

[3] These forecasts are discussed in Portney (1981).

[4] A third problem, argued by Dorfman (1973) in response to Evans (1973) but not treated in the current study, is that there will never be a true general-equilibrium model until pollution emitted by all sources is included in the production functions of all industries. Given this point, a partial equilibrium framework is the only alternative.

terms are the j, ith coefficient in the total requirements matrix.[5] Thus, exports of industry i are indirectly affected by a change in the price of good j.

The abatement-induced price increase for industry j can increase or decrease export revenues depending on the export price elasticities, export volume, the coefficients of the jth row of the total requirements matrix, and industry prices. The impact of a change in the price of industry j on the export revenues of each industry i is found by multiplying exports of i times the j, ith element of the total requirements matrix times the price ratio times one plus $\eta_i E$. Revenues will rise, remain constant, or fall if $\eta_i E$ is greater than, equal to, or less than -1, respectively.

The dollar value of imports will rise for two reasons. First, with a higher domestic price more goods will be imported. Second, higher domestic prices will permit foreign producers to charge higher prices for the larger quantity they are shipping to the U.S. Obviously, the lower the import price elasticity is, the smaller the impact on the balance of trade.

In order for an abatement cost-induced change in the price of industry j to have a positive net effect on the balance of trade, the export impact must be positive and greater than the import impact. While not the case in the current study, this condition might occur if the import and export price elasticities for all sectors were small in absolute value. It might also occur if the U.S. exports several times more of product j than it imports (a strong net exporter), with low export and import price elasticities.

The assumption of full-cost pass-through into prices is reasonable in the long run. Alternate assumptions and their implications are discussed in detail in Lake, Hanneman, and Oster (1979), but a summary is warranted. Greater than full-cost pass-through is possible in oligopolistic industries. This non-competitive behavior has the effect of encouraging more foreign firms to compete for domestic markets than would simple cost pass-through. Conversely, foreign firms would be discouraged from competing in U.S. markets by price increases smaller than cost increases, as might occur in highly competitive industries. Use of the full-cost pass-through assumption helps ensure that the estimates are, in fact, upper bounds.

Most studies using input-output to measure price changes caused by environmental regulations simply multiply the abatement cost vector by the standard total requirements matrix (see, e.g., Pasurka, 1984; Mutti and Richardson, 1977; Walter, 1973). This approach misses a small portion of abatement costs, those implicit in capital goods used in the production process. To account for these additional costs, a "capital-included" total requirements matrix is formed following the method described in Robison (1985). The general procedure is to add to the standard coefficient "A" matrix a capital flows coefficient, or "B" matrix that has been scaled so that the columns sum to the depreciation to output ratio of that industry. Given this modified A matrix, called A^*, the capital-included total requirements matrix is formed $(1 - A^*)^{-1}$. Each j, i element of the total requirements shows the percentage price increase for industry i caused by a one percent increase in the price of input j.

C. DATA

All economic data and the estimated import and export price elasticities are provided by the INFORUM project.[6] A discussion of the general form of the merchandise import and export equations can be found in Almon et al. (1974), while modifications, recent estimation results, and the non-merchandise equations are described in the INFORUM staff paper (1986).[7]

---

[5] This simple result comes from the input-output price determination equation: $p = pA + v$, where: p = vector of 78 sector prices; v = vector of 78 sector value added; A = 78 by 78 input-output coefficient matrix. Solving this equation for prices: $p = (1 - A)^{-1}v$. A change in industrial pollution abatement for a sector changes the value added for that sector and prices of all sectors according to the coefficients of the total requirements matrix, $(I - A)^{-1}$.

[6] INFORUM is the Interindustry Forecasting Project of The University of Maryland.

[7] The general forms of the merchandise trade equations are $M_t = (a + bD_t)P_t^\eta$ and $E_t = (a + bF_t)P_t^\eta$ where: M_t = imports in domestic-port prices; D_t = domestic demand for the product; P_t = the effective ratio of the foreign price to the domestic price; η = the price elasticity; E_t = exports; F_t = foreign demand index. The non-merchandise equations are too sector specific to present a general form. See Almon (1974) and INFORUM (1986) for details.

TABLE 12.1
Average Abatement Content of U.S. Trade
(1.00 Means 1 cent per dollar)

	1973	1977	1982
Total U.S. imports	0.428	0.628	0.993
Total U.S. exports	0.372	0.538	0.715
Total U.S. output	0.499	0.692	0.927
Ratio of import content to export content	1.151	1.167	1.389
U.S. imports from Canada[a]	0.652	0.971	1.512
U.S. exports to Canada[a]	0.586	0.893	1.319
Ratio of import content to export content	1.113	1.087	1.146

[a] U.S.–Canada trade data is available only for the merchandise sectors. For this reason, the results are not directly comparable to the total trade figures.

Abatement costs for each firm are defined as including only the costs of abating pollution generated by the firm, not the cost of abatement equipment built into the goods the firm produces. For example, the costs of abatement equipment built in U.S. automobiles are classified as being incurred by the auto's purchaser, not the producer.

The abatement data for the manufacturing sectors are taken from *Pollution Abatement Costs and Expenditures* for 1973, 1977, and 1983 (1976, 1979, 1984). No annual operating cost data are available for the non-manufacturing sectors. Capital expenditures on abatement equipment are, however, available for non-manufacturing in *The Survey of Current Business* (1984, 1978, 1980). Operating costs are approximated by assuming that the ratio of capital expenditures for abatement equipment to annual operating costs is the same for non-manufacturing as for manufacturing.[8]

D. RESULTS

The two major findings of this study are much as expected. First, there is some evidence that U.S. pollution control programs have changed the U.S. comparative advantage such that more high-abatement-cost goods will be imported and more low-abatement-cost goods exported. Second, the impact on the U.S. balance of trade of marginal changes in IPA is negative for most industries, growing with trade volume, and small relative to domestic consumption.

Table 12.1 presents the average abatement content of U.S. total and U.S.–Canadian trade for the years 1973, 1977, and 1982 as well as the average abatement content of U.S. output.[9] No detailed abatement data are available for countries (including Canada) for which similar input output tables are available. Hence, what is reported as abatement content of imports is, in fact, the abatement content of the goods had they been produced in the U.S. Because of this data restriction, a rise in the abatement content of imports relative to exports indicates that the United States is importing goods with high U.S. abatement costs.

While the level of abatement costs implicit in both imports and exports rose rapidly over the period, they rose relatively more quickly for total imports than total exports. Relative shifts can be seen by looking at the ratio of import to export abatement content, which rose from 1.17 to 1.39 between 1977 and 1982. The rise in the abatement content ratio clearly indicates a shift towards importing goods with high U.S. abatement costs. Further evidence of a shift in comparative

---

[8] More detailed information about the data is available from the author upon request.
[9] A 1972 input-output table is used for the 1973 computations. The 1977 input-output table is also used for the 1982 computations, given that no 1982 table is available.

advantage is found in comparing the abatement content of U.S. exports, imports, and output. Over the 1973 to 1982 period the abatement costs implicit in U.S. exports and output grew proportionately. The abatement content of imports grew more quickly, rising above that of output in 1982. Thus, in 1982, the average dollar's worth of imports had more abatement costs implicit in it than the average dollar's worth of U.S. output. Stated another way, if the U.S. had produced the goods it imported, the average abatement content of U.S. output would have been higher.

Table 12.1 shows virtually no change in the abatement content ratio for U.S.–Canadian trade, thus implying no major shift in trading pattern. Given Canada's adoption of environmental quality regulations, this result is not surprising. In total trade, rising U.S. abatement costs increased the price differential for some products relative to countries with little or no environmental regulation and induced a change in the trading pattern. Canada, by adopting environmental regulations, prevented the price differential and corresponding shift in trading pattern from occurring. The Canadian results also suggest that the trading pattern shift is stronger for all other countries than is indicated in Table 12.1. No shift is apparent for trade with Canada, which accounts for approximately 15 percent of all U.S. trade. Thus, to get the total trade shift of Table 12.1, the change in trading patterns with all other countries must be greater than the average indicated.

Table 12.2 presents the results of Equation (12.5) for selected sectors under the assumption that additional abatement costs raised the sectoral price by 1 percent. Thus, the value of –245.1 for the agricultural products sector indicates that, if additional abatement costs raised the price of agricultural products holding everything else constant, the U.S. balance of trade would have been reduced by at most $245.1 million. Of the $245.1 million impact, $95.8 million is attributed to reduced agricultural exports and increased agricultural imports, while the remainder is the result of indirect effects of agricultural prices on other sectoral prices.

As can be seen in Table 12.2, the direct impacts are large relative to indirect impacts in most manufacturing sectors such as food and tobacco, ferrous metals, and motor vehicles. The indirect effects are relatively large for service sectors such as gas utilities, retail trade, and business services. The importance of the indirect effects is best exemplified by electrical utilities (sector 56), which has large abatement costs but a small trade volume. A 1 percent increase in the price of electrical utilities would have only a $–1.5 million direct impact on the balance of trade, yet an $–80.9 million total impact.

Table 12.2 and Appendix Tables 12.4 and 12.5 strongly support most economists' intuitive position that higher abatement costs will reduce the U.S. balance of trade. The only sectors with positive impacts on the trade balance are coal mining (1973 and 1982 for all trade, and all years for trade with Canada), transportation services (1973 for trade), and special industry machinery (1973 and 1982 for trade with Canada). All three sectors have low export and import elasticities, and the U.S. is a strong exporter of coal and transportation services. In trade with Canada, the U.S. is a strong net exporter of special industry machinery, whereas in total trade the U.S. is a net importer.

Relative to domestic consumption of the respective sector, the impacts on the balance of trade are quite small, with a range of 0.006 percent (retail trade) to 0.8 percent (ferrous metals). When measured as a percentage of total U.S. trade (exports plus imports) of individual sectors, the impacts range from –0.12 percent (special industry machinery) to –7.08 percent (copper) for the merchandise sectors, with an average of –2.69 percent. For trade with Canada, the impacts range from 0.2 (coal mining) to –8.9 (petroleum refining) percent, with an average of –2.53 percent.

Table 12.2 also provides a measure of a side benefit derived from the U.S. governmental policy of paying a large portion of farmers' abatement costs. If this program held agricultural prices down by 1 percent, the 1977 U.S. balance of trade was improved by $245.1million.

Little can be said about the growth in the effects shown in Appendix Tables 12.4 and 12.5. Although additional abatement costs will have an impact on trade patterns, other influences such as macro-economic policies and influences of changing exchange rates make it impossible to determine exactly what portion of the shifts can be attributed to increased abatement costs.

TABLE 12.2
Impacts on U.S. Balance of Trade for 1 Percent Price Increases
(in millions of 1997 dollars)

		Total U.S.		U.S.–Canada		Total U.S.
		Direct Impact	Total Impact	Direct Impact	Total Impact	Domestic Consumption
1	Agriculture	–95.8	–245.1	–52.3	–71.9	133,273
2	Iron ore mining	–12.8	–41.0	–7.4	–12.7	3,567
4	Coal mining	22.1	–6.3	8.1	2.1	16,701
6	Crude petroleum	–316.4	–434.5	–16.7	–36.0	64,920
9	Food and tobacco	–249.4	–285.8	–17.3	–28.2	210,733
13	Paper	–111.0	–180.0	–75.4	–86.0	52,294
15	Agricultural fertilizers	–5.6	–25.2	–8.3	–13.7	10,523
16	Other chemicals	–132.5	–314.6	–26.9	–63.3	99,808
17	Petroleum refining	–82.0	–201.5	–5.2	–32.3	104,290
20	Plastic products	–24.3	–78.6	–6.1	–16.2	24,206
24	Stone, clay, and glass	–46.5	–83.1	–5.6	–12.9	35,462
25	Ferrous metals	–376.4	–566.5	–54.5	–106.7	71,719
26	Copper	–28.2	–64.4	–4.9	–14.0	9,300
27	Other nonferrous metals	–76.0	–172.5	–26.8	–47.4	35,033
28	Metal products	–91.4	–219.8	–22.1	–54.8	86,228
29	Engines & turbines	–13.3	–28.9	–9.1	–13.4	9,454
32	Metalworking machinery	–15.3	–42.3	–1.6	–7.4	13,734
33	Special industry machinery	4.9	–4.5	2.9	1.1	7,547
35	Computers	–29.5	–34.4	–6.6	–7.5	10,424
40	Household appliances	–20.1	–22.7	–1.2	–1.8	10,728
42	TV, radio, and phonographs	–50.9	–52.5	–4.0	–4.6	9,492
43	Motor vehicles	–285.5	–330.2	–119.2	–126.5	128,208
54	Transportation services	3.3	–0.1	a	–0.5	2,025
55	Communication services	–7.5	–50.4	a	–7.9	60,105
56	Electrical utilities	–1.5	–80.9	a	–17.2	63,568
57	Gas utilities	–3.0	–62.0	a	–13.4	46,708
60	Retail trade	–0.4	–11.2	a	–2.0	194,613
61	Eating and drinking places	0.0	–33.4	a	–6.5	87,819
62	Finance and insurance	–9.0	–57.0	a	–9.5	127,952
66	Business services	–17.7	–210.9	a	–37.1	161,049

a No direct effects can be measured, owing to data limitations.

Table 12.3 presents estimates of the total impact of IPA on the balance of trade under the assumption that the INFORUM import and export price elasticities hold for the full price change. In addition, all general equilibrium mitigating effects that might occur through changes in income or exchange rates are ignored. Therefore, these estimates should be the maximum potential effects. Given these caveats, the net reduction in the balance of trade for 1977 is $2,392.3 million for all trade and $544 million for trade with Canada. These figures are 0.67 and 1.02 percent of the respective trade volumes.

The total impacts found in this study are above those forecasted in Evans (1973) and the Chase Econometrics (1981) studies but below those in the Data Resources Study (1981). The most striking difference between the current study and previous work is the growth in the size of IPA's impact over time. This study finds the impact grew by 218.9 percent in constant dollars between 1973 and

TABLE 12.3
Total Impact of Industrial Abatement on Trade
(in millions of 1977 dollars)

	1973	1977	1982
Total trade			
U.S. export impact	–133.5	–258.8	–426.3
U.S. import impact	–1,247.9	–2,133.5	–3,978.9
Total trade impact	–1,381.4	–2,392.3	–4,405.3
Total U.S. trade volume	256,405.0	354,895.0	453,825.0
U.S.–Canadian trade			
Export impact	–33.7	–63.2	–64.0
Import impact	–303.7	–480.8	–786.0
Total trade impact	–337.4	–544.0	–850.0
U.S.–Canadian trade volume	48,467.0	53,413.0	51,733.0

1982, compared with a high of 28 percent (the Data Resources study) among the other studies. The tremendous difference may in part be due to general equilibrium effects in the forecasting studies, but it is likely that these studies mis-estimated both the size of IPA costs and the length of the adjustment period.

E. CONCLUDING REMARKS

By providing a framework for examining the impact industrial pollution abatement has on the U.S. balance of trade, this chapter has moved one step closer to a more accurate assessment of U.S. environmental policy. General conditions for positive and negative impacts are derived in an interindustry framework under the assumption of full cost pass-through. This framework establishes an upper bound for the impact by explicitly ignoring all offsetting general-equilibrium effects such as changes in exchange rates. The statistical results indicate that marginal changes in IPA will reduce the U.S. balance of trade for all but a few industries. In addition, empirical support is found for the proposition that IPA is inducing changes in the U.S. comparative advantage such that the abatement content of imported goods is rising relative to that of exported goods.

The implications of this work for U.S. policy makers are fairly obvious. First, marginal changes in abatement costs, such as those resulting from the recent acid-rain talks with Canada, will result in a reduction in the U.S. balance of trade. Second, the inefficiency of the current regulatory system contributes substantially to the impact on the balance of trade. A 1982 Government Accounting Office report (1982) suggests that if a more efficient regulatory scheme were used, air pollution abatement costs could be reduced 40 to 90 percent, which would in turn improve the U.S. balance of trade. Finally, the adoption of similar environmental regulations by other countries will reduce the impact on the trade balance, although high-abatement-cost industries are likely to move to countries that do not adopt an environmental policy.

ACKNOWLEDGMENTS

I would like to thank Wallace Oates, Clopper Almon, Matt Hyle, Ralph Monaco, Margaret McCarthy, and Don Meyer for their helpful comments on earlier drafts. I would also like to thank Douglas Nyhus and Lorraine Sullivan of the INFORUM staff for their assistance with some of the data work.

APPENDIX

TABLE 12.4
Impacts on U.S. Balance of Trade for 1 Percent Price Increases
(in millions of 1977 dollars)

		1973	1977	1982
1	Agriculture	−258.3	−245.1	−295.0
2	Iron ore mining	−27.4	−41.0	−41.5
4	Coal mining	4.6	−6.3	5.8
6	Crude petroleum	−225.0	−434.5	−469.7
9	Food and tobacco	−249.4	−285.8	−319.7
13	Paper	−151.6	−180.0	−221.8
15	Agricultural fertilizers	−8.4	−25.2	−25.7
16	Other chemicals	−210.5	−314.6	−445.7
17	Petroleum refining	−149.3	−201.5	−335.3
20	Plastic products	−43.5	−78.6	−111.3
24	Stone, clay, and glass	−74.1	−83.1	−108.7
25	Ferrous metals	−429.2	−566.5	−651.1
26	Copper	−65.9	−64.4	−78.0
27	Other nonferrous metals	−127.1	−172.5	−251.1
28	Metal products	−167.7	−219.8	−284.3
29	Engines & turbines	−22.0	−28.9	−46.3
32	Metalworking machinery	−33.0	−42.3	−71.8
33	Special industry machinery	−6.9	−4.5	−32.3
35	Computers	−29.4	−34.4	−97.2
40	Household appliances	−17.7	−22.7	−27.0
42	TV, radio, and phonographs	−47.0	−52.5	−70.1
43	Motor vehicles	−269.0	−330.2	−438.9
54	Transportation services	0.7	−0.1	−1.8
55	Communication services	−47.7	−50.4	−65.1
56	Electrical utilities	−43.1	−80.9	−126.9
57	Gas utilities	−25.9	−62.0	−108.6
60	Retail trade	−10.0	−11.2	−13.0
61	Eating and drinking places	−29.4	−33.4	−50.1
62	Finance and insurance	−60.8	−57.0	−72.7
66	Business services	−173.2	−210.9	−284.7

TABLE 12.5
Impacts on U.S. Balance of Trade with Canada for 1 Percent Price Increases (in millions of 1977 dollars)

		1973	1977	1982
1	Agriculture	−76.6	−71.9	−82.3
2	Iron ore mining	−8.4	−12.7	−11.9
4	Coal mining	3.1	2.1	0.8
6	Crude petroleum	−49.1	−36.0	−58.1
9	Food and tobacco	−26.8	−28.2	−31.1
13	Paper	−78.2	−86.0	−90.7
15	Agricultural fertilizers	−7.1	−13.7	−13.0
16	Other chemicals	−47.2	−63.3	−86.3
17	Petroleum refining	−23.5	−32.3	−56.0
20	Plastic products	−11.6	−16.2	−18.1
24	Stone, clay, and glass	−12.8	−12.9	−13.4
25	Ferrous metals	−78.7	−106.7	−104.4
26	Copper	−19.8	−14.0	−13.6
27	Other nonferrous metals	−45.5	−47.4	−50.0
28	Metal products	−45.1	−54.8	−60.8
29	Engines & turbines	−11.1	−13.4	−10.6
32	Metalworking machinery	−7.0	−7.4	−8.9
33	Special industry machinery	1.2	1.1	−0.4
35	Computers	−6.5	−7.5	−7.6
40	Household appliances	−1.7	−1.8	−2.1
42	TV, radio, and phonographs	−4.3	−4.6	−5.6
43	Motor vehicles	−96.3	−126.5	−136.1
54	Transportation services[a]	−0.3	−0.5	−0.7
55	Communication services[a]	−9.2	−7.9	−8.2
56	Electrical utilities[a]	−10.2	−17.2	−22.6
57	Gas utilities[a]	−5.3	−13.4	−20.5
60	Retail trade[a]	−2.3	−2.0	−2.1
61	Eating and drinking places[a]	−6.8	−6.5	−7.8
62	Finance and insurance[a]	−11.9	−9.5	−10.1
66	Business services[a]	−35.9	−37.1	−42.8

[a] Includes only the indirect impacts on merchandise sectors due to data limitations.

REFERENCES

Almon, C., Jr., Buckler, M., Horwitz, L., and Reimbold, T., *1985: Interindustry Forecasts of the American Economy,* Lexington Books, Lexington, MA, 1974.

Baumol, W. J. and Oates, W. E., *The Theory of Environmental Policy,* Prentice–Hall, Englewood Cliffs, NJ, 1975.

D'Arge, R. C., International trade, domestic income, and environmental controls: some empirical estimates, in *Managing the Environment,* Kneese, A.V., Rolfe, S. E., and Harned, J. W., Eds., Praeger Publishers, New York, 1971.

Dorfman, R., Discussion of presented papers, *Am. Econ. Rev.,* 63, 253, 1973.

Evans, M. K., A forecasting model applied to pollution control costs, *Am. Econ. Rev.,* 63, 244, 1973.

INFORUM Staff, The INFORUM models: a closer look, *INFORUM Research Report,* 75, 1986

Lake, E. E., Hanneman, W. M., and Oster, S. M., *Who Pays for Clean Water? The Distribution of Water Pollution Control Costs,* Westview Press, Boulder, CO, 1979.

Leontief, W., Environmental repercussions and the economic structure: an input-output approach, *Rev. Econ. Stat.,* 52, 262, 1970.

Magee, S. P. and Ford, W. F., Environmental pollution, the terms of trade and the balance of payments, *Kyklos,* 25, 101, 1972.

Mutti, J. H. and Richardson, D. J., International competitive displacement from environmental control: the quantitative gains from methodological refinement, *J. Environ. Econ. Manage.,* 4, 135, 1977.

Pasurka, C. A., The short-run impact of environmental protection costs on U.S. product prices, *J. Environ. Econ. Manage.,* 11, 380, 1984.

Portney, P. R., The macro-economic impacts of federal environmental regulation, in *Environmental Regulation and the U.S. Economy,* Peskin, H. M., Portney, P. R., and Kneese, A. V., Eds., Johns Hopkins University Press, Baltimore, 1981.

David, R. H., Who pays for industrial pollution abatement?, *Rev. Econ. Stat.,* 67, 702, 1985.

Russo, W. J. and Rutledge, G. L., Plant and equipment expenditures by business for pollution abatement, 1983 and planned 1984, *Surv. Curr. Bus.,* 64, 31, 1984.

Rutledge, G. L., Dreiling, F. J., and Dunlap, B. C., Capital expenditures by business for pollution abatement 1973–7 and planned 1978, *Surv. Curr. Bus.,* 58, 33, 1978.

Rutledge, G. L. and O'Connor, B., Capital expenditures by business for pollution abatement, 1977, 1978, and planned 1979, *Surv. Curr. Bus.,* 60, 19, 1980.

U.S. Bureau of the Census, *Pollution Abatement Costs and Expenditures 1973,* U.S. Government Printing Office, Washington, D.C., 1976.

U.S. Bureau of the Census, *Pollution Abatement Costs and Expenditures 1977,* U.S. Government Printing Office, Washington, D.C., 1979.

U.S. Bureau of the Census, *Pollution Abatement Costs and Expenditures 1982,* U.S. Government Printing Office, Washington, D.C., 1984.

U.S. General Accounting Office, *A Market Approach to Air Pollution Control Could Reduce Compliance Costs Without Jeopardizing Clean Air Goals,* U.S. Government Printing Office, Washington, D.C., 1982.

Walter, I., The pollution content of American trade, *West. Econ. J.,* 11, 61, 1973.

13 The Effects of Domestic Environmental Policies on Patterns of World Trade: An Empirical Test

James A. Tobey[1]

A. INTRODUCTION

In theory, environmental control costs encourage reduced specialization in the production of polluting outputs in countries with stringent environmental regulations (Pethig, 1976; Siebert, 1977; McGuire, 1982). In contrast, countries that fail to undertake an environmental protection program presumably increase their comparative advantage in the production of items that damage the environment. This relationship between trade and environmental policy receives considerable attention whenever countries are in the process of passing new pollution control measures. Groups who oppose existing measures or the implementation of stiffer measures argue that they reduce the ability of polluting industries to compete internationally.[2] With foreign trade an increasingly important sector in many of the world's economies, the arguments of such groups are now frequently weighted very heavily.

The premise that trade suffers from the imposition of environmental policy has a strong element of a priori plausibility but, surprisingly, has little empirical support. Several macroeconometric models (D'Arge, 1974; Robison, 1986; OECD, 1985) have predicted that pollution control measures should lead to a small but discernible effect on the balance of trade, but there are few studies to confirm this prediction.

The location-of-industry studies (Leonard, 1988; Pearson, 1987, 1985; Walter, 1985) have explored the related ideas that stringent pollution control measures push industries out of the U.S. (the "industrial-flight" hypothesis), and that less-developed countries compete to attract multinational industries by minimizing their own environmental policies (the "pollution-haven" hypothesis). Their investigations, however, have been unable to find evidence in support of either hypothesis.

This chapter complements the results of the less rigorous location of industry studies by providing an empirical test of the hypothesis that stringent environmental policy has caused trade patterns to deviate in commodities produced by the world's "dirty" industries.

The empirical tests are conducted using the cross-section Hechscher–Ohlin–Vanek (HOV) model of international trade. Leamer (1984), Bowen (1983) and others (Murrell, 1990 and Ryterman, 1988) have popularized the application of this model to empirical tests of the sources of international comparative advantage. The basic HOV model is extended, in addition, to allow for non-homothetic

---

[1] This chapter does not necessarily represent the views of the Department of Agriculture. I am very grateful for the thoughtful comments of Professors Peter Murrell and Wallace Oates, and the support of the Bureau of Business and Economic Research, University of Maryland.

[2] In addition, the imposition of environmental regulations has often been suggested as a partial explanation for inflation in the United States and declines in productivity growth experienced by most industrialized countries in the 1970s.

---

preferences (NHP) and scale economies/product differentiation (EOS). Under all three specifications, two different approaches are taken to conduct the empirical tests. The first involves inclusion of a qualitative variable in the estimated equation to represent the stringency of pollution control measures. The second examines the signs of the estimated error terms when the variable measuring the stringency of a country's environmental policy is not included in the estimated equation.

It is found that the stringent environmental regulations imposed on industries in the late 1960s and early 1970s by most industrialized countries have not measurably affected international trade patterns in the most polluting industries. This result is consistent with and reinforces the results of the location-of-industry studies which use a very different empirical approach, but it may nevertheless be intuitively surprising to those who still fear the undesirable competitive effects of environmental policy.

Before proceeding with the empirical test, commodities whose production draws heavily on environmental resources, termed "pollution-intensive" commodities, are defined.

B. DEFINITION OF POLLUTION-INTENSIVE COMMODITIES

A commodity's relative pollution intensity is defined in this study by the pollution-abatement costs incurred in its production in the United States. Such costs are reported by the U.S. Department of Commerce (1975) and the Environmental Protection Agency (1984). Kalt (1985) has taken these data for 1977 and matched them to sixty-four agricultural and manufacturing 2-digit I–O industries. These direct pollution abatement costs are then multiplied by the 1977 total expenditures I–O table to generate an estimate of total (direct and indirect) PACs per dollar of industrial output.

Commodities termed pollution-intensive are defined as the products of those industries whose direct and indirect abatement costs in the U.S. are equal to or greater than 1.85 percent of total costs. The cut-off of 1.85 percent is chosen because it results in a set of industries that are generally considered the most polluting (metals, chemicals, and paper industries) throughout the world.[3] Moreover, there is a considerable difference between the pollution-abatement costs in these industries and in those of the remaining group of industries.

In Table 13.1, these input-output industries which I have defined as pollution-intensive are matched to 3-digit SITC commodities and aggregated into five commodity groups including paper and pulp products (paper), mining of ores (mining), primary iron and steel (steel), primary non-ferrous metals (nfmetals), and chemicals (chems). It is interesting to note at the outset the relatively small size of pollution abatement costs as a percentage of total costs in even the most highly regulated industries.

C. HECKSCHER–OHLIN–VANEK MODEL OF INTERNATIONAL TRADE

The HOV equations are a multi-factor, multi-commodity extension of the Heckscher–Ohlin (H–O) model of international trade. They have been used in three different ways. The factor content studies and cross-commodity regressions use measures of factor intensities and trade to infer factor endowments. The third methodology, and the approach taken in this study, regresses trade in a specific commodity across countries on country characteristics, i.e., resource endowments. In that resource endowments are the explanatory variables, such regressions reveal the direct influence of resources on trade in a specific commodity. Since this study seeks to reveal information on the most pollution-intensive commodities across countries, the cross-country analysis is chosen as the most appropriate approach.

---

[3] This cut-off does not include the petroleum industry. Petroleum is excluded because it is believed that the dynamics of this industry during this time (1975) were heavily influenced by extraordinary circumstances affecting the availability and processing of crude oil. Leonard (1988) supports this view.

TABLE 13.1

I–O Industry	SITC	Description	Pollution Abatement Costs as Percentage of Total Costs
Mining			
5	281	Iron ore, concentrates	2.03
6	283	Ores of nonferrous base metals	1.92
Primary Nonferrous Metals			
38	681	Silver, platinum, etc.	2.05
38	682	Copper	2.05
38	683	Nickel	2.05
38	685	Lead	2.05
38	686	Zinc	2.05
38	687	Tin	2.05
38	689	Nonferrous base metals, n.e.s.	2.05
Paper and Pulp			
24	251	Pulp and waste paper	2.40
24	641	Paper and paperboard	2.40
24	642	Articles of paper	2.40
Primary Iron and Steel			
37	671	Pig iron	2.38
37	672	Ingots	2.38
37	673	Iron and steel bars	2.38
37	674	Universals, plates	2.38
37	675	Hoops and strips	2.38
37	676	Railway material	2.38
37	677	Iron and steel wire	2.38
37	678	Tubes and fittings	2.38
37	679	Iron, steel castings	2.38
Chemicals			
27	513	Inorganic elements	2.89
27	514	Other inorganic chemicals	2.89
28	581	Plastic materials	2.36

A set of eleven resource endowments for the year 1975 is used to explain net exports of the most polluting industries under the HOV model.[4] These endowments are provided by Leamer (1984) and include:

1. CAPITAL (CAP) Accumulated and discounted gross domestic investment flows since 1948, assuming an average life of 15 years.
2. LABOR 1 (LAB1) Number of workers classified as professional or technical.
3. LABOR 2 (LAB2) Number of literate nonprofessional workers.
4. LABOR 3 (LAB3) Number of illiterate workers.
5. LAND 1 (LND1) Land area in tropical rainy climate zone.
6. LAND 2 (LND2) Land area in dry climate zone.

---

[4] Trade data for 1975 comes from United Nations' trade tapes.

7. LAND 3 (LND3) Land area in humid mesothermal climate zone (for example, California).
8. LAND 4 (LND4) Land area in humid microthermal climate (for example, Michigan).
9. COAL (COAL) Value of production of primary solid fuels (coal, lignite, and brown coal).
10. MINERALS (MINLS) Value of production of minerals: bauxite, copper, flourspar, ironore, lead, manganese, nickel, potash, pyrite, salt, tin, and zinc.
11. OIL (OIL) Value of oil and gas production.

The HOV model can be summarized by the following equations which explicitly incorporate this list of endowments:

$$N_{it} = CST_{i0} + b_{i1}V_{1t} + b_{i2}V_{2t} + \ldots + b_{i11}V_{11t} + \mu_{it} \tag{13.1}$$

where N_{it} are net exports of commodity i by country t; V_{kt} are endowments of resource k ($k = 1, \ldots, 11$) in country t; b_{ik} are the coefficients which indicate the total effect (production and consumption) of an increase in a resource on net trade of a specific commodity; μ_{it} is the disturbance term, and; CST_{i0} is the equation's constant term. The constant term embodies one resource endowment or country characteristic, which all countries are assumed to possess identically and which has a non-zero mean value.

If the environmental endowment, measured by the stringency of environmental regulations,[5] has an effect on trade patterns, then the set of eleven endowments in Equation (13.1) is incomplete. In this case, estimation of the HOV trade equations implies a specification error involving an omitted variable. Two approaches are taken to test the effect of the environmental endowment on trade patterns under the HOV model when cross-country quantitative data on the environmental endowment are not available. In the first, a qualitative variable is included in Equation (13.1) to represent the omitted variable, and in the second, an omitted variable test is conducted.

D. THE TESTING EFFECTS OF POLLUTION CONTROL MEASURES

1. THE HOV MODEL WITH ENVIRONMENTAL STRINGENCY

To test the pollution-haven hypothesis under the first approach, the following equation is estimated under ordinary least squares (OLS):

$$N_{it} = CST_{i0} + b_{i1}V_{1t} + b_{i2}V_{2t} + \ldots + b_{i11}V_{11t} + b_{iE}D_{Et} + \mu_{it} \tag{13.2}$$

where D_{Et} is qualitative variable measuring the stringency of pollution control measures in country t based on a 1976 UNCTAD survey (Walter and Ugelow, 1979).

The degree of environmental stringency is measured on a scale from one (tolerant) to seven (strict); the mean score for developed countries is 6.1, while for developing countries it is 3.1. There are observations for 23 countries: 13 industrialized and 10 developing countries (see Table 13.2).

The results are presented in Table 13.3 (absolute value of the t radio shown in parentheses beside the estimated regression coefficient). In no instance is the t ratio found to be statistically significant on the measure for the stringency of environmental policy in the five regressions of net exports of polluting industries.

---

[5] Although pollution emissions are a joint product of the production process, they can also be interpreted as an input in the production function because they can be viewed as one of the various uses of the environment (Baumol and Oates 1988). Since use of the environment is typically a public good, the environmental endowment has no price attached to it and will be used freely by industries until pollution control measures are instituted. Thus, in this study, a country's environmental endowment is measured by their stringency of pollution control measures.

TABLE 13.2
Index of the Degree of Stringency of Environmental Policy
(7 = strict, 1 = tolerant)

	Group A	Index		Group B	Index		Group C	Index
1	Austria	4	1	Chile	4	1	Colombia	5
2	Australia	5	2	Cyprus	1	2	Liberia	1
3	Belux	3	3	Israel	4	3	Nigeria	2
4	Denmark	5	4	Malta	1			
5	Finland	6	5	Singapore	6			
6	Germany	5	6	Spain	4			
7	Japan	7	7	Panama	4			
8	New Zealand	5						
9	Nether.	5						
10	Norway	6						
11	Sweden	7						
12	U.K.	4						
13	U.S.	7						

Source: Walter and Ugelow (1979).

TABLE 13.3

	Equations (D.F. = 10)				
Variable Name	Mining ($R^2 = 0.99$)	Paper ($R^2 = 0.96$)	Chems ($R^2 = 0.93$)	Steel ($R^2 = 0.89$)	NFMetals ($R^2 = 0.92$)
CAP	−192 (2.4)	177 (1.6)	583 (5.6)	1537 (2.6)	−89 (1.0)
LAB1	735 (1.9)	−267 (0.5)	981 (1.9)	−1434 (0.5)	−550 (1.2)
LAB2	−111 (3.2)	−25 (0.5)	−154 (3.5)	54 (0.2)	44 (1.1)
LAB3	−15 (0.6)	50 (1.5)	−49 (1.6)	84 (0.5)	69 (2.5)
LND1	385 (1.5)	278 (0.8)	521 (1.6)	237 (0.1)	−254 (0.9)
LND2	−104 (0.7)	−192 (1.0)	−31 (0.2)	503 (0.5)	−247 (1.5)
LND3	1295 (2.8)	100 (0.2)	−268 (0.5)	−2898 (0.9)	−414 (0.8)
LND4	435 (0.9)	6089 (9.2)	−2003 (3.2)	−1374 (0.4)	−589 (1.1)
COAL	−78 (0.6)	−110 (0.6)	−283 (1.6)	−83 (0.1)	88 (0.6)
MINLS	338 (2.0)	330 (1.4)	88 (0.4)	26 (0.1)	715 (3.7)
OIL	−30 (1.6)	−110 (4.3)	−20 (0.8)	−142 (1.0)	17 (0.8)
D	−10314 (0.3)	2454 (0.1)	−1531 (0.1)	98844 (0.4)	48658 (1.3)
CST	−5669 (0.1)	−168370 (1.0)	−107110 (0.7)	−697020 (0.8)	−122980 (0.9)

2. OMITTED VARIABLE TEST

A second approach to testing the effect of pollution control measures on trade patterns investigates the bias in the regression residuals when the variable representing countries' environmental endowments are not included in the HOV equations.[6]

Consider first a simple HOV equation with one known and one unknown independent variable. Let x_{t2} represent a factor endowment for country t. Under the null hypothesis that the environmental

---

[6] The idea for this methodology is attributable to Ryterman (1988). Ryterman was able to show that technological flexibility is a significant variable in determining trade patterns between market economies and centrally planned economies.

factor (x_{t3}) has no effect on the pattern of trade, the equation specifying net exports (N_t) may be written as:

$$N_t = \beta_1 + \beta_2 \chi_{t2} + \tilde{\mu}_t \qquad (13.3)$$

The alternative to the null hypothesis is represented by the following equation:

$$N_t = \beta_1 + \beta_2 x_{t2} + \beta_3 x_{t3} + \tilde{\mu}_t \qquad (13.4)$$

If Equation (13.3) is correct, the least squares estimators of β_1 and β_2 using Equation (13.3) will be unbiased and efficient for all sample sizes. If Equation (13.4) is correct, then estimation of Equation (13.3) will still generate an unbiased estimator of β_2 given the following assumption:[7]
   A1: The omitted variable is not correlated with any of the included independent variables.
   Given assumption A1, estimation of Equation (13.3) when the omitted variable (x_{t3}) does not equal zero will not affect $\hat{\beta}_2$. Its presence will, however, be embodied in the constant and disturbance term. Solving for $\tilde{\mu}_t$, the following equation can be derived:

$$= \beta_3 \left(x_{t3} - \overline{X}_3 \right) + \mu_t \qquad (13.5)$$

Under the null hypothesis that x_{t3} has no effect on the pattern of trade so that $\beta_3 = 0$, $\tilde{\mu}_t$ is a consistent estimator of μ_t. Under the alternative case where pollution control measures are expected to have an effect on the pattern of trade, so that $\beta_3 \neq 0$, then (given assumption A1) $\tilde{\mu}_t$ provides a consistent estimate of Equation (13.5).
   A methodology to test the effect of pollution control measures on the pattern of trade may now be presented. Under the alternative hypothesis that Equation (13.4) is correctly specified and assuming it also has all the properties of the classical regression model, then the sign of μ_t is expected to be random. Therefore, the expected sign of $\tilde{\mu}_t$ in Equation (13.5) is the same as that of $\beta_3(x_{t3} - \overline{X}_3)$. β_3 is expected to be negative if pollution control measures reduce net exports of pollution-intensive commodities. To determine the sign of $(x_{t3} - \overline{X}_3)$, consider the distribution of the stringency of environmental regulations, x_{t3}, over the world. Industrialized, high-income countries have environmental endowments greater than the population mean $\overline{X}_3$, and less-developed countries have environmental endowments less than the population mean. Thus, the pattern of signs of $\tilde{\mu}_t$ under the alternative hypothesis depends on the distribution of x_{t3} over countries. Given the distribution above, the proportion of error terms that are positive for developing countries is expected to be significantly greater than the proportion of error terms that are positive for industrialized countries.
   Let T_n represent the true proportion of errors for countries in group n (where $n = 1$ corresponds to industrialized and $n = 2$ corresponds to developing countries). The null hypothesis (H_0) states that the proportion of errors that are positive is the same for both industrialized and developing countries. The alternative hypothesis (H_1) states that the proportion of such errors is greater for developing countries than for industrialized countries.

$$H_0 : T_2 = T_1$$

$$H_1 : T_2 > T_1$$

---

[7] Kmenta (1971) shows that while $\hat{\beta}_2$ is unbiased, $\text{var}(\hat{\beta}_2)$ is positively biased. Therefore, the usual tests of significance concerning $\hat{\beta}_2$ are not valid since they will tend to accept the null hypothesis more frequently than is justified by the given level of significance.

A nonparametric statistical procedure was chosen to conduct the statistical test because it requires few assumptions regarding the distribution of the error terms. Under the null hypothesis, the test statistic may be given as:[8]

$$A = \frac{R_2 - R_1}{\left[T \times (1-T)\left[\left(1/I \times J_2\right)+\left(1/I \times J_1\right)\right]\right]^{1/2}}$$

where $R_n = S_n/(i \times J_n)$ represents the proportion of estimated errors that are positive; "T" equals the total number of commodity groups (= 5), J_n equals the total number of countries in country group n; and S_n equals the number of estimated error terms for countries in group n that are positive.

$\hat{T}$ is an estimate of the true proportion under the null hypothesis. The best estimate of the true population proportion is constructed by combining the observations for both industrialized and developing countries as follows:

$$\hat{T} = \left(S_1 + S_2\right)\big/\left(I \times \left(J_1 + J_2\right)\right)$$

To perform the omitted variable test a set of 58 countries is arranged in three groups (see Table 13.4). Group one consists of industrialized, high-income countries. Environmental regulatory costs in this group are predicted to generate a comparative disadvantage in the production of polluting commodities. Group two is composed of upper-income developing countries and semi-industrialized nations without a stringent environmental program in 1975.[9] Group three is composed of middle to low-income developing countries.

A summary of information on the estimated residuals when Equation (13.1) is estimated using this set of 58 countries is shown in Table 13.5. It is not possible to reject the null hypothesis that $T_2 = T_1$ in the comparison of industrialized countries with any combination of the developing country groups. These results reinforce the earlier finding which used a qualitative variable to represent the environmental endowment and which also found no effect of pollution control measures on HOV trade patterns.

E. EXTENSIONS TO THE HOV MODEL

This section specifies alternative trade models when two of the HOV assumptions are relaxed (see Leamer, 1984 for a complete list of necessary assumptions) and reports the empirical results when the above empirical tests are again performed on these models.

The first extension of the HOV model allows for non-homothetic preferences. The HOV model assumes identical homothetic tastes, meaning that individuals facing identical commodity prices will consume commodities in the same proportions. In this cross-section study, with countries at various levels of development, this assumption may not be completely reasonable. To allow for NHP, I assume that consumption across countries is a linear function of population and national income. When the HOV model is modified to incorporate this assumption, per capita net exports (n_{it}) become a linear function of per capita resource endowments (v_{kt}) as given by the following equation:

$$n_{it} = b_{i0}^* + \left(\sum_{k=1}^{K} b_{ik} v_{kt}\right) \tag{13.6}$$

---

[8] See Yamane (1967).

[9] A review of international environmental policy (Tobey, 1989) is used to specify those countries with and without enforced pollution control measures as of the mid 1970s.

TABLE 13.4
Country Observations

Country	Gdp Per Capita[1]	Country	Gdp Per Capita[1]	Country	GDP Per Capita[1]
			(1975 Constant U.S. $)		
Group One		**Group Two**		**Group Three**	
Australia	5919	Argentina	3159	Afghanistan	380
Austria	4994	Brazil	1978	Burma	312
Belux	5569	Chile	1834	Colombia	1596
Canada	6788	Costa Rica	1835	Dom Rep	1443
Denmark	5969	Cyprus	1811	Ecuador	1300
Finland	5192	Greece	3360	Egypt	929
France	5864	Hong Kong	2559	El Salvador	1005
Germany	5758	Ireland	3067	Ghana	952
Iceland	5201	Israel	4154	Honduras	871
Japan	4904	Italy	3870	India	472
Netherlands	5321	Malta	2154	Indonesia	536
New Zealand	4769	Mexico	2276	Jamaica	1763
Norway	5419	Panama	2026	Korea	1530
Sweden	6749	Peru	1860	Liberia	830
Switzerland	6082	Portugal	2397	Lybia	6680[2]
U.K.	4601	Singapore	2875	Malaysia	1532
U.S.	7132	Spain	4032	Mauritius	1260
AVERAGE	5661	Turkey	1738	Nigeria	1179
		Yugoslavia	1960[2]	Paraguay	1186
		AVERAGE	2567	Philippines	912
				Sri Lanka	661
				Thailand	930
				AVERAGE	1002[3]

[1] Heston and Summers (1984).
[2] 1977 GNP per capita, from World Bank (1979), *World Development Report*.
[3] Excluding Libya.

TABLE 13.5

Country Group	Positive Residuals								
	Paper	Steel	Chems	NFMetals	Mining	S_n[1]	J_n[2]	R_n[3]	A[4]
One	7	9	5	8	11	40	17	0.47	—
Two	5	10	9	5	6	35	19	0.37	−1.32
Three	10	18	15	5	11	59	22	0.54	0.96
2+3	15	28	24	10	17	94	41	0.45	−0.15

[1] S_n is the number of errors for group n that are positive.
[2] J_n is the number of countries in group n.
[3] $R_n = S_n/(I \times J_n)$.
[4] "A" is the test statistic computed using group one as the sample of industrialized countries compared against groups of developing countries (T = 0.46). An absolute value of 1.65 for the test statistic in the normal distribution corresponds to a probability of 95 percent.

where $b_{iO}^* = -c_{iO}$, and c_{iO} is a parameter that relates consumption of commodity i in country t to country t's population. As before, b_{ik} indicates the total effect of an increase in a resource on net trade of a specific commodity.

The same set of empirical tests is conducted using this equation. Without burdening the reader with additional tables, we simply report that Walter and Ugelow's variable is not significant in any of the OLS regressions. Similarly, the statistical comparison of error terms under the omitted variable test does not support the hypothesis being tested.

The second extension of the HOV model emphasizes scale economies and product differentiation. All of the pollution-intensive commodity groups are associated with what are generally considered very large-scale production processes. Moreover, Hufbauer (1970) has found the production of paper products to be subject to large economies of scale and the production of nonferrous metals to be subject to diseconomies of scale. Scale economies may therefore be important in determining international comparative advantage for these goods. If this is the case, developing countries with small economies and limited infrastructure outside of major cities may be at a disadvantage in the production of these commodities. A model which allows for the fact that the production function for polluting commodities exhibits economies of scale is therefore tested.

I use a model which Murrell (1990) derives and which follows Helpman and Krugman (1985) closely. The set of assumptions used to generate this model is essentially the same as that of the previous HOV model except that product differentiation and economies of scale are introduced. Specifically, it is assumed that each good can be produced in an infinite number of varieties. Each variety of a single good has the same production function, but such functions vary between goods. Furthermore, the production function for each variety exhibits economies of scale, at least at low levels of output.

With the introduction of economies of scale and product differentiation, the export of good i by country t, x_{it} is specified as follows (Murrell, 1990):

$$x_{it} = \sum_{k=1}^{K} b_{ik}^* V_{kt}\left(1 - G_t/G_w\right) \tag{13.7}$$

where G_t is the national income of country t and G_w is total world income.

Equation (13.7) cannot be derived from the H–O theory and the asterisks on the coefficients of the equation are a reminder that these coefficients are not equivalent to b_{ik} in the previous two models.

Again, the same set of empirical tests is conducted, this time using Equation (13.7). Once again, Walter and Ugelow's variable is not found significant in any of the five OLS regressions. In contrast, the test statistics comparing the sign of the estimated residuals of industrialized and developing countries are significant at the 95 percent level of confidence, but, because they are negative, they also fail to support the hypothesis being tested.

F. A FIXED-EFFECTS EMPIRICAL TEST

A reasonable explanation for the empirical results above may be that the magnitude of environmental expenditures incurred by the industrialized countries in the late 1960s and early 1970s were not sufficiently large to cause a noticeable effect on trade patterns between countries with and without environmental protection programs. The cross-section HOV model may not be sufficiently precise to identify these small changes in factor abundance and comparative advantage. Thus, the effect of domestic environmental policy on trade may be getting lost in the "noise." By examining the change in trade patterns before and after the introduction of environmental measures in the industrialized countries, one might be able to detect the hypothesized shifts in trade patterns in response to environmental policies that do not show up in the equations using data from a specific point in time. Such a methodology would also be effective in capturing the effect of environmental policy even if there was a significant lag in the impact of pollution controls on international competitiveness.

TABLE 13.6

Variable Name	Equations (D.F. = 22)				
	Mining (R^2 = 0.03)	Paper (R^2 = 0.0+)	Chems (R^2 = 0.05)	Steel (R^2 = 0.0+)	NFMetals (R^2 = 0.04)
E	−54155 (1.1)	−2298 (0.1)	78007 (1.9)	49437 (0.4)	−65593 (1.1)

Although endowment data are only available for the year 1975, most resource endowments change little over time. At least for the most polluting industries, one might argue that the most important endowment change over the period 1970 to 1984 was the increase in environmental regulations. Consider, then, a HOV model where the change in net exports over 1970–84 is linearly related to the change in factor endowments over the same period. Under a 'fixed-effects' specification, assume that, except for the environmental endowment, the change in factor endowments equals zero. In this case, one is left with the following equation:

$$\Delta N_{it} = E_t + \mu_{it}$$

where ΔN_{it} are 1984 minus 1970 net exports of commodity i by country t. E_t is the Walter and Ugelow (1979) measure of the degree of the stringency of environmental policy in 23 countries in 1977. Since these countries generally did not have enforced environmental programs in place by 1970, the level of environmental policy given by this index is a reasonable proxy for the change in environmental policy. Finally, μ_{it} are the random error terms.

The results of the OLS estimation of this model are shown in Table 13.6. If environmental policies reduce countries' international comparative advantage in the most pollution-intensive commodities, then the sign on the environmental endowment coefficient should be negative and significant. Only in the chemicals group does the significance of the coefficient approach a conventionally accepted level of confidence. The sign, however, is positive, and once again does not support the hypothesized impact of pollution control measures on trade patterns.

G. CONCLUDING REMARKS

Several tests are undertaken under a variety of specifications, but in no case is there any evidence that the introduction of environmental control measures has caused trade patterns to deviate from the HOV predictions. This result is important in that it casts serious doubt on the balance-of-trade argument against environment control which, as mentioned above, can be a strong deterrent to the implementation of new or stronger pollution control measures. The test results also lend support to the less empirically rigorous location-of-industry studies which maintain that the world distribution of dirty industries has not been affected by differing country levels of environmental stringency.

As already noted, a reasonable explanation for these empirical results may simply be that the magnitude of environmental expenditures in countries with stringent environmental policies is not sufficiently large to cause a noticeable effect. This seems like the most likely explanation. Other explanations can, however, be postulated, and we offer two possibilities below.

The first relates to the validity of the HOV equations themselves. The accuracy of the cross-section HOV model was recently questioned by Bowen et al. (1987), who show that since this model makes no reference to factor input intensities, it is only a weak test of the H–O hypothesis. The stricter H–O hypothesis requires that factor supplies, magnitude of the difference in robustness, and accuracy of the weak version vis-à-vis the strict version is not clear.

The second observation relates to the nature of the pollution-intensive commodity groups and violations of the assumptions of the HOV model. This model assumes that commodities move internationally at zero cost of transportation, and that there are no other impediments to trade. However, transportation costs, tariffs, and subsidies[10] are important considerations in these industries and may affect a country's composition of trade. These trade impediments are difficult to deal with at an empirical level, since they are very difficult to measure for large sets of countries. However, if they are distributed across countries in the same pattern as environmental controls, then they would interfere with the ability of the HOV model to pick up the effect of environmental policy.

SUMMARY

The premise that trade suffers from the imposition of environmental policy has a strong element of a priori plausibility, but, surprisingly, little empirical support. The present chapter provides an empirical test of the hypothesis that stringent environmental policy has caused trade patterns to deviate in commodities produced by the world's dirty industries. The empirical tests are conducted using the cross-section Heckscher–Ohlin–Vanek (HOV) model of international trade. It is not found that the introduction of stringent environmental control measures has caused trade patterns to deviate from the HOV predictions. This result casts serious doubt on the balance of trade argument against the imposition of stronger environmental control.

REFERENCES

Baumol, W. and Oates, W., *The Theory of Environmental Policy*, 2nd Ed., Cambridge University Press, New York, 1988.

Bowen, H., Changes in the international distribution of resources and their impact on U.S. comparative advantages, *Rev. of Econ. and Stat.*, 65, 402, 1983.

Bowen, H., Leamer, E., and Sveikauskas, L., Multicountry, multifactor tests of the factor abundance theorem, *Am. Econ. Rev.*, 77, 791, 1987.

D'Arge, R., International trade, domestic income, and environmental controls: some empirical estimates, in *Managing the Environment: International Economic Cooperation for Pollution Control*, Kneese, A., Ed., Praeger, NY, 1974, 289.

Krugman, P. and Helpman, E., *Market Structure and Foreign Trade*, MIT Press, Cambridge, 1986.

Heston, A. and Summers, R., Improved international comparisons of real product and its composition: 1950–1980, *Rev. of Income and Wealth*, 30, 207, 1984.

Hufbauer, G. C., The impact of national characteristics and technology on the commodity composition of trade in manufactured goods, in *The Technological Factor in International Trade*, Vernon, R., Ed., Columbia University Press, New York, 1970.

Kalt, J. P., The impact of domestic environmental regulatory policies on U.S. international competitiveness, Energy and Environmental Policy Center, Discussion Paper Series, John F. Kennedy School of Government, Harvard University, 1985.

Leamer, E. E., *Sources of International Comparative Advantage: Theory and Evidence*, MIT Press, Cambridge, 1984.

Leonard, H. J., *Pollution and the Struggle for World Product*, Cambridge University Press, Cambridge, 1988.

Mcguire, M., Regulation, factor rewards, and international trade, *J. of Public Econ.*, 17, 335, 1982.

Murrell, P., *The Nature of Socialist Economies: Lessons from East European Foreign Trade*, Princeton University Press, 1990.

OECD, *The Macro-Economic Impact of Environmental Expenditures*, Organization for Economic Cooperation and Development, Paris, 1985.

---

[10] Many countries with stringent pollution control measures have offered various subsidies to polluting industries which lessen the impact of pollution control measures on world trade patterns in these industries.

Pearson, C., *Down to Business: Multinational Corporations, the Environment and Development*, World Resources Institute, Washington, D.C., 1985.

Pearson, C., Ed., *Multinational Corporation, the Environment and Development*, World Resources Institute, Washington, D.C., 1987.

Pethig, R., Pollution, welfare, and environmental policy in the theory of comparative advantage, *J. of Environ. Econ. and Manage.*, 2, 160, 1976.

Robison, D. H., Industrial Pollution Abatement: The Impact on Balance of Trade, unpublished manuscript, University of Maryland, 1986.

Ryterman, R., Technological Flexibility and the Pattern of East–West Trade, Ph.D. thesis, University of Maryland, 1988.

Siebert, H., Environmental quality and the gains from trade, *Kyklos*, 30, 657, 1977.

Tobey, J., The Impact of Domestic Environmental Policies on International Trade, Ph.D. thesis, University of Maryland, 1989.

U.S. Department of Commerce, Bureau of the Census, *Pollution Abatement Costs and Expenditures*, Current Industrial Reports, Washington, D.C., 1975.

Walter, I., Environmentally induced industrial relocation to developing countries, in *Environment and Trade*, Rubin, S. J., and Graham, T. R., Eds., Allanheld, Osmum, and Publishers, 1982, 67–101.

Walter, I. and Ugelow, J., Environmental policies in developing countries, *Ambio*, 8, 102, 1979.

Yamane, T., *Statistics: An Introductory Analysis*, Harper & Row, New York, 1967, 584.

14 Unilateral CO$_2$ Reductions and Carbon Leakage: The Consequences of International Trade in Oil and Basic Materials[1]*

Stefan Felder and Thomas F. Rutherford

A recursively dynamic general equilibrium model featuring six world regions with trade in energy and non-energy goods is used to simulate the period from 1990 through 2100 in 10-year intervals. The simulations explore the effect of unilateral action by the OECD to curb global CO$_2$ emissions. Unilateral cuts create incentives for free-riding by non-participating regions, so that global emissions are reduced by less than the amount that the OECD cuts back its regional emissions. Carbon "leakage" occurs through two channels. First, basic materials production increases in unconstrained regions, resulting in increased carbon intensity of GDP. Second, reductions in OECD oil imports cause the world oil price to fall, leading to an increased energy intensity in the non-participant regions. In this chapter, we use a general equilibrium model to assess the extent to which these two mechanisms reduce the effectiveness of OECD reductions in curbing global CO$_2$ concentrations.

© 1993 Academic Press, Inc.

A. INTRODUCTION

Two international conferences (Toronto, 1988 and Cairo, 1990) have called for significant reductions in worldwide carbon dioxide emissions associated with the combustion of fossil fuels. Europe, Japan, and Australia have already made commitments to lower their CO$_2$ emission levels. At the last UN conference on global warming held in Rio (June, 1992), the European Community proposed a carbon tax for the developed world to curb global CO$_2$ emissions.

The purpose of this chapter is to explore the economic consequences of unilateral cutbacks of CO$_2$ emissions. The framework for this analysis is a recursively dynamic general equilibrium model which is designed to identify the economic channels through which restrictions on CO$_2$ emissions affect international trade and the pattern of comparative advantage. Emission limits produce price effects which spill over into other markets on both domestic and international levels. The effects are likely to be significant for goods with a high energy content. On the international level, our model incorporates trade in oil, an aggregate non-energy good, and basic intermediate materials

[1] We thank Alan Manne and an anonymous referee for helpful comments on this chapter. We remain responsible for the views expressed here. The research reported in this chapter was supported by the Electric Power Research Institute, the Swiss National Science Foundation under Grant 8210–028322, and the Canadian Natural Science and Engineering Research Council under Operating Grant T306A1. Earlier versions of this chapter have been presented at the Conference of the Applied Econometrics Association, Geneva, January 1992, and at the Conference on Computational Economics, Austin, TX, May 1992.

(BMAT). BMAT is a category which includes relatively energy-intensive goods such as chemicals, steel, plastic, and glass. The model distinguishes six regions (USA, other OECD countries, USSR, China, MOPEC [Mexico and OPEC], and ROW [rest of the world]). The regional submodel features two macro sectors (BMAT and aggregate output) and a disaggregated energy sector in which final products are electric and non-electric energy.

The model is a "general economic equilibrium" in that all economic activities are summarized in a consistent, albeit highly aggregate, fashion. In this model, prices adjust so that all domestic and international markets clear while producers and consumers make optimal decisions taking market prices as given. When there are region-specific restrictions on carbon emissions associated with energy consumption, we presume that crude oil imports count against the importing region's emission quota. At the same time, we assume that there can be no effective tax on carbon embodied in other imported goods. For this reason, trade in energy-intensive goods such as basic materials can offset the effects of CO_2 cutbacks, particularly if one or more regions take unilateral measures. "Leakage" can also occur when carbon taxes depress the world-market price for oil and thereby change the energy consumption in non-signatory countries.

In the simulations reported here, we "benchmark" our model to quantities and prices for 1990. The projections cover eleven 10-year intervals extending from 1990 through 2100. In the baseline scenario "business as usual," an equilibrium trajectory is simulated when no carbon emission limits apply in any region. The counterfactual experiments reduce CO_2 emissions in the OECD between 1 and 4% per annum. We are concerned with the order of magnitude of economic variables; hence, for simplicity, we focus on cuts of specified percentage per annum from the business as usual case rather than on a particular path of emissions reductions.[2]

In a second set of counterfactual cases, we assess the effects of policies which could be adopted to restrict leakage through trade in basic materials. The OECD regions are presently net exporters of basic materials, and the model suggests that these export will continue into the next century. As a consequence, trade policies to reduce leakage through BMAT trade will involve subsidies for OECD exports rather than taxes on BMAT imports. A policy measure which combines carbon taxes and export subsidies parallels the energy tax proposal by the European Community's commissioner for the environment, Carlo Ripa de Meana. The commissioner's tax design includes exemptions intended to moderate the impact on the competitive position of domestic industry. Notably, the list of sectors which the EEC wants to exempt from the carbon tax are those which are included in our BMAT category.[3]

We conduct a calculation in which OECD regions subsidize exports of basic materials so that exports remain at business-as-usual levels through the model horizon. This experiment is also motivated by recent developments in the area of trade and environment. Demands by ecological groups to restrict international trade in environmentally sensitive commodities are being reflected by new trade legislation in the United States and other countries.[4] Our scenario in which BMAT exports by the OECD are subsidized in the presence of carbon taxes exemplifies the idea of this legislation.

Our numerical experiments provide insight into the consequences of unilateral OECD emission restrictions and the extent to which increased emissions by non-OECD countries offset OECD reductions. Our model suggests that the welfare cost of a 2% per annum (compounded) unilateral reduction in carbon emissions by the OECD would have a cost of roughly 2% of GDP. We find that

---

[2] This approach was recommended by the OECD Public Economics Division in their model comparison project. (See Dean and Hoeller, 1992.)

[3] We do not provide an explicit calculation of the economic effects of the EC proposal in this chapter, but we feel that our analysis captures some of the effects. The current EC proposal actually exempts a number of energy-intensive industries from carbon taxes. A precise calculation of the effects of this policy would require a substantial revision of our model structure in view of the possibility that exemptions would induce fuel switching in the affected industries.

[4] For a survey, see Subramanian (1992).

at the margin this reduction would be offset by 25% through increased emission in non-signatory countries. Trade policies to offset leakage through subsidization of energy-intensive industries have at best limited effectiveness. We conclude that for reducing carbon emission, trade restrictions are a poor substitute for a comprehensive global agreement.

The body of this chapter is organized a follows. Section B describes the model structure and key parameters underlying technologies and preferences. Section C formalizes the connection between costs of unilateral carbon abatement and leakage. Section D presents an overview of the central business as usual scenario. Section E presents the results of the counterfactual cases with unilateral carbon restrictions in the OECD. Sections D and E employ a series of graphs to illustrate the results. Section F summarizes our findings and makes suggestions for future research. An algebraic description of the model is provided in Appendix A, and a description of the key input data is presented in Appendix B. Both appendices are available from the authors upon request.

B. MODEL STRUCTURE

The present model is based on an earlier model constructed to analyze the welfare effect of carbon taxes in a recursively dynamic five-region framework (Rutherford, 1992b).[5] Both of these models are based on the Global 2100 model and dataset from Manne and Richels (1990, 1992). The Global 2100 ROW region is an aggregate of heterogenous countries including OPEC and other oil exporters as well as oil-importing LDCs. For the present study, which is focused on unilateral carbon reductions and leakage, we separated OPEC and Mexico from the Global 2100 ROW region. In the new region (MOPEC), shares of oil and gas production of GNP are significantly higher than in ROW so that one can expect different impacts from OECD carbon taxes on the two regions.

Our model shares a number of structural features with Global 2100. These similarities include a world economy distinguishing five regions, 10-year time intervals beginning in 1990 and extending to 2100, a process submodel describing the energy sector, carbon constraints applied on a region-specific basis, and no interregional capital flows (each region operates with balance of payments in every period.)

There are, however, important differences between these models. First, our model is based on a recursive rather than a fully intertemporal dynamic structure. Savings fractions of final consumption are input data which are unaffected by changes of the real interest rate. Second, the model distinguishes two non-energy goods: basic materials and an aggregate output. The presence of traded commodities with differing energy intensities provides insights into the extent to which changes in carbon taxes lead to the reallocation of comparative advantage. Third, in the recursive model, the international oil market clears in every period. Finally, Global 2100 incorporates forward-looking resource policies, whereas in our model, the timing of energy extraction and the resulting changing composition of energy supplies are based on static expectations. The treatment of expectations has important consequences for the analysis.

The detailed structure of a single period in each of the periods is shown in Figure 14.1. Primary factors (capital and labor) are employed together with electric and non-electric (E,N) and basic intermediate inputs (B) to produce the domestic region's macro output. This output is in turn used for final demand (consumption and investment, C+I), as well as for exports (X), and as inputs to the production of energy (EC) and basic materials (BC), respectively. These flows are represented by arrows in the diagram. Aggregate output, basic intermediate materials, and oil are traded in the international market.

---

[5] In Rutherford (1992b), Rutherford extended an earlier static general equilibrium model developed with Perroni and described in Perroni and Rutherford (1993). In the static model, the authors focused on the year 2020, in which energy supplies arise from upward-sloping marginal cost curves.

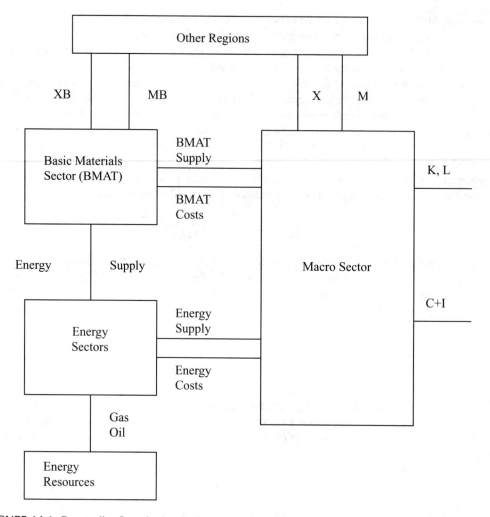

FIGURE 14.1 Commodity flows in the single-region submodel

Within each single-period model, the following classes of constraints apply:

International markets apply to macro output, crude oil, and basic intermediate materials. There are import and export quotas for oil.

Regional markets apply for primary factors (labor and capital), primary energy supply (oil and natural gas), and secondary energy supplies (electric and non-electric).

Energy sectors submodels describe current and future sources of energy supplies in different regions. Constant cost and carbon coefficients and upper and lower bounds apply to output of all technologies. The rate of introduction of new technologies is limited by marginal costs, which rise steeply as production levels exceed the baseline introduction rate.

Low- and high-cost oil and gas supplies arise from a constant ratio depletion model.[6] The extraction profile for low-cost supply is exogenous. The initiation date for tapping high-cost supplies is endogenous but the subsequent production profile is exogenous.

---

[6] The energy submodel corresponds to the ETA model in Global 2100. The electric production technologies include hydro-electricity, remaining coal-fired plants, remaining nuclear plants, new-vintage gas-fired generation, new-vintage coal-fired plants, and "advanced technology electricity." The non-electric supply sectors include coal for direct use, natural gas used for non-electric purposes, synthetic fuels, renewable non-electric power, and low-cost and high-cost oil and gas.

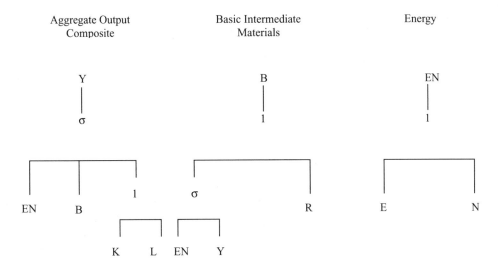

FIGURE 14.2 Nested CES production structures, σ, elasticity of substitution for energy, non-energy, and basic materials inputs to production

Nested separable constant elasticity production functions are employed to characterize substitution possibilities in the production of basic materials and aggregate output. The basic structure of these functions is illustrated in Figure 14.2.

Both Y and B sectors use the same secondary energy composite, in which there is a unitary elasticity of substitution between electric and non-electric energy. In basic materials, the energy composite trades off with inputs of aggregate output according to a constant elasticity of substitution. The macro production function features a constant elasticity of substitution between the secondary energy composite, basic materials, and a Cobb–Douglas composite of primary factors (labor and capital). Finally, a distinction is made between new vintage and extant vintage production. The composition of inputs in both extant macro and basis material production function in 1990 is fixed for successive periods.

The model has been implemented using GAMS for model generation, updating, and report writing and MPS/GE for model solution.[7] In the single-period submodel, a simultaneous system of 300 non-linear inequalities is solved. One simulation of the 11-period horizon requires roughly 15 minutes with a 25-MHz 80486 microcomputer.

C. LEAKAGE AND THE MARGINAL COST OF ABATEMENT

In this section, we present an analytical framework in which to illustrate how the marginal leakage rate affects the optimal level of emissions for a region undertaking unilateral action. We begin with a static model in which there are two final goods and a single representative agent. Let X represent environmental quality, and let Y represent a composite of other commodities. In this stylized economy, the representative agent chooses a level of carbon emissions C which solves

$$\text{Maximize } W(X,Y) \quad \text{s.t.} \quad X = f(C), \quad Y = g(C)$$

---

[7] GAMS (general algebraic modeling system) is described in Brooke et al. (1988). MPS/GE stands for mathematical programming system for general equilibrium analysis (see Rutherford, 1989). The interface between these modeling environments is described in Rutherford (1992a).

where function f captures the relationship between carbon emissions and environmental quality as emission rise, environmental quality declines ($f'(C) < 0 \ \forall \ C$) and function g captures the relationship between emissions and consumption possibilities for other goods ($g'(C) > 0 \ \forall \ C$).

The first-order conditions for this single-agent problem are

$$-\frac{\partial W}{\partial X}\frac{df}{dC} = \frac{\partial W}{\partial Y}\frac{dg}{dC}.$$

This equation can be interpreted as equating marginal benefits of carbon emissions with the marginal cost. Marginal benefits, on the left-hand side, are the product of a term representing the marginal utility of environmental quality and a term representing the extent to which carbon emissions affect environmental quality (recall that $f' < 0$). On the right-hand side of this equation we have the product of a term representing the marginal utility of consumption for goods and a term which represents the extent to which increases in carbon emissions make possible additional units of other goods ($g' > 0$). In an intertemporal framework, this model extends into an optimal control setting. The algebraic relations are somewhat more complicated, but the optimality conditions are effectively unchanged. At any point in time, the marginal benefits (present and future) of increased emissions are balanced with the present and future costs.

Now consider a model in which there are two regions, denoted n and s. Suppose that total emissions are given by $C = C_n + C_s(C_n)$. That is, global emissions are the sum of both regions' emissions. We consider the optimal level of emissions for the first region, n, taking the resulting level of emissions by the second region, s, as a function of C_n. Due to international energy market effects, we expect that $dC_s/dC_n < 0$. (As region n's fossil carbon emissions rise, this increases the international price of oil and, ceteris paribus, causes C_s to decline.)

Suppose that region n seeks to maximize its own welfare, taking into account the response of region s. That is, region n solves

$$\text{maximize } W_n(X,Y) \quad \text{s.t.} \quad X = f\big[C_n + C_s(C_n)\big], \qquad Y = g(C_n).$$

For this model, the first-order conditions read

$$-\frac{\partial W}{\partial X}\frac{df}{dC}\left(1 + \frac{dC_s}{dC_n}\right) = \frac{\partial W}{\partial Y}\frac{dg}{dC_n}.$$

In this model we interpret $l = -dC_s/dC_n$ as the *marginal* leakage rate, l. When leakage is zero, we see that the optimality condition for region n acting unilaterally is the same as that for the single-agent model. The optimal level of abatement (when $l = 0$) equates benefits ($b = (\partial W/\partial X)(df/dC)$) with marginal cost ($c = (\partial W/\partial Y)(dg/dC_n)$). If, on the other hand, the leakage rate is positive, then the first-order condition can be interpreted as equating marginal benefits with marginal cost *adjusted for leakage*; i.e.,

$$b = \frac{c}{1-l}.$$

For example, if the marginal leakage rate is 50%, then a marginal cost calculation based on a single-region model underestimates the true marginal cost of global abatement through unilateral action by a factor of 2.

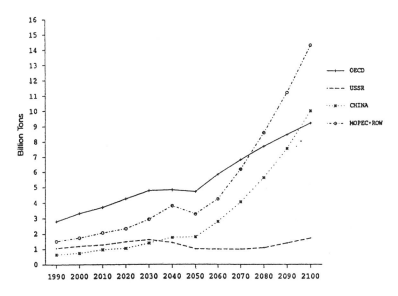

FIGURE 14.3 Total carbon emissions for BAU

We conclude by simply pointing out that marginal and average leakage rates are quite distinct. We define the average leakage rate based on a reference level of emissions, $\bar{c}_n$ and $\bar{c}_s$. For a given level of emissions by n, c_n, we have

$$l^a = \frac{c_s - \bar{c}_s}{\bar{c}_n - c_n} \, .$$

In our numerical model, we find that it is common that the average leakage rate may be small while that marginal rate is quite large.

D. BUSINESS AS USUAL SCENARIO

The business as usual (BAU) assumptions are based on Energy Modeling Forum (EMF 12) study guidelines (Energy Modeling Forum, 1991). The reference case assumes that world population will double and world GDP will increase by more than ninefold over the next century. In our BAU calculations, the growth factors lead to a rise in carbon emissions from 6 billion tons in 1990 to roughly 36 billions tons in 2100. Figure 3 presents projections for carbon emissions on a regional basis. In all regions but the USSR, which has large natural gas reserves, carbon accumulation accelerates after 2050 when oil and gas are replaced by coal-based synthetic fuels. The rate of emissions growth is lower in the OECD, where labor growth rates projections are smaller. The non-OECD regions increase carbon emissions dramatically as their per capita income levels approach current levels in the OECD. These projections underscore the importance of having a comprehensive global agreement on carbon reduction.

The slight decline in emissions in 2030 and 2040 (in Figure 14.3) is explained by the oil price paths which are shown in Figure 14.4. Oil prices overshoot their backstop price in 2030 and 2040 in all regions but the USSR. In the Global 2100 dataset, China and the USSR are subject to oil export limits which in the BAU scenario cause the domestic prices in these regions to fall below the prices in the other regions. The oil prices are low in the early decades due to high oil production. Consequently, there is no production of synthetic fuels in this period. Expansion constraints on synfuels production and an early exhaustion of resources cause oil prices to overshoot in 2030

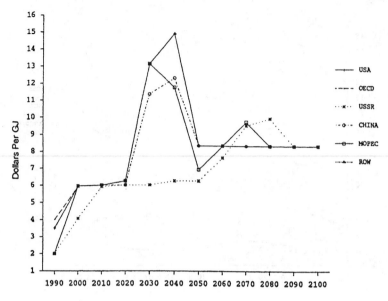

FIGURE 14.4 Oil prices for BAU

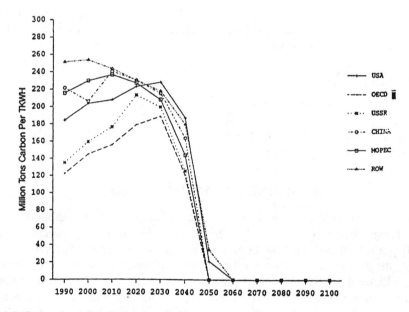

FIGURE 14.5 Carbon intensity of electric energy

and 2040. In the USSR and MOPEC, the introduction of backstop technologies occurs later, resulting in overshooting oil prices for a second time. Note that the backstop price of oil equals the cost of synthetic fuels, $8.33 per GJ.

Figures 14.5 and 14.6 provide some insights into the nature of the carbon emissions through the next century. Figure 14.5 shows the carbon intensity of electric energy consumption and Figure 14.6 the carbon intensity of non-electric energy consumption. In 1990, the carbon intensity of electric energy is lower in the OECD and the USSR, reflecting the relatively larger nuclear share in those regions. The carbon intensity is higher in regions such as ROW and China, where coal has a higher share of the electric capacity. After 2030, the carbon intensity of electric power drops off sharply with the introduction of cost-effective carbon-free electric technologies (solar or nuclear power).

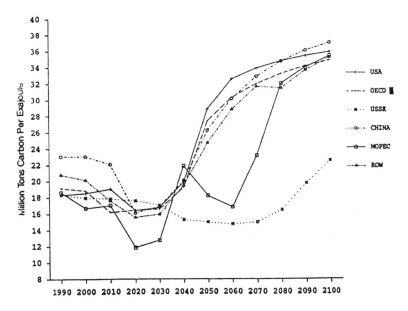

FIGURE 14.6 Carbon intensity of non-electric energy

The 1990 carbon intensity of non-electric energy is higher in China due to the larger share of coal direct use as compared with the other four regions. In all regions except the USSR, carbon intensity of non-electric energy rises sharply after 2040 as oil and gas supplies are exhausted and synthetic fuels are introduced. The USSR carbon coefficient remains lower than those of other regions due to the persistence of natural gas supplies.

E. THE SCOPE OF UNILATERAL ACTION

The counterfactual cases study the impacts of unilateral OECD carbon emission restrictions in two alternative model structures. First we analyze OECD cutbacks ranging from 1 to 4% per annum (pa) from the 1990 level. In an alternative experiment, we fix OECD BMAT trade at the BAU levels throughout the model horizon and study its impact on global carbon emissions. In all counterfactual cases, we assume that emission rights are freely traded among the OECD countries.

Figure 14.7 displays OECD carbon taxes as a function of unilateral cutbacks. Severe (4%) reductions result in an early introduction date for high-cost supplies. Consequently, tax rates are high and relatively stable. For less restrictive cutbacks, tax rates jump on a higher level when tapping of high-cost natural gas reserves occurs. After that, the availability of larger quantities of natural gas softens the impact of the constraints.

Figure 14.8 presents marginal leakage rates for emission cutbacks between 1 and 4% per annum. One percent carbon reductions produce significantly positive leakage rates from 2020 onward. Marginal leakage peaks in 2030 to 2040 at roughly 45% and decreases slowly over the rest of the century. Marginal leakage rates equal to 45% translate to a premium of 82% on the direct cost of carbon restrictions expressed in the revenues of the carbon tax. Two percent and higher cutbacks increase marginal leakage rates substantially in the first three decades but result in negative marginal rates in the second half of the century. This happens because OECD carbon reductions result in lower levels of fossil fuels consumption and delayed extractions of relatively low carbon oil and gas supplies in the region ROW. As a result, there is significantly lower output of carbon-intensive synthetic fuels in the years 2050 to 2080 when the OECD drastically reduces carbon emissions.

Figure 14.9 combines the information of the last two figures and presents the welfare costs of unilateral reductions of carbon emission by the OECD. The costs are calculated according to the

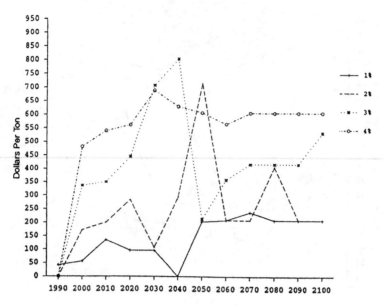

FIGURE 14.7 Carbon taxes in the OECD

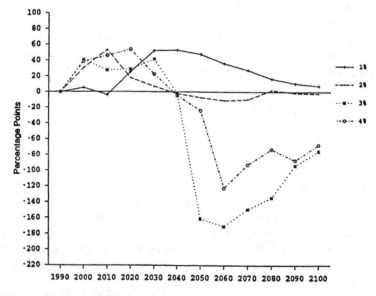

FIGURE 14.8 Marginal leakage rates

equation $\partial W/\partial c$ displayed in Section C. The figure indicates the costs as percentages of OECD GDP. For the first half of the century, we observe that rising OECD emission reductions produce increasing welfare costs. Drastic (4%) cuts produce welfare costs in the range of 3 to 9% of GDP over this period. In the second half of the century, welfare costs are low and decrease with a rise in OECD carbon cutbacks. This happens because of the marginal leakage rates figures which are explained above.

Figure 14.10 displays a regional decomposition of the global average leakage rate for a 2% pa unilateral emission cut in the OECD. Global leakage rates rise in the early decades, peak in 2030 to 2040 at 35%, and steadily decline over the remaining period. Carbon taxes in the OECD reduce aggregate world demand for oil and put a downward pressure on world energy prices. As a

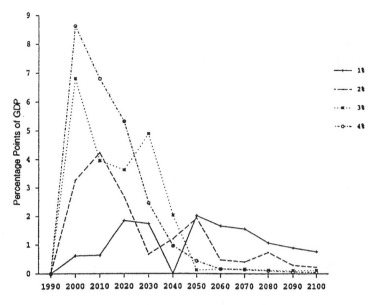

FIGURE 14.9 Welfare costs of unilateral action

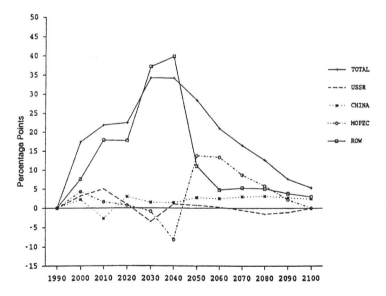

FIGURE 14.10 Global and regional average leakage rates — 2% cutbacks

consequence, ROW increases its oil imports and delays the extraction of high-cost energy resources and the introduction of backstop technologies. This leads to high leakage rates in the transition to the use of new vintage capacities. Interestingly, for MOPEC we do not observe a delayed tapping of high-cost resources in the counterfactual cases. Instead we see that MOPEC reduces reserve accumulation of high-cost supplies in 2010 to 2030. Consequently, there is a larger supply of natural gas in the following decades, and hence less demand for synfuel production. This explains the negative leakage rates for MOPEC in 2040 as well as the subsequent jump in the leakage rate. Leakage rates in China and the USSR are very small. This is because their energy markets are virtually decoupled from the world market due to export and import quotas. For more severe unilateral cutbacks by the OECD the pattern of regional contributions to global leakage remains virtually unchanged. It is ROW which determines predominantly the size of global leakage rates.

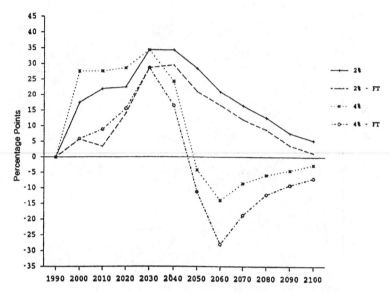

FIGURE 14.11 Leakage rates for alternative model structures

Leakage can arise from two channels. The first is leakage through trade, which occurs when the production of carbon-intensive goods migrates from carbon-constrained to unconstrained locations. The second source of leakage is price-induced substitution, which leads to an increased carbon intensity of production outside the OECD. A decomposition of these two effects is displayed in Figure 14.11. This figure shows the average leakage rates which result from two models. The first is the standard model, with a 2 and 4% emission cut in the OECD. The second model applies an alternative trade structure, which maintains (by export subsidies) OECD trade in basic materials at exactly the baseline levels throughout the model horizon. In the alternative model structure, carbon leakage arises solely through a price-induced substitution effect. The curves show that both types of leakage are important. This result implies that trade policies designed to prevent migration of basic material production from carbon-constrained to -unconstrained regions do have a negative effect on global carbon emissions. However, the policies must include a higher carbon tax in order to reduce carbon consumption in the sectors which do not produce basic materials.

F. SUMMARY AND CONCLUSION

In this study, we have investigated the implications of international trade linkages in determining the economic effects of unilateral CO_2 emission restrictions. We have analyzed these issues in the context of a recursively dynamic general equilibrium model. The insights which emerge from the simulations are as follows:

1. The time profiles of carbon emissions from 1990 to 2100 show that electric energy is relatively more carbon intensive than non-electric energy up to the time that a carbon-free electric energy becomes cost effective. The trend in carbon intensity of non-electric energy is the reverse. Through the period when traditional fossil fuels are dominant, the carbon coefficient for non-electric energy is relatively low. There is a sharp increase in carbon content of primary energy when coal-based synthetics are used to replace oil and natural gas.
2. Over the period 2000 to 2050, marginal leakage rates are positive and increase with carbon emission cutbacks in the OECD. Severe (3 to 4%) cuts produce marginal leakage rates in the range of 40%. Over the period 2060 to 2100, leakage rates are negative for cuts greater

than 1%. This happens because carbon taxes in the OECD delay tapping of high-cost energy supplies and the introduction of carbon-intensive production of synthetic fuels.

3. The welfare costs of unilateral OECD carbon abatement are substantial for the first half of the century. For this period, welfare costs range between 2 and 8% of GDP for emission cuts between 2 and 4% per annum. Due to negative marginal leakage rates, welfare costs in the second half of the century are below 1% of GDP.

4. A regional decomposition of leakage reveals that change in carbon emissions in the ROW is the primary factor for leakage. This increase results from an economy-wide increase in energy intensity as well as from a shift toward the production of carbon-intensive goods.

5. Restrictions on trade in basic materials produce lower leakage rates. In this sense, OECD trade policy can be used to affect emissions in other regions, but this instrument has only limited effectiveness. Trade restrictions to reduce carbon emissions are a poor substitute for a comprehensive global agreement.

These results should be viewed as purely exploratory, given the high degree of uncertainly surrounding the key parameters describing technologies and preferences. Finally, this chapter did not consider the benefits of reduced greenhouse gases for current and future generations. The benefits of carbon emission reductions are much more uncertain than the costs. This notwithstanding, it would be interesting to undertake further research along the line of Nordhaus (1991), who incorporates estimated damage curves from global warming.

REFERENCES

Brooke, A., Kendrick, D., and Meeraus, A., *GAMS: A User's Guide*, Scientific Press, San Francisco, 1988.

Dean, A. and Hoeller, P., Costs of reducing CO$_2$ emissions: evidence from six global models, OECD working paper 122, OECD Secretariat, Paris, 1992.

Energy Modeling Forum, *Study Design for EMF 12 Global Climate Change: Energy Sector Impacts of Greenhouse Gas Emission Control Strategies*, Terman Engineering Center, Stanford University, Stanford, 1991.

Manne, A. S. and Richels, R. G., Global CO$_2$ emission reductions: the impact of rising energy costs, *Energy J.*, 12, 87, 1990.

Manne, A. S. and Richels, R. G., *Buying Greenhouse Insurance — The Economic Costs of CO$_2$ Emission Limits*, MIT Press, Cambridge, 1992.

Nordhaus, W. D., To slow or not to slow: the economics of the greenhouse effect, *Econom. J.*, 101, 920, 1991.

Perroni, C. and Rutherford, T. F., International trade in carbon emission rights and basic materials: general equilibrium calculations for 2020, *Scand. J. Econom.*, 95, 257–278, 1993.

Rutherford, T. F., General Equilibrium Modeling With MPS/GE, mimeo, University of Western Ontario, 1989.

Rutherford, T. F., Applied general equilibrium modeling using MPS/GE as a GAMS subsystem, working paper 92–15, University of Colorado, 1992a.

Rutherford, T. F., The welfare effects of fossil carbon restrictions: results from a recursively dynamic trade model, OECD working paper 112, OECD Secretariat, Paris, 1992b.

Subramanian, A., Trade measures for environment: a nearly empty box?, *World Economy*, 135, 1992.

15 International Trade and Environmental Quality: How Important Are the Linkages?

Carlo Perroni and Randall M. Wigle

This chapter develops a numerical general equilibrium model of the world economy with local and global environmental externalities. The model is then used to investigate the relationship between trade and the environment. Our results suggest that international trade has little impact on environmental quality. Furthermore, the magnitude of the welfare effects of environmental policies is not significantly affected by changes in trade policies. At the same time, the size and distribution of the gains from trade liberalization appear to be little affected by changes in environmental policies.

A. INTRODUCTION

The last few years have seen a remarkable increase in environmental concerns, both in the public at large and among policy makers. One particular focus of these concerns is the interrelationship between international trade and the environment. A widely held view is that international trade is a primary contributor to environmental degradation (Shrybman, 1990). Trade liberalization has also come under criticism on the grounds that domestic environmental policies can be significantly undercut through trade diversion and relocation of dirty industries to "pollution havens," particularly when governments are restricted by trade treaties.[1]

The implications of the environmental externalities for the theory of comparative advantage have been extensively explored in the literature. Environmental emissions affect the pattern of comparative advantage: countries with comparatively higher levels of environmental assimilative capacity, or with comparatively lower valuations of environmental quality have a comparative advantage in the production of pollution-intensive goods (Sierbert, 1977; Blackhurst, 1977). Less than full internalization of environmental externalities may have implications both for equity (if the victims of the externalities are not compensated) and for efficiency (emission levels will be above optimal). With respect to international trade, the lack of internalization distorts the pattern of comparative advantage, resulting in non-optimal levels of trade. In addition to these efficiency effects, the lack of internalization affects the distribution of the gains from trade. In this situation, a country might find autarky to be preferable to free trade (Pethig, 1976).

The thrust of the literature on trade and the environment is mainly theoretical, and few of the results are even qualitatively robust. Nevertheless, there appears to be some consensus among economists that trade barriers are a very poor substitute for domestic environmental policies (Bhagwati, 1981a), and that trade liberalization and environmental protection may be treated as

---

[1] Specific GATT rules have come under attack, especially those concerning subsidies and "National Treatment" (see Whalley, 1991).

separable objectives, but to our knowledge no previous study has attempted to provide quantitative support for this conjecture.

This chapter represents an attempt to assess the relative contribution of international trade to environmental degradation, and the extent to which environmental policies affect the size and distribution of the gains from trade liberalization. To this end we formulate an operational general equilibrium model encompassing international trade and environmental externalities, which is calibrated to a consistent world data set for 1986. We employ the model to explore the quantitative significance of the links between international trade and the environment.

The main obstacle to performing such an exercise is simply our limited information on how economic activities affect the natural environment, and on how individuals value environmental quality. Although there is a large literature on the valuation of environmental damages, there are many parameters we use for our model for which there are no estimates. In this respect our results are of necessity speculative. Nevertheless, we feel that the urgency of the debate on trade and environment warrants a first attempt to assess the importance of these linkages.

Given our parameterization, numerical simulations indicate that although trade liberalization appears to have a discernible impact on the environment, its relative contribution to environmental degradation is very limited, and that this impact is likely to be virtually zero when appropriate environmental policies are in place. Furthermore, we find that environmental policies and trade policies are not as interdependent as some appear to believe. Specifically, environmental policy seems to have little effect on the size and distribution of the gains from trade liberalization. Similarly, we find that trade policy has a very limited impact on the welfare effects of environmental policies.

The plan of the chapter is as follows: Section B outlines the general methodology employed to model environmental externalities, while Section C details the structure of our model. Section D describes the data set and the calibration procedure. Our experiments and numerical results are presented in Section E.

B. ENVIRONMENTAL EXTERNALITIES IN THE GLOBAL ECONOMY

The processes that link environmental externalities to both the damage caused by them and the costs of reducing them can be grouped into four categories.

- *Institutional processes.* While there is normally a market mechanism to mediate transactions in private goods, there may be no private market for environmental quality. In the absence of a market or some other corrective policy, polluters do not pay for their "use" of environmental resources.
- *Technological processes.* Production activities generate emissions that reduce environmental quality. These emissions can be reduced by switching to cleaner technologies or through the cleaning up of existing technologies ("end-of-pipe treatment"). Just as there are different technical processes for production, there are alternative processes of abatement.
- *Natural and geographical processes.* The impact of emissions on environmental quality varies across regions, depending on the nature of the emissions (local/transnational/global) and on the assimilative capacity of the environment. The later will depend on the geographical pattern of emissions and on the local characteristics of the natural environment.
- *Subjective processes.* The value placed by consumers on environmental quality depends on consumers' preferences and there is a strong presumption among economists that the willingness to pay for environmental quality is positively correlated with lifetime wealth.[2]

---

[2] There is limited evidence on the matter. Braden and Kolstad (1991) give a good discussion of the associated problems. The general conclusion of Radetzki (1991) is consistent with the view that valuation rises with income.

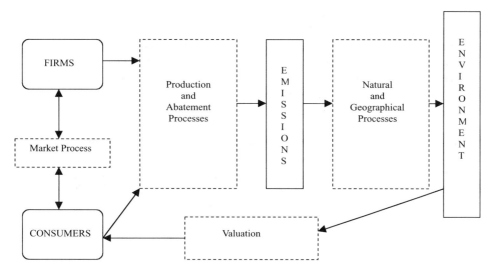

FIGURE 15.1 Economy-environment linkages

The connections between these processes, the economy's agents, and the environment are summarized in Figure 15.1.

These categories may become quite blurred in practice. For instance, we may observe that the valuation of environmental externalities appears to be higher in rich countries than it is in poor ones, but it is hard to tell whether this effect is mainly due to differences in assimilative capacity or differences in income levels. Also, the distinction between technological and physical processes is largely conventional and depends on how we define the pollutant being emitted. Thus, unless detailed information on the various aspects of environmental externalities is available, one may be tempted to abandon some of the above distinctions. In our view, however, maintaining these distinctions is important *precisely because of* the lack of detailed information: a careful categorization provides structure to the analysis and order to the policy debate. Also, maintaining such distinctions is essential for an interdisciplinary approach to the problem of environmental externalities.

The fact that these distinctions are often subtle means that they are often ignored in quantitative models; the need for preserving these distinctions in a model with fully integrated international trade spurred our modeling efforts. In order to explore the relationship between trade and the environment, in the next section we develop a model of the global economy that attempts explicitly to include all of the above aspects of environmental externalities.

C. MODEL STRUCTURE

This section describes the structure of our model, with a special emphasis on the modeling of environmental externalities. A complete description of the model is available from the authors by request.

1. EMISSIONS AND ABATEMENT

As mentioned above, we distinguish between externalities arising from production and from consumption activities, as well as between local and global externalities.[3] Gross environmental emissions, in the absence of any regulation, represent a fixed proportion of activity levels. To illustrate,

---

[3] There is nothing in the model structure to prevent us from having "regional" externalities that affect only geographically close countries. Since our country aggregation is not based on geography, such a categorization becomes meaningless.

let us denote by m^L and m^G, respectively, the local and global emission coefficients of a given production or consumption sector.[4] Then the sector's gross emissions are simply equal to $m^L X$ and $m^G X$, where X represents the level of output or consumption. Net emissions depend on gross emissions and on the degree of emission abatement. Denoting by A^L and A^G, respectively, the proportional abatement of local and global emissions by this sector, its net emissions can be expressed as $m^L X(1 - A^L)$ and $m^G X(1 - A^G)$.

Abatement activities are sector specific. We assume that there exists some degree of output complementarity in the "production" of abatement of local and global emissions. This is achieved by postulating a constant-elasticity transformation (CET) function linking "aggregate abatement", A, with the proportional abatement levels for local and global emissions with elasticity of transformation τ between the two types of abatement. Unit abatement costs are assumed to be independent of production levels; as sectors increase the level of abatement as a proportion of their emissions (i.e., as they decrease their per-unit emissions), however, unit emission abatement costs rise (Ballard and Medema, 1988). We represent this cost structure by means of a constant-elasticity unit cost function of the form

$$\frac{F(\vec{p}, A)}{A} = \frac{f(\vec{p})}{1 + \eta} A^\eta, \tag{15.1}$$

where $F(\vec{p}, A)$ is the cost of abating a given proportion, A, of emissions for each unit of output or consumption, $\vec{p}$ is a vector of input prices, and η (which is positive) is the elasticity of unit abatement costs with respect to A. Marginal abatement costs as a function of A are then

$$f(\vec{p}) A^\eta. \tag{15.2}$$

The government levies unit local and global emission fees. The resulting emission charges, $z^j L$ and $z^j G$, paid on each physical unit of the "unabated" portion of output or consumption, are equal to the emission fee times the amount of emissions per unit of output or consumption (i.e., the emission coefficient). For given emission fees, producers and consumers select cost-minimizing abatement levels for which marginal abatement costs equal private marginal abatement benefits. The latter are equal to a CET aggregation of unit emission charges. Notice that we assume abatement technologies to be identical in all regions, which implies that differences in abatement choices across regions arise exclusively from differences in emission charges or in factor prices.[5]

2. Damage and Valuation

Total local emissions experienced by a certain region, E^L, are the sum of net local emissions by all sectors in that region. Total global emissions experienced by a region, E^G, are the sum of net global emissions by all sectors in all regions. The relationship between emissions experienced and environmental damage D is modelled by means of convex, constant-elasticity damage functions:

$$D^L(E^L) = k^L(E^L)^{\rho^L} \tag{15.3}$$

$$D^G(E^G) = k^G(E^G)^{\rho^G}, \tag{15.4}$$

---

[4] By consumption sector, we mean merely the domestic consumption of the good in question.
[5] We assume that no region is inherently "dirtier" than others.

where k^L and k^G are constants, and ρ^L and ρ^G represent elasticities of damage with respect to emissions (assumed to be greater than unity) (Dewees, 1992). These elasticities are likely to be different for different types of emissions and to vary across regions to reflect differences in assimilative capacity. Further discussion of the parameters of the damage functions follows below.

We also distinguish between a local dimension and a global dimension of environmental quality on the consumption side of the economy. Environmental quality is equal to the difference between endowments of environmental quality and damage:

$$Q^L = \overline{Q}^L - D^L\!\left(E^L\right) \tag{15.5}$$

$$Q^G = \overline{Q}^G - D^G\!\left(E^G\right) \tag{15.6}$$

It follows that the marginal valuation of emission experienced by a region is simply equal to

$$p_E^L = N p_Q^L \frac{dD^L}{dE^L} \tag{15.7}$$

$$p_E^G = N p_Q^G \frac{dD^G}{dE^G} \tag{15.8}$$

where N is population, and p_Q^L and p_Q^G are individual marginal valuations for local and global environmental quality, respectively.

The individual valuations of environmental quality in turn depend on the level of environmental quality and on per capita consumption of goods. We model this relationship via a utility function defined over consumption goods and environmental quality: First we aggregate the two types of environmental quality by means of a constant-elasticity of substitution (CES) function with an elasticity of substitution of σ. Combined environmental quality then enters a CES utility function together with per capita consumption, with an elasticity of substitution equal to γ.[6] It can be easily shown that the reciprocal of γ equals the income elasticity of environmental quality valuation, ω.[7] Also notice that the degree of responsiveness of the marginal valuation of environmental quality to an increase in damage depends on the size of environmental quality endowments. Thus, we can choose these endowments so as to reflect an exogenously selected value for the elasticity of marginal valuation with respect to damage, ς.

Because of the static nature of the model, we implicitly assume that environmental quality is a renewable resource. If all pollutants were "fund" pollutants (i.e., if the environment has some assimilative capacity for them) this inaccuracy may not be too serious. For "stock" pollutants (those for which the assimilative capacity is very small or zero), this assumption leads to an underestimation of the social cost of environmental emissions.[8]

3. Environmental Policies

Environmental controls are modelled as emission fees that force the emitting party to internalize part or all of the external cost. For local emissions generated by a sector in region j, emission charges per unit of unabated output or consumption are

---

[6] We assume strong separability between consumption and environmental quality, which enables us to treat environmental quality as a single consumption aggregate (Foster, 1981).

[7] Notice that since individuals are rationed in their consumption of environmental quality, and since observable income does not include the value of environmental quality consumed, it is possible to obtain a positive income elasticity of the marginal valuation of environmental quality from a homothetic formulation of preferences (Lucas et al., 1992).

[8] The terms "stock" and "fund" pollutants are suggested in Tietenberg (1991).

$$z^j L = \pi_D^j p_E^{jL} m^L ,\tag{15.9}$$

where π_D^j represents the rate at which the emitter is forced to internalize the externality. For global emissions, charges are

$$z^j G = \pi_D^j p_E^{jG} m^G + \pi_X^j \sum_{r \neq j} p_E^{rG} m^G ,\tag{15.10}$$

where π_D^j is an internalization rate for the domestic component of the externality, and π_X^j is the rate of internalization for international spillovers. Even when π_X^j is greater than zero, we assume that the revenue of emission charges is transferred not to the affected parties but to the residents of the country where the emissions originate. In our model structure this approach, unlike real-world environmental regulatory schemes is inherently efficient, though efficient, though not necessarily equitable.

4. KEY PARAMETERS

The respective shapes of the marginal cost and benefit functions for emission abatement are crucial determinants of optimal abatement choices and their responsiveness to policy changes. This sub-section summarizes how the key parameters and variables in our environmental submodel affect the shape of these functions.

The level of marginal abatement costs depends on factor prices and is increasing in the proportion of emissions abated. The marginal cost schedule is steeper the larger is the elasticity of unit abatement costs with respect to proportional abatement, η. This relation merely reflects the fact that a given incremental reduction in emissions is easier to achieve when emissions are high than when almost all emissions have been eliminated.

The marginal benefits of abatement depend on internalization rates, population size, and the marginal valuation of environmental quality, which in turn depends directly on per capita income and inversely on the level of environmental quality. The marginal benefit schedule is decreasing in the level of abatement, since an increase in environmental quality results in a reduced marginal valuation of environmental quality and thus in lower emission charges for a given rate of internalization. The marginal benefit schedule is steeper the larger are the elasticity of environmental damage with respect to emissions, ρ, and the elasticity of damage valuation with respect to damage, ς.

5. PRODUCTION AND TRADE

The environmental submodel described above is embedded in a multisectoral-multiregional trade model with non-homogeneous traded goods. Primary factors, labor and capital, are domestically mobile but internationally immobile, and factor endowments differ across regions. Because of capital immobility, and the static nature of the model, dirty industries are prevented from relocating to regions with low emission charges, a possibility that some authors indicate as a primary source of environmental "leakage" (Lucas et al., 1992). As we shall mention below, we view this as a main limitation to our analysis. Production technologies are identical across regions and are modeled by means of constant returns to scale CES production functions. Preferences are modeled through homothetic nested CES functions, as discussed above. Substitution elasticities are identical across regions, but demand share parameters are region specific. In order to accommodate cross-hauling in trade flows, we adopt an Armington structure: goods of the same type produced in different regions are less than perfect substitutes.[9]

---

[9] From a theoretical point of view an Armington structure is inconsistent with the assumption that technologies are identical across regions. Nevertheless, in a model with aggregated flows, the assumption of imperfect substitutability between imports and domestically produced goods may be justified by the existence of discrepancies in the commodity compositions of trade and production flows.

D. DATA AND PARAMETER CALIBRATION

In this section we describe the basic data used to calibrate the model. The commodities were chosen with the idea of identifying goods by their factor intensities in labor, skills, capital, and environmental inputs. As a result, there are six goods (and sectors) chosen to separate industries with more emissions from those with less and also to separate higher-technology industries from lower-technology industries.

Countries were grouped according to their per capita incomes (since it is an important determinant of the valuation of environmental quality) and the relative levels of environmental quality they enjoy. The following abbreviations are used:

NA   United States and Canada (assumed to have relatively low unit emissions and limited assimilative capacity)

OD   Other Developed Countries (assumed to have relatively low unit emissions and limited assimilative capacity)

LI   Low- and Middle-Income Countries (assumed to have relatively high unit emissions and abundant assimilative capacity)

For standard CGE models the calibration procedure is straightforward, but for the environmental portion of our model, calibration is complicated by the inadequacy of data, and specifically by the absence of markets for environmental quality.

Population, trade, demand, and value-added data were used to calibrate the model to 1986.[10]

Sectoral abatement costs as a percentage of total costs in NA were obtained from Lucas et al. (1992) and Low (1992b). Since data on abatement costs for consumption emissions were not available to us, we simply assumed that these costs are one-half of the abatement costs in the corresponding production sectors. For this specification, total abatement costs represent less than 1 percent of GNP, which is in line with typical estimates (Low, 1992a).

Although the choice of a reference abatement level is somewhat arbitrary, if we assume a 50 percent abatement level in conjunction with a value of 1/3 for η (the elasticity of unit abatement costs with respect to proportional abatement), the resulting abatement cost schedules behave consistently with empirical findings.[11]

The parameters of damage functions are specified as follows. For global emissions we assume identical parameters for all regions. With respect to local emissions, we specify different scale coefficients in the various regions. These coefficients, k^{jL}, reflect the degree of dispersion of local emissions in the local environment and are likely to be different for different pollutants and to depend on the size and physical characteristics of the local environment. Since we were not able to find aggregate estimates for these coefficients, as a first approximation they were specified according to the formula $k^{jL} = k^L(N^j)^{-\rho^{jL}}$, where N^j is the population of region j, and k^L is a scaling parameter that is endogenously determined by our calibration procedure. Given this specification, if two regions of different size (and with an identical ρ^L) experienced local emissions that are in proportion to their respective population sizes, then they would experience the same level of local environmental damage. For the elasticities of damage with respect to emissions, we employ educated guesses based on a comparison of emission data with concentration levels for selected air and water pollutants in different regions (World Resources Institute, 1990).

To calculate emission coefficients, we first compute marginal abatement costs, given our choice of η (see above) and for given unit abatement costs. Notice that our choice of functional form implies that marginal abatement cost is simply equal to average cost times $(1 + \eta)$. These marginal

---

[10] Further details are available from the authors by request.

[11] Maloney and Yandle (1984) focus on the price elasticity of demand for emission permits, which, given our specification of abatement costs, can be expressed as $\Omega = A/[\eta(1 - A)]$. For $A = 0.5$ and $\eta = 1/3$ one obtains $\Omega = 3$, which is consistent with their findings (they report values of 3 and above).

abatement costs provide an estimate of the implied emission charges (which equal the marginal benefit of abatement). We then employ an exogenous imputation scheme to obtain separate local and global emission charges.[12] For given domestic internalization rates, and assuming zero internalization rates for international spillovers, we can compute net emissions values per unit of output. Given reference abatement levels, we obtain gross emission values. We then arbitrarily choose units for physical emission so that, in the benchmark equilibrium, the marginal valuation in NA is equal to unity. We thus obtain physical sectoral unit coefficients of gross emissions, which we adopt for all regions.

Clearly, the absolute size of environmental externalities that we obtain through this process is crucially dependent on the choice of benchmark internalization rates: the higher the benchmark rate of internalization, the smaller the implied external effects in the benchmark equilibrium. Since no reliable estimate is available for the value of environmental damages generated by economic activities, any choice of internalization rates must by necessity be purely speculative. In order to reflect this indeterminacy, we choose a base case value of 0.33 for the domestic internalization rate in all regions and subsequently perform sensitivity analysis around this central value. Notice that assuming identical internalization rates for all regions does *not* contradict the idea that environmental controls are tighter in high-income countries, since, for a given internalization rate, emission charges and abatement levels will be higher in high-income regions, where the marginal valuation for environmental quality is higher.[13]

For the income elasticity of the marginal valuation of environmental quality, ω, the elasticity of marginal valuation of environmental quality with respect to environmental damage, ς, and the elasticity of substitution between local and global environmental quality, σ, we select, respectively, central-case values of 1.0, 0.5, and 1. Although the available evidence on the parameters characterizing preferences for environmental quality is rather scant,[14] there seems to be an agreement that, given a certain level of environmental quality, its marginal valuation would be substantially higher in rich countries than in low- and middle-income countries. Given our choice of parameters, we obtain equilibrium values for the social marginal valuation of environmental quality in different regions that do not conflict with our prior beliefs: 1.0 for NA, 0.63 for OD, and 0.12 for LI.

For all production and abatement functions, as well as for the aggregation of domestic demand, we employ a Cobb–Douglas specification, which implies a unitary elasticity of substitution. Finally, for the elasticity of substitution between domestically produced goods and imports we use a value of 2.5, while the elasticity of substitution between imports of different origin is made equal to 5 (see discussion in Jomini et al., 1991).

When the model is calibrated, a direct computation of benchmark emissions and abatement levels in all regions would require a previous knowledge of the values of the marginal valuations of environmental quality in all regions, which in turn depend on net emissions. This requirement forces us to resort to a simultaneous calibration procedure: all parameters are endogenized, and the model is augmented by a set of calibration restrictions. This augmented model, together with our assumed data and parameters, is sufficient to identify all the parameters of the model. The final step of our calibration procedure consists of solving the augmented model numerically to obtain a set of parameters that is consistent with our data set.[15] The calibration procedure is detailed in model documentation available from the authors.

---

[12] For production emissions, we assume that local emissions charges account for 80 percent of total emission charges. For consumption charges, 100 percent of emission charges are assumed to account for local emission charges in all sectors except EH, where this share is reduced to 80 percent (to reflect the presence of taxes on consumption of fossil fuels).

[13] This assumption, however, contradicts the observation that internalization rates appear to differ across countries by type of emission. For example, taxes on fossil fuels in the EEC are substantially higher than they are in North America, suggesting that the internalization rate for global emissions is higher in the EEC than in North America.

[14] "Economists are as far now from placing a single dollar value on fishable water, unimpaired scenery, or other environmental phenomena as they were two decades ago, perhaps farther." (Braden and Kolstad, 1991, 324).

[15] The model is conceived and solved using MPS/GE (Rutherford, 1988).

TABLE 15.1
Percent Changes in Combined Environmental Quality — Central Case

Region	NA	OD	LI
Benchmark Internalization			
Benchmark	0.0	0.0	0.0
Free trade	−1.3	−0.9	−0.7
Trade wars	+0.8	+0.5	+0.4
Global Internalization			
Benchmark	+95.9	+68.1	+39.4
Free trade	+95.5	+67.5	+39.5
Trade wars	+96.1	+68.6	+39.5
Unilateral Action by NA			
Benchmark	+45.7	+2.0	+2.3
Free trade	+44.5	+1.2	+1.7
Trade wars	+46.5	+2.5	+2.7

E. SIMULATIONS AND RESULTS

We performed simulations with our model to assess the relative contribution of international trade (including trade policy changes) to environmental degradation and to explore the implications of environmental externalities for the size and distribution of the gains from trade liberalization. Specifically, we explore the effects on environmental quality and welfare of all combinations of the following trade-policy configurations:

1. Business as usual (current levels of environmental protection)
2. A move to full, global internationalization (i.e., $\pi_D^j = 1$ and $\pi_X^j = 1$ in all regions)
3. Unilateral domestic environmental action by NA (i.e., $\pi_D^j = 1$ in NA); and of the following three trade-policy scenarios:
   a. Benchmark trade barriers (benchmark)
   b. A removal of all trade barriers (free trade)
   c. A three-fold increase in trade barriers (trade wars).

The results of our simulations for all cases are summarized in Tables 15.1 and 15.2.[16] The rest of this section is devoted to a discussion of our findings. Some limited sensitivity analysis is presented in subsection E1.

At the benchmark levels of internalization, international trade does appear to have some negative impact on environmental quality. A move to free trade slightly worsens the environment, while an increase in trade barriers has a small but positive impact on the environment. Trade liberalization produces sizable welfare gains only for NA and OD. On the other hand, an increase in trade barriers has a substantial negative welfare impact on all regions.

---

[16] The numbers in these tables should all be interpreted as percentage changes compared with the initial situation (benchmark trade barriers and benchmark internalization in all countries). Given our parameterization, in the benchmark equilibrium a 1 percent increase in net emissions produces approximately a 1 percent decrease in the level of environmental quality in the NA bloc. Admittedly, the figures representing changes in environmental quality cannot be given any cardinal interpretation, since in our model environmental quality is defined only in a conventional fashion; these figures, however, may provide information on the *relative* effects of different policies. Percentage welfare changes are computed as equivalent variations as a proportion of benchmark full income (income plus the value of environment quality enjoyed by consumers).

TABLE 15.2
Percent Welfare Changes — Central Case

Region	NA	OD	LI
Benchmark Internalization			
Benchmark	0.0	0.0	0.0
Free trade	+0.4	+1.3	0.0
Trade wars	−1.1	−2.6	−1.6
Global Internalization			
Benchmark	+1.3	+0.8	+0.8
Free trade	+1.7	+2.2	+0.8
Trade wars	+0.2	−1.8	−0.8
Unilateral Action by NA			
Benchmark	+0.4	+0.1	+0.2
Free trade	+0.8	+1.4	+0.1
Trade wars	−0.7	−2.5	−1.4

Full internalization of all externalities causes marked improvement in environmental quality in all regions. All countries that fully internalize externalities experience an improvement in environmental quality of more than 39 percent, regardless of the trade policy regime. By contrast, the improvement in environmental quality brought about by a move from free trade to a trade war is never more than 2 percent. Full internalization also yields substantial welfare gains for all regions.

The most striking message of the results is that compared with environmental policies, trade policies have very little impact on environmental quality. This finding suggests that the relative contribution of international trade to environmental degradation is quite modest. Notice also that the size and the distribution of the gains from trade liberalization are not substantially affected by the change in environmental policies.

Unilateral domestic internalization by NA causes a marked improvement in NA environmental quality, but the addition of trade restrictions has virtually no added effect on environmental quality. However, trade restrictions still cause a large welfare loss. The idea that appropriate environmental policies, when implemented unilaterally, must be buttressed by trade restrictions is not supported by our results. Even though NA internalizes only the domestic impact of the emissions, such a policy still produces a reduction in global emissions. As a result, the welfare of non-participants rises and their environmental quality improves.

The *qualitative* effect of trade policy on environmental quality is unrelated to the environmental policy regime. Liberalization causes small quality losses, and trade restriction causes small gains in environmental quality. In most cases, environmental quality is highest for all environmental policies when trade is restricted. In the LI region, virtually the same environmental quality is achieved when global internalization is combined with free trade as it is when it is combined with trade wars. In the other cases, the reduction in environmental quality caused by free trade is smallest (sometimes almost zero), however, when appropriate environmental policies are in place. Stated differently, the indirect negative impact of trade liberalization on environmental quality is likely to be smaller the higher the level of internalization of environmental externalities.

1. SENSITIVITY ANALYSIS

Given the acknowledged uncertainty about many of our parameters, we performed some limited sensitivity runs on two experiments (Global Internalization and Trade Wars). The central case elasticities and alternate values used are reported in Table 15.3. The choice of values was based on

TABLE 15.3
Parameter Values Used in Sensitivity Analysis

Parameter	LOW	CC	HIGH
Domestic internalization rate (π_D^i)	0.10	0.33	0.50
MAC w.r.t. degree of abatement (η)	0.25	0.33	0.50
Valuation w.r.t. income (ω)	0.95	1.00	1.05
Valuation w.r.t. damage (ς)	0.25	0.50	0.75
Local-global substitution (σ)	0.85	1.00	1.15

TABLE 15.4
Sensitivity: Global Internalization

Environmental Quality (percent changes)

Region	CC	π_D^i L	π_D^i H	η L	η H	ω L	ω H	ς L	ς H	σ L	σ H	$\pi_D^{NA,OD} = 0.5$ $\pi_D^{LL} = 0.33$
NA	95.9	100.0	77.2	98.5	91.7	97.4	94.2	49.9	132.3	95.9	95.8	81.7
OD	68.1	106.2	54.0	72.5	60.9	64.9	69.8	37.3	95.1	68.4	67.9	63.1
LL	39.4	42.8	38.1	43.0	34.1	32.3	43.9	21.7	54.5	39.9	39.1	50.5

Welfare (percent changes)

Region	CC	π_D^i L	π_D^i H	η L	η H	ω L	ω H	ς L	ς H	σ L	σ H	$\pi_D^{NA,OD} = 0.5$ $\pi_D^{LL} = 0.33$
NA	1.3	5.6	0.6	1.4	1.2	1.0	1.5	1.6	1.1	1.3	1.3	1.0
OD	0.8	5.4	0.3	0.9	0.7	0.6	1.0	1.0	0.7	0.8	0.8	0.7
LL	0.8	2.8	0.5	0.8	0.8	0.8	0.8	0.9	0.7	0.8	0.8	0.6

a number of considerations. First, where we had some evidence (as in the case of the elasticity of marginal abatement cost with respect to the degree of abatement), we used this evidence to choose the midpoint. Where there was no directly applicable evidence, we adopted values that seemed plausible. We also did a run corresponding to the case where the internalization rates in the developed regions are higher (0.5) than the low-middle income region (0.33). This run is reported in the last column of our sensitivity results. (See Tables 15.4 and 15.5.)

The main message from the sensitivity runs is that the main conclusion of our analysis is robust to the range of parameters we tried. In all configurations, trade restrictions have relatively minor effects on environmental quality and cause significant welfare losses to all regions, whereas following appropriate environmental policies significantly improves environmental quality and raises economic welfare.

The analysis also identifies two parameters that are crucial to the analysis. The first is the domestic internalization rate (π_D^i). The lower the benchmark internalization rate, the higher the (welfare and environmental quality) returns to internalization. The other important parameter is the elasticity of marginal valuation with respect to damages. The higher this value is, the higher are the returns to liberalization as well.

2. SUMMARY OF RESULTS

Our numerical findings must be viewed as exploratory for a number of reasons, not least of which is our limited knowledge of precisely how environmental emissions affect the environment. The main qualifications to our results fall into the following categories.

TABLE 15.5
Sensitivity: Trade Wars

Region	CC	Environmental Quality (percent changes)										$\pi_D^{NA,OD} = 0.5$
		π_D^i		η		ω		ς		σ		$\pi_D^{LI} = 0.33$
		L	H	L	H	L	H	L	H	L	H	
NA	0.8	0.8	0.8	0.8	0.8	0.8	0.8	0.6	0.9	0.8	0.8	0.6
OD	0.5	0.9	0.4	0.5	0.5	0.5	0.5	0.4	0.6	0.5	0.5	0.4
LL	0.4	0.4	0.4	0.4	0.4	0.3	0.4	0.3	0.5	0.4	0.4	0.3

Region	CC	Welfare (percent changes)										$\pi_D^{NA,OD} = 0.5$
		π_D^i		η		ω		ς		σ		$\pi_D^{LI} = 0.33$
		L	H	L	H	L	H	L	H	L	H	
NA	−1.1	−1.0	−1.1	−1.1	−1.1	−1.1	−1.1	−1.0	−1.1	−1.1	−1.1	−1.1
OD	−2.6	−2.4	−2.6	−2.6	−2.6	−2.6	−2.6	−2.5	−2.6	−2.6	−2.6	−2.6
LL	−1.6	−1.5	−1.6	−1.6	−1.6	−1.6	−1.6	−1.5	−1.6	−1.6	−1.6	−1.6

- *Parameters.* Our reading did not yield consensus values for a number of key parameters in the model, particularly those describing individual and social valuation of environmental quality, which crucially determine the size of the benefits of internalization.
- *Data.* Data on policy regimes, emissions, and environmental quality are inadequate, particularly for the LI bloc.
- *Model structure.* Because of the static nature of our model, our results do not capture the potential effects of trade and environmental policies on industrial location and on technological change. The model is very highly aggregated.
- *Policy regime.* Both benchmark and counterfactual environmental policies are implemented as emission fees, which are at least cost effective and can be efficient. This regime contrasts with the actual policy regimes in many countries that employ inefficient regulatory schemes.

F. SUMMARY AND CONCLUSION

The model developed here represents an attempt to achieve a comprehensive treatment of local and global environmental externalities in a static framework and of their linkage with international trade. In spite of our efforts carefully to separate the various aspects of environmental externalities, our numerical findings must be viewed as exploratory, owing to our limited knowledge of precisely how environmental emissions impact on the environment. We also realize that two features of the model, perfect competition and capital immobility, are likely to dampen the impact of both trade and environmental policies. Even with these limitations, we believe that the results of our simulations shed some light on some fundamental questions concerning the linkages between environmental and trade policies.

Our results suggest that, although free trade may have a negative impact on environmental quality, its relative contribution to environmental degradation appears to be limited. Further, achieving significant environmental quality improvements via trade policy is likely to be extremely costly. This finding can be attributed to the following circumstances.

1. Trade accounts for a relatively small share of world production.
2. A large share of trade involves trade in environmentally clean goods.
3. Estimates suggest that, even for dirty traded goods, abatement costs account for a relatively modest share of total costs.

These facts contribute to our conclusion that trade restrictions are a very poor substitute for direct environmental protection.

Our experiments also indicate that the welfare gains from environmental protection are not significantly affected by changes in trade policies; at the same time, the size and distribution of the gains from trade liberalization are little affected by the presence of environmental externalities. These findings suggest that trade restrictions are not necessary companions to effective environmental policy, and that well-conceived environmental policies are unlikely to be disastrous in an open economy, as some appear to believe.

Protection of the environment is becoming a dominant concern in the design of economic policies, and our numerical results indeed suggest that the gains from increased environmental action may be very substantial. Nevertheless, the importance of coordinating trade and environmental policies should not be overstated. Even in the absence of international cooperation on environmental policies, the process of trade liberalization may promote a more efficient use of world resources.

REFERENCES

Ballard, C. L. and Medema, S. G., The efficiency effects of pigouvian taxes and subsidies: a computational general equilibrium approach, paper presented at the NBER Applied General Equilibrium Workshop, 1988.

Bhagwati, J., The generalized theory of distortions and welfare, in *International Trade: Selected Readings*, 2nd ed., Bhagwati, J., Ed., MIT Press, Cambridge, 1981a.

Bhagwati, J., *International Trade: Selected Readings*, 2nd ed., MIT Press, Cambridge, 1981b.

Blackhurst, R., International trade and domestic environmental policies in a growing economy, in *International Relations in a Changing World*, Blackhurst, R., et al., Eds., Sijthoff, Leiden, 1977.

Braden, J. B. and Kolstad, C. D., Eds., *Measuring the Demand for Environmental Quality*, Elsevier Science, New York, 1991.

Dewees, D. N., The efficiency of pursuing environmental quality objectives: the shape of damage functions, mimeo, Department of Economics, University of Toronto, 1992.

Forster, B. A., Separability, functional structure and aggregation for a class of models in environmental economics, *J. Environ. Econ. Manage.*, 8, 118, 1981.

Jomini, P., et al., SALTER: A General Equilibrium Model of the World Economy — Volume 1: Model Structure, Database and Parameters, Study undertaken by the Industry Commission for the Department of Foreign Affairs and Trade, 1991.

Low, P., Ed., *International Trade and the Environment*, International Bank for Reconstruction and Development, Washington, D.C., 1992a.

Low, P., Trade measures and environmental quality: the implications for Mexico's exports, in *International Trade and the Environment*, Low, P., Ed., International Bank for Reconstruction and Development, Washington, D.C., 1992b.

Lucas, Robert, E. B., Wheeler, D., and Hettig, H., Economic development, environmental regulation and the international migration of toxic industrial pollution: 1960–88, in *International Trade and the Environment*, Low, P., Ed., International Bank for Reconstruction and Development, Washington, D.C., 1992.

Maloney, M. T. and Yandle, B., Estimation of the cost of air pollution control regulation, *J. Environ. Econ. Manage.*, 11, 244, 1984.

Pethig, R., Pollution, welfare, and environmental policy in the theory of comparative advantage, *J. Environ. Econ. Manage.*, 3, 160, 1976.

Radetzki, M., Economic growth and the environment, in *International Trade and the Environment*, Low, P., Ed., International Bank for Reconstruction and Development, Washington, D.C., 1992.

Rutherford, T. F., General equilibrium modeling with MPS/GE, paper, Department of Economics, University of Western Ontario, 1988.

Shrybman, S., 1990, International trade and the environment: an environmental assessment of the General Agreement on tariffs and trade, *Ecologist*, 20, 30, 1990.

Siebert, H., Environmental quality and the gains from trade, *KYKLOS*, 20, 657, 1977.

Tietenberg, T., *Environmental and Natural Resource Economics*, 3rd ed., Harper Collins, New York, 1991.
Whally, J., The interface between environmental and trade policies, *Econ. J.*, 101, 180, 1991.
World Resource Institute, *World Resour, 1990–1991*, Oxford University Press, London, 1990.

16 Environmental and Trade Policies: Some Methodological Lessons[1]

V. Kerry Smith and J. Andrès Espinosa

This chapter provides an overview of recent research on the interconnections between environmental and trade policies. It describes how the assumptions of these models can be important to the conclusions derived on the gains from coordination. The Harrison–Rutherford–Wooton CGE model of the European Union is used to illustrate the importance of these issues. This model was extended to include three air pollutants and their health effects as non-separable influences on household preferences in each region described by the model. The results suggest that the conventional assumption of separability in preferences between marketed and non-marketed goods is central to conclusions about the importance of coordination of these policies.

A. INTRODUCTION

Trade and the environment has been a front page issue for most of the nineties. Both popular and professional publications have offered lessons about how policies should be modified to harmonize both sets of objectives.[2] This interest has, in turn, stimulated a considerable body of new conceptual research (see, e.g., Krutilla, 1991; Chichilnisky, 1994; Copeland, 1994; Copeland and Taylor, 1994, 1995a; Lopez, 1994). We have learned, as a result of the research to date, that domestic environmental regulations will *either increase or decrease* the international competitiveness of the products traded by the country undertaking them. By contrast, we know that removing trade barriers will *either improve or reduce* the level of well-being of the typical household in the country reducing trade barriers. Finally, while the verdict here is not as clear-cut as in these first two areas, it seems likely that research currently under way will soon establish that efforts to link trade and environmental policies will *either enhance or degrade* the current levels of well-being experienced in developed economies from the levels reached using the current independent approaches to these problems.

While this summary is presented in part to amuse the reader, it is nonetheless a reasonably accurate description of where we stand. Moreover, it is not an indictment of the research to date. Rather, it reflects what is learned from conceptual research in most areas: *the devil is in the details*. That is, the specific circumstances of each country's economy in relation to the world markets relevant for its products must be considered before clear-cut conclusions can be offered. Thus, we can agree that

> The international market transmits and enlarges the externalities of the global commons. No policy that ignores this connection can work (Chichilnisky, 1994, p. 865).

[1] Thanks are due to Glenn Harrison for providing the data and code to implement HRWI as well as for helping us to use it; to an anonymous referee, Edward Barbier and Charles Perrings for most helpful comments on an earlier draft; and to Paula Rubio for preparing and editing several versions of this chapter. Partial support for Smith's research was provided by the U.S. National Oceanic and Atmospheric Administration through grant No. NA46GP0466. Espinosa's research was supported by Resources for the Future. The conclusions presented are those of the authors, not of their affiliated organizations.

[2] For a good summary of this discussion as well as citations of empirical literature bearing on this topic see Repetto (1995).

Nonetheless, this agreement does not prevent policy makers from rejecting most proposals for greater linkage in these policies. Moreover, their caution may well be warranted based on the available empirical evidence. As Repetto (1995) observes, there is no solid empirical evidence that increasing the stringency of environmental regulations necessarily has detrimental effects on trade and investment. Of course, this does not mean that coordination should be ruled out. Rather, it heightens the importance of understanding the features of each case before attempting to offer generalizations about when there will be advantage in linking trade and environmental policies.

Research on these questions is also likely to have payoffs in other areas. Indeed, we will argue that important methodological issues have emerged from the intermediate results of the research to date on trade and the environment. In what follows, we summarize these insights and illustrate their importance with a new CGE model that incorporates local and transboundary externalities. Using Krutilla (1991) and Anderson's (1992) adaptation of the conventional applied welfare framework, Section B illustrates how environmental and trade policies become intertwined. This graphical analysis also provides a platform for our description of how assumptions of recent theoretical and computational models influence these studies' conclusions about environmental and trade policy. Section C describes the Espinosa and Smith (1995) extension to the Harrison et al. (1989) model of most of the economies currently comprising the European Union, to include environmental externalities and how that framework can be used to gauge the importance of several of the simplifying details of earlier research. While most of the empirical detail in this model considers developed economies, the generic lessons derived from examining the implications of incorporating environmental resources within preferences are equally relevant to the developing context. These relationships are discussed in the closing section of the chapter as well as throughout our description of the findings. The last section also uses the results from our analysis to suggest a corollary to Krugman's (1991) arguments about the importance of geography for international economic interactions.

B. A SIMPLIFIED STORY OF INTERDEPENDENCE

Using the Krutilla and Anderson demand and supply format, it is possible to develop a straight-forward categorization of the current literature's description of trade and the environment. Consider first, the case of a small, open economy exporting a product to the world market. This is presented in simple terms in Figure 16.1. $D\overline{D}$ and $S\overline{S}$ describe the domestic demand and supply, and OP_W the world price. Small implies this country's actions alone will not affect the world market. Under autarky, the market equilibrium would be at point X with domestic price, OP_X, and quantity OQ_W. If SS_E defines the supply function reflecting the full social cost of producing Q (i.e., including both incremental private and any incremental social costs arising from externalities from the production of Q),[3] then the welfare loss associated with failing to adopt efficient environment policies is given by XEH. SHX is the added social costs due to producing Q_X. Most of these losses count against the economic surplus we would have attributed to producing OQ_X (i.e., the triangle SXD is economic surplus and SXE the amount subtracted due to added environmental costs).

If this economy is opened to the world market, producers now face the world price, OP_W and would export (i.e., $OP_W > OP_X$). They now produce Q_W, selling OQ_A on the domestic market and Q_AQ_W on the world market. Social costs due to environmental externalities increase by $HXCG$. Some

---

[3] This is the framework often used in describing environmental costing. It is important to acknowledge that the 'cost' added by private incremental costs represent the aggregate incremental willingness to pay to avoid the externalities associated with each level of production. Thus, this arises from the affected people's preferences for pollution abatement. To derive the link between $S\overline{S}$ and SS_E would require understanding how each sector responds to control as required to realize the net pollution arising from each level of output. See Smith (1992) for a discussion of some of these assumptions in the context of applying environmental costing to agriculture.

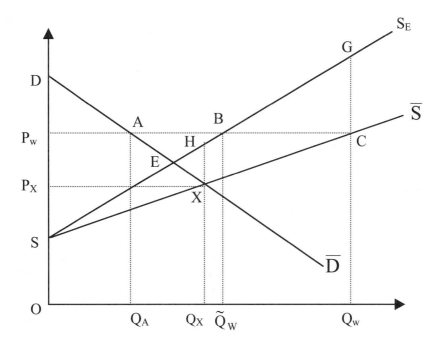

FIGURE 16.1 Krutilla–Anderson framework for a small exporting country

of this increase displaces the increased rent (*AXC*) earned by producers, and the net outcome depends on the relative size of *ABE* (the remaining rent) and *BGC* (the incremental environmental cost that does not displace the rent change). Any export restrictions intended to reduce incentives to participate in the world market that fail to reflect the presence of environmental costs may alter the judgment on rent losses versus environmental cost, depending on the incremental private and social costs.

There are a number of issues that cannot be addressed in this type of simple, one-market evaluation. For example, this diagram does not take account of the effect of each policy on the position of the private supply, social cost, and private demand functions. The treatment of these general equilibrium effects is what distinguishes most of the new conceptual literature from this simple approach. There are also the details about the factors underlying these functions that have implications relevant to complex domestic environmental policies. As a result, we first consider the factors that can be important to the results with small open models and then turn to the effects of general equilibrium feedbacks.

1. Extensions to the Krutilla–Anderson Framework

Copeland (1994) and Lopez (1994) consider the potential for multiple effects influencing either welfare or the pollution/income relationship in a single economy. General equilibrium adjustments are not considered. Copeland proposes a small, open economy where the primary focus of his evaluation is a generalization of the analysis to include several markets and programmes. Thus, he uses a framework that provides a proper accounting of the price, policy-induced rent, and amenity changes required to properly measure Hicksian gains and losses for *exogenous* effects. There are no feedback effects of domestic policy through either international markets or the global commons, because the domestic economy is small relative to the world market, and pollution is localized.

While it is not described in these terms, Copeland's evaluation of policies uses a Hicksian compensating variation measure. He bases his evaluation on marginal changes in the Anderson and Neary (1992) balance-of-trade function with changes in either environmental (represented as effluent charges) or trade (import tariff and export subsidies) policies. This function reduces to the Hicksian compensating variation with adjustments for changes in income (or rents) due to existing policies.[4]

In terms of our diagram for the case of a small open economy, Copeland's model generalizes the analysis in three ways. It allows for multiple commodities (consistently accounting for the effects of trade and environmental policy for each); expresses the welfare measures in Hicksian terms; and, perhaps most importantly from the perspective of our later discussion of general equilibrium effects of trade and environmental policy, allows for indirect consumer effects between pollution and marketed commodities. Thus, his framework allows a marginal change in pollution to shift the demand function for the marketed good. In terms of our diagram, this approach recognizes that the introduction of policies taking account of the externalities inherent in the SS_E function could, in principle, lead to shifts in $D\overline{D}$. This effect is ignored in virtually all the other analysis of trade and environment.

Lopez also considers a small open economy and provides a more detailed description of domestic pollution generation, as the equivalent of using an environmental input to produce marketed output along with labor and capital. However, the environmental input is assumed to be weakly separable from conventional inputs. His focus is on how growth in income, whether due to technological change or trade liberalization, will influence environmental quality. In the absence of stock effects on the environment, in his scheme income growth (and greater conventional factor usage) will initially increase pollution and then cause it to decline as income grows. This result allows for a role for the environment in preferences, but requires that Frisch's index of the importance of income effects be separate from pollution.[5] This latter point is important because income changes do not influence the rate of substitution between marketed goods and amenities. His model does allow income to increase willingness to pay for pollution reduction (or equivalently for amenities). In terms of Figure 16.1, the primary focus of his model is on the details underlying $S\overline{S}$ and the connection to SS_E.

These restrictions are also relevant to using the model to understand the role of coordination between trade and environmental policies in the developing as well as developed context. For example, an important by-product of Lopez's model is the demonstration that income effects can lead to differences in the levels of environmental quality selected at different income levels. Developed and developing countries can appear as different points on a parabolic relationship between pollution and income, with pollution initially increasing with income and then declining (comparable to the Grossman and Krueger (1993) empirical link between pollution and GDP). In the Lopez analysis, this result follows from a comparison of the importance of substitution in two contexts as the level of income grows. The first of these involves the environment's importance to people's preferences, and the second considers its role in production at different levels of output.

These relationships become more complex if the environment influences people's demands for marketed goods and, simultaneously, affects some marketed inputs differently from others. Earlier research, by Lopez and Niklitschek (1991), for example, illustrates how the positive contributions of environmental resources (e.g., natural biomass) to a traditional sector can have these types of effects on production. They found that direct production taxes to take account of trade's detrimental effects on the environment would be necessary to enhance welfare with trade liberalization.

---

[4] In practice, one would also need to include differential domestic taxation and subsidy policies as influences on the expenditure, GNP and rent changes. Our analysis incorporates these effects as well in the welfare measures described for the amended Harrison–Rutherford–Woton model described below. All taxed collected in each region are redistributed to the individual household representing all consumers in that region.

[5] This condition relies on a partial equilibrium evaluation of a small open economy and assumes a CES (constant elasticity of substitution) production function for the single index of all market output in terms of a single marketed input and pollution services. With the assumption that the Frisch money flexibility of income is independent of the effects of pollution on preferences, Lopez shows that the relationship between pollution and income will depend on the inverse of the elasticity of substitution in production in comparison to the Frisch index.

With the Krutilla–Anderson framework, the inclusion of social costs requires considering the added incremental costs due to externalities, so that they are proportional to increases in marketed outputs. Lopez's model focuses on the simplifications underlying the Krutilla–Anderson type of analysis and how they influence what is likely to be observed (across countries) from a process that views pollution levels as outcomes of the trade-off between marketed goods and amenities as households experience growth in real income.

2. Incorporating General Equilibrium Feedbacks

The remaining papers in this area consider different approaches for treating the general equilibrium interactions between market and non-market outcomes. Chichilnisky's (1994) analysis uses a conventional general equilibrium framework with two countries (designated the North and South); the North is intended to represent a developed economy with a well-established system of property rights to all resources, while the South designates a developing economy with a different set of conditions describing the availability of factor inputs. Her model relies on the differences in access conditions between private and open access resources (as factor inputs) to alter the supply functions for tradable goods. Because endowments are not fixed and respond to prices, the supply of marketed outputs produced using the environmental resource's services will be greater under open access conditions. The model allows for the general equilibrium interaction between product and factor market conditions in both economies. Establishing private property rights to the environmental resource offers the only feasible policy response to the overexploitation induced by the access conditions to the environmental resources. Trade simply increases the problem, even when policies designed to raise other factor costs are considered.

In terms of Figure 16.1 this framework is an analysis of how access conditions to environmental resources can influence $S\overline{S}$. There are no public goods or externalities, aside from the inefficiencies due to open access. While Chichilnisky's examples introduce amenities, the model does not. Indeed, private property rights succeed *because* there are no preference-related effects served by protecting the environmental resources. Thus, SS_E does not exist in this framework.

Copeland and Taylor (1994) adopt a similar two-country (North–South) setting but introduce externalities, local to each country's decisions. Thus, this model adds general equilibrium effects to the social cost issues identified in Figure 16.1 with SS_E. It allows for interactions between the two countries; endogenous (to each country) pollution taxes; and a continuum of goods, differing in both their propensity to generate pollution and their importance to consumers. The model assumes Cobb–Douglas technology and Cobb–Douglas preferences in marketed goods. Domestic pollution is treated as strongly separable from marketed goods.

As with Lopez, income effects are the key source of differences in environmental policy and, in turn, lead to the motivation for trade. Increased pollution in the South provides the way it can increase the gains to be realized from trade. Lower incomes in the South ensure that this trade is possible, because pollution control is less highly valued. If the differences in the South's ability to absorb pollution correspond to income differences (e.g., the less developed country has both lower income levels and lower existing pollution), these differences in initial conditions would simply reinforce what the model displays. The Chichilnisky and the Copeland–Taylor models either ignore or assume separability between marketed goods and the environment. As a result, they fail to take account of the same substitution effects (arising through preferences) that were noted in our discussion of Lopez. Pollution itself may alter the composition of goods demanded.

The Copeland–Taylor framework shifts the focus of attention from the goods' markets and the effects of trade on them, as illustrated in Figure 16.1 for the one-commodity case, to the market for pollution. In Copeland and Taylor's (1994) first paper, the general equilibrium effects of trade are seen through the domestic market for pollution. Trade both influences the supply of pollution and exerts effects on income to shift the incremental value of pollution reduction. In a subsequent paper, Copeland and Taylor (1995a) allow for transboundary pollution and multiple countries.

Separability (of preferences in aggregate pollution) implies that the pollution-induced effects of greater production in one country now are not limited to that country. They act as an offset to the increased income and lower costs for marketed goods from trade. All pollution in any country is a pure public good (bad). But these effects apply with equal force to *all* countries. As the authors note, the results parallel what we find in the literature on charitable contributions (see Bergstrom et al., 1986): policy changes that expand aggregate output do not reduce worldwide pollution.

Barbier and Rauscher (1994) consider a framework that relates more specifically to the problems of developing economies by evaluating the effects of trade on incentives for tropical deforestation. Using an optimal control model, they allow for stock externalities to influence trade policy for both small and large economies. As with Chichilnisky and Copeland and Taylor, the model assumes separable preferences, so there are no preference-related substitution effects influencing how the externality in each model influences optimal trade policy.

3. Computable General Equilibrium (CGE) Approaches

A framework with pure public goods has been the primary specification used for those computable general equilibrium (CGE) models that introduced non-market environmental resources (see Piggott et al., 1992; Ballard and Medema, 1993). Environmental resources are allowed to influence the level of household well-being but not the preference-related substitutions between goods due to changes in these resources. Equally important, these models do not have an explicit spatial dimension. Their structure treats the environmental resource as a uniform public good. This specification is comparable to the form used in Copeland and Taylor's multiple-country model with global pollution.

To our knowledge, the only model that has attempted to develop a more explicit treatment of local and transboundary pollution is Perroni and Wigle (in press). Their framework allows for both local and global effects. It incorporates separate regional abatement decisions, but assumes the same technology. A constant elasticity of transformation (CET) function is used to aggregate net local and net global pollution abatement activities. Within each region, local environmental quality is determined by the level of output of the sectors generating emissions, decisions about abatement activities, a region-specific constant elasticity damage function linking aggregate net (post-abatement) emissions to quality, and the initial stock of quality assumed present in the region. Control over the mix of global versus local environmental quality is realized through the abatement function. The emissions generation, damage functions, and environment quality endowment differ across regions. Global environmental quality is described in a similar way, but the parameters of damage functions are assumed to be the same across regions. It is difficult to evaluate how the model's implied response in the final mix of local and global damages to policy accords with the assumptions generally used in domestic policy evaluations.

Preferences are specified as a nested CES function, with the pollution composite and goods aggregate in the top nest. This implies that substitution between goods will *not* be influenced by the level of pollution (local or global). Nonetheless, the structure is an interesting step toward more realism.

4. Overview

Our review of the recent research on trade and the environment adopted the modeling perspective used in environmental economics to evaluate what details should be judged as important in evaluating the work to date. First, the models, both theoretical and computational, have consistently assumed separable preferences. This approach separates the effects of the environment on well-being from people's decisions made about different mixes of private goods. As we will discuss in more detail below there are no preference-related feedback effects of externalities transmitted outside markets on the outcomes observed in markets. This approach is *inconsistent* with all revealed preference methods for non-market valuation. It is *the absence of separability*, and indeed the specific assumptions linking private goods to public goods, that have been the most important aspect of the preferred methods to estimate the economic values of non-marketed commodities.

Second, in actual applications, we generally find that environmental externalities arise through some type of spatial interaction. That is, activities in one location have effects on other activities in that location, as well as on others outside the immediate area. Virtually all policy responses suggest that the details of both the activity and the transmission mechanism matter.[6] Recently, Copeland and Taylor (1995b) have demonstrated how spatial differences in the effects of externalities on production can give rise to non-convexities akin to those due to increasing returns (see Helpman and Krugman, 1985). From an applied CGE perspective, only Perroni and Wigle (in press) acknowledge this type of connection. However, as we noted, their specification of these connections may oversimplify matters in important ways.

Finally, attempts to link trade and environmental policies must be evaluated within a framework that acknowledges the complexity of trade restrictions present in most countries. Equally important, they must allow welfare judgments to take account of the general equilibrium effects that changes both relative prices and incomes (rents) that can arise from existing policies. As Chichilnisky (and others before her) acknowledge, these types of increases in complexity are difficult to incorporate in a meaningful way in theoretical models. While they can be accommodated in CGE analyses, they must be introduced so that the specification matches some existing baseline.

The reason for this approach is direct. It is relatively easy to acknowledge that complexity in the nature of regulations and policy-induced price wedges will matter. Because each specific form will change outcomes, it makes little sense to catalogue all possible distortions, many of which do not exist. Instead, the most productive route for model design would call for selecting a specific realistic description of a policy regime and calibrate the analysis to deal with it. Clearly, the specific calibration influences both the parameters and the conclusions drawn from that model. Nonetheless, by selecting an existing model and incorporating environmental effects within it, there is a ready benchmark for comparison. For this reason, we selected a re-specified version of an established model for the major economies in the European Union (EU). The model was re-calibrated to include local and transboundary pollutants as well as the health effects they cause, using damage functions based on those currently used in regulating these pollutants in the U.S.

C. DEVELOPING A NON-MARKET CGE MODEL

In this section we describe the Harrison–Rutherford–Wooton model (HRW1) and the adaptations to it to incorporate three air pollutants: sulphur oxides (SO_X) as a transboundary externality, and particulate matter (PM) and nitrogen oxides (NO_X) as sources of local externalities. The model links the pollutants to production in relevant sectors of each country; incorporates a simple diffusion system for the transboundary pollutant; and introduces the morbidity-related health effects from air pollution as non-separable effects on preferences. Mortality impacts are treated as separable effects on individual well-being. The abatements were made to incorporate current estimates of health damages, both morbidity and mortality from these air pollutants. The design also permits an evaluation of the implications of: (a) omitting the environment from evaluations of changes in trade policy; (b) ignoring the joint effects of local and transboundary pollutants; and (c) restricting preferences to be separable in environmental resources.

After describing the structure of the model and our amendments, we outline the welfare measure used to evaluate a scenario involving both trade liberalization and environmental degradation. Our analysis considers three different ways that environmental damage might have been introduced into the model. The first of these directly calibrates the preference and production structure to meet

---

[6] One example of this role for environmental resources is found in evaluating the net benefits of command and control versus incentive-based regulations. See Oates et al. (1989).

base year conditions, including health effects as translating parameters on a Stone–Geary specification of preferences. The second treats the analysis of environmental damages as an *exogenous* valuation task, with computation after the CGE model is solved for a market equilibrium in marketed goods. This follows the strategy recently proposed by Boyd et al. (1995). The last considers health-damage impacts on the source for household income that endogenously reduce the effective labor endowment available to households.

1. THE REVISED HARRISON–RUTHERFORD–WOOTON MODEL, HRW1

HRW1 was calibrated to represent conditions in the EU in 1985, and it includes eleven regions. Eight include the EU countries: Germany, France, Italy, the Netherlands, Belgium, the U.K., Denmark, and Ireland, as well as the U.S., Japan, and an aggregate region that includes all other countries. The model distinguishes six aggregate commodities (agriculture and food; mining and energy; non-durable manufacturing; durable manufacturing; construction and transportation; and services) and three factor inputs (land, labor, and capital). Factors are not traded outside the home region. Consumption goods trade using a nested Armington structure.[7]

Production is characterized using constant returns to scale, nested, production functions. These functions assume a constant elasticity of substitution in the primary factors (land enters only the agricultural sector) and a linear technology in intermediate inputs. Each region's demand is represented by a single consumer with a Cobb–Douglas utility function in the six consumption goods. The budget constraint is specified by the aggregate consumer's exogenous endowment of primary factors, net revenues from taxes, and level of foreign investment (assumed to equal a lump-sum transfer in the model). The model includes a detailed representation of domestic taxes, trade barriers, government and EU programs. For example, all regions (except the U.S.) have value-added taxes on the primary factors represented as a fixed *ad valorem* tax rate. Producers in the EU countries also receive subsidies on production aimed at fostering exports. The net EU transfers are equal to the value of receipts from intervention purchases and variable agricultural import levies (due to the Common Agricultural Policy) less the cost of export subsidies and the net contributions to the EU.

Our model uses the HRW1 structure in exactly the form developed and reported in various publications (e.g., Harrison et al., 1989, 1991). We have made three modifications to the model: (a) replaced the Cobb–Douglas with Stone–Geary utility functions for the aggregate consumer in each region (the Cobb–Douglas specification takes the form $U = \Pi_i x_i^{\beta_i}$, while a Stone–Geary displaces each commodity by a translation or subsistence parameter, γ_i, so $U = \Pi_i (\chi_i - \gamma_i)^{\beta_i}$); (b) introduced nine air-pollution-induced morbidity effects as translating effects on each household's subsistence parameter for services (this is the aggregate that HRW1 designates for health expenditures); and (c) introduced an explicit set of pollution generation and dispersion models for each of the three air pollutants identified earlier. As a result, air pollution will affect the marginal rates of substitution for consumption goods relevant to these services. Our specification for air pollution effects also allows for mortality effects as a separable shift in preferences. The magnitude of this effect is also directly linked to existing estimates of mortality-air-pollution-dose response models. Finally, monetary loss measures use estimates of the value of statistical lives. Each of these modifications has been introduced so that the model's initial calibration is maintained.

There are four important features of our adaptation to HRW1 to incorporate air pollution: (a) developing benchmark air-pollution measures for each sector in each country; (b) specifying the air diffusion system for transfrontier pollutants (e.g., SO_X); (c) linking the ambient concentrations of pollutants to morbidity and mortality effects; and (d) monetizing these health impacts. We will summarize the highlights associated with each step.

---

[7] With the nested Armington assumption, importers are assumed to minimize import costs subject to an import aggregation function linking the first nest domestic consumption of the commodity with a CES aggregate of the imports of the same commodity from all regions.

TABLE 16.1
Benchmark Air-Pollution Levels and the Morbidity and Mortality Losses as Shares of Service and Total Expenditures

Region	Benchmark Air Pollution				Benchmark Morbidity Effects		Mortality Effects
	PM (µg/m³)	SO$_x$ (ppm)	NO$_x$ (µg/m³)	Own Deposition Rate	Share of Total Expenditures	Share of Services Expenditures	Share of Total Expenditures
Germany	59.55	0.024	49.50	0.42	0.028	0.112	0.172
France	24.32	0.010	22.34	0.42	0.013	0.049	0.045
Italy	83.85	0.029	46.98	0.65	0.031	0.146	0.157
Netherlands	60.28	0.010	46.05	0.15	0.025	0.131	0.110
Belgium	91.40	0.015	49.59	0.31	0.040	0.216	0.194
U.K.	19.85	0.014	31.98	0.83	0.013	0.046	0.051
Ireland	16.32	0.009	25.16	0.51	0.006	0.023	0.014
Denmark	60.61	0.009	49.33	0.15	0.047	0.255	0.275
U.S.	32.78	0.006	53.46	1.00	0.012	0.031	0.047
Japan	35.02	0.008	39.15	1.00	0.023	0.069	0.058
Rest of World	48.40	0.013	46.19	0.95	0.047	0.128	0.337

Emissions of PM, NO$_X$, and SO$_X$ are based on the *OECD Environmental Data Compendium 1989*. Each region was assigned the population-weighted estimate of the average annual concentration for urban and rural locations. Because SO$_X$ is assumed to be a transfrontier pollutant, we used 1991 estimates of the relationship between emissions and depositions to each country from each source (see Markandya and Rhodes, 1992). The first three columns of Table 16.1 report the benchmark concentrations used for all three pollutants. The fourth column reports the share of SO$_X$ deposition due to domestic sources. For local pollutants (PN and NO$_X$), the deposition is assumed to equal emissions, and ambient concentration coefficients are constant multiples converting deposition units to the appropriate ambient measures.

The allocation of emissions to specific production sectors was based on the U.S. emissions by production sector. These emissions per unit of output rates were applied to each sector's output in each region, and the resulting estimates of aggregate emissions were re-normalized by country, using the ratio of actual to predicted aggregate emissions.

The third component of the process of incorporating environmental resources into the HRW1 involves computation of the health effects. The nine morbidity effects are given by pollutants as follows: PM (bronchitis, chronic cough, croup, emphysema, upper respiratory disorders, and cough episodes), NO$_X$ (lower respiratory disorders, eye irritation), and SO$_X$ (chest discomfort). The dose-response functions were based on the Desvousges et al. (1994, Vol.1) study of morbidity effects. Implementing these models requires the correct pollution measure (and we used conventional conversion ratios in these adaptations), along with a base population to apply the revised incidence rate, relevant for each health effect. In many cases these populations were specific demographic groups (often either older populations or young children). In general, it was not possible to identify populations at this level of detail. We used the overall population of each region to estimate a base incidence level. For NO$_X$, it was possible to adjust for the fraction of the population under fifteen years of age in developing the health impact.

Monetization of the health effects requires estimates for the incremental willingness to pay for each health event. These were taken from Desvousges et al. (1994, Vol. 3). Because all of the valuation estimates were in 1993 U.S. dollars, they were re-scaled to 1980 (the reference year used by HRW to deflate prices in their benchmark data set) and converted to each country's currency. Mortality effects were assumed due to particulate matter, and a $3.5 million estimate was used as the value of a statistical life (in 1989 dollars).

Changes in pollution are assumed to change the level of the subsistence parameter for the service aggregate (Pollak and Wales 1981). This translation was calibrated to meet the benchmark solution for HRW1 given our estimates of the ambient concentrations of each pollutant in each country and the dose-response functions. The subsistence parameters for the Stone–Geary specification in each country were developed using Gamaletsos (1971) and Lluch and Powell (1975). Based on these estimates, adjusted to reflect real GDP in the benchmark year for HRW1, total demand for each product was reduced by the value of its subsistence parameter for benchmark conditions (i.e., including estimates of air pollution). The remaining parameters of the Stone–Geary functions were derived using HRW1's calibration method.

By allowing the subsistence parameter for services to be a linear function of estimated morbidity losses in the benchmark year, the intercept in the translating equation was used to adjust the impact of pollutants so that the benchmark estimate corresponded for each subsistence parameter. Mortality effects were treated as a separable adjustment to utility.

2. Measuring the Welfare Consequences of Trade and Environmental Policies

To measure the welfare effects of changes in trade or environmental policies, it is important to acknowledge that existing trade policies (e.g., tariff, quota, and subsidy) generate income for some economic agents. Policy changes can lead to income changes. Thus, efficiency gains arising from efforts to remove distortions in product or factor prices or reduce environmental externalities can, in a general equilibrium framework, change these rents. To acknowledge them in applied welfare analysis implies we accept a second-best situation in evaluating policy changes.

Following Anderson and Neary (1992) we adopt a balance-of-trade function (BT) to evaluate the welfare effects of policy change. This is consistent with Copeland, except he assumed a balance-of-trade equilibrium and we allow for trade surplus and deficits (β). Equation (16.1) defines the BT function in terms of the Hicksian expenditure function, $e(.)$, the GNP or aggregate revenue function, $g(.)$, and tariff and quota rents.

$$BT(q,\phi,\bar{u}) = e\big(p(q,\pi,\bar{u}),\pi,\bar{u}\big) - g\big(p(q,\pi,\bar{u}),\pi\big) - (\phi-1)\pi^* m - (1-w)\big(p-p^*\big)^T q - \beta \quad (16.1)$$

where q is quotas (vector); ϕ is the proportional mark-ups due to tariffs; $\bar{u}$ is the level of well-being for a single household (assumed to represent all households in the economy); π is domestic prices (i.e., $\pi_i = \pi^* + t_i$, so $\pi = \pi^*\phi$, with t_i = tariff on ith commodity, π^* a diagonal matrix and $\phi_i = 1 + (t_i \pi_i^*)$); π^* is the diagonal matrix of world prices on diagonal and zeros elsewhere; m is the vector of import demands ($m = e_\pi - g_\pi$); p^* is the vector of foreign prices on quota-restricted commodities (p = domestic price vector); $p(q,\pi,\bar{u})$ is virtual prices of quota-restricted goods; $(1-w)$ is the fraction of quota rents that returns to domestic residents (w is the share that goes to foreigners); β is the trade surplus or deficit; and the superscript "T" identifies the transpose of a vector.

Introducing environmental externalities regulated at some level Z_d for domestic and Z_f trans-boundary sources adds these two terms to $e(.)$, $g(.)$, and, therefore, to the virtual prices functions, $p(.)$.

Consider now a simultaneous change in tariff restrictions from ϕ to $\bar{\phi}$ with $\phi < \bar{\phi}$) and from Z_d^0 to Z_d^*, Z_f^0 to Z_f^*. Using the balance-of-trade functions to evaluate the welfare implications of this policy change we have Equation (16.2):

$$\Delta BT = e\big(p(q,\pi,Z_d^0,Z_f^0,\bar{u}),\pi,Z_d^0,Z_f^0,\bar{u}\big) - e\big(p(q,\bar{\pi},Z_d^*,Z_f^*,\bar{u}),\bar{\pi},Z_d^*,Z_f^*,\bar{u}\big)$$

(conventional Hicksian compensating variation)

$$-g\Big(p,\big(q,\pi,Z_d^0,Z_f^0,\bar{u}\big),Z_d^0,Z_j^0,\bar{\pi}\Big)-g\Big(p\big(q,\bar{\pi},Z_d^*,Z_f^*,\bar{u}\big),Z_d^*,Z_f^*,\bar{\pi}\Big)$$

(change in the value of factor endowments − household income)

$$-\Big[(\phi-1)\pi^* m-\big(\bar{\phi}-1\big)\pi^*\bar{m}\Big]$$

(change in tariff income)

$$-(1-w)\Big[\big(p-p^*\big)^T q-\big(\bar{p}-p^*\big)^T q\Big]$$

(change in quota rents). (16.2)

Reorganizing terms we se that the Anderson–Neary criteria for the restrictiveness of trade policy are completely consistent with the Hicksian welfare theory in the presence of multiple restrictions. The first term in Equation (16.2) is the Hicksian willingness to pay for the composite change, and the remaining terms adjust for income and rent changes due to the policy change, The zero profit condition along with an assumption that factors are immobile (a characteristic of the HRW model) ensures that changes in the GNP function correspond to the change in value of domestic endowments.

This formulation also allows consideration of the effects of separability. With separable preferences, environmental resources would be yet another adjustment to the income changes. They would not influence the virtual price of quotas or the Hicksian WTP for the policy change independent of this income adjustment. In this general formulation they can, in principle, influence the pure substitution effect captured in the first term as well as the other three income-related changes.

A final issue that this expression concerns the role of the measures of environmental resources. We assumed in Equation (16.2) that Z_d and Z_f enter both household expenditure and GNP functions in the same way. This specification implicitly embeds the role of the environment, as a mechanism converting emissions to ambient concentrations, in one of these two behavioral functions. Restrictions on domestic emissions are relevant to Z_d as well as to any contribution domestic emissions make to the level of the pollutants influenced by transboundary sources, and thus to the Z_f in the GNP function. The ambient environment and the transboundary transfer function influence what is relevant for the level of Z_f entering the expenditure function.

In practice the level of Z_d in the GNP function is not the same as that affecting the expenditure function. There is some conversion relationship. Because there is only one household in each country, the approach simply scales emissions in the current version of the model. It is nonetheless one of the ways spatial differences within each region could be introduced in a model without requiring preference differences. The diffusion functions serve as the equivalent of preference differences.

As we noted earlier, the Perroni–Wigle use of conventional CES functions to capture these differences is one of the ways of recognizing their importance. Unfortunately, we know very little about the correspondence between the use of this type of averaging function and the types of diffusion models used by environmental scientists to link ambient concentrations to emissions of different types of pollutants. Both influence excess demands and virtual prices. Thus, the simplifying assumptions highlighted earlier can in principle influence all aspects of a welfare evaluation of policy change. Separability assumptions incorrectly confine their impacts to the terms adjusting for changes in income.

3. Do the Assumptions Matter?

To evaluate whether the assumptions of the recent research on trade and the environment influence conclusions about welfare effects of liberalization, we have considered one composite scenario in four different ways. Our policy change introduces a reciprocal reduction in non-tariff barriers by

50 percent (from 0.20 to 0.10) for all trade between the U.K. and each of its EU trading regions in the goods from the durable manufacturing sector. This is combined with a 25 percent increase in the emission rates for all three pollutants for this sector in the U.K. We might consider the second component of the scenario as an approximate method for describing a relaxation in environmental standards to represent the entry of marginal plants in response to the trade liberalization.

The first approach for evaluating this change ignores the environmental effects completely. Labelled NE (no-environment), it focuses on the market effects of the reduction in trade restrictiveness. The second approach adopts the Boyd et al. (1995) framework by keeping track of the changes in emissions due to the composite policy *after* the model is solved for a new vector of general equilibrium prices with the reciprocal reduction in non-tariff barriers in this sector for EU regions. The increment to the balance-of-trade function with environment effects is computed based on the scenario in comparison to the baseline. The changes in emissions and the corresponding health effects (morbidity and mortality) are computed and valued at baseline incremental WTPs. These are treated as the equivalent of an exogenous income change using the fixed prices for the health effects. This approach (labeled BKV) imposes a stronger restriction than separability of preferences (in environmental and marketed commodities) because reductions in income due to the damages are not allowed to influence the composition of final demands for marketed goods and implicitly the general equilibrium levels of emissions. The third approach (labeled ENDOW) treats morbidity effects as reducing the available labor endowment. It might seem plausible to interpret this approach as a pure income reduction. This would not be correct. The reduced endowments alter factor and product prices. As a result, substitution effects do accompany the change in pollution. They arise in this case from the incremental cost increases due to increased labor scarcity rather than from a specified role of the environment in preferences.

Finally, the last evaluation framework (labeled ES for Espinosa–Smith) recognizes the effects of morbidity on threshold demands for services, and as a result reflects the substitution effects induced for other marketed goods as well as any income effects through re-valuation of endowments of changes in rents (as given in Equation 16.2).

Tables 16.2 and 16.3 present the results for the trade/environment scenario evaluated with each of the four different strategies. Table 16.2 compares their impacts on producer and consumer prices and trade. Qualitatively, the results are similar, with the direction of change in all the sectoral prices and trade similar regardless of how the policy is implemented (i.e., with and without environmental consequences as well as with and without various forms of separability imposed). Table 16.3 offers the most direct illustration of the effects of these later assumptions. To gauge the implications of separability, consider first the implications for taking account of the air-pollution-induced effects on morbidity, as given in the first row of Table 16.3 (row IA.1). Taking the ES approach as the 'correct' basis for evaluating the composite trade and environment scenario, we find that ignoring this aspect of the environment would cause a 12 percent overstatement of the gain relative to GDP. It should be noted that both the ΔBT and GDP (i.e., the numerator and denominator in the ratio reported in the table) are changing in these comparisons.

General equilibrium adjustments affect the composition of final demand, the level of emissions, and the value of factor endowments. The BKV approach would *overstate* the losses: this error is 6 percent, as seen in the bracketed term in Table 16.3, row IA.1. This outcome is what should be expected, because their approach ignores the prospects for substitution.

The results from treating morbidity effects as an endogenous adjustment to the labor endowment may at first seem surprising, but they should not be. The measure of welfare change as a percentage of GNP is comparable to the ES case, in part because of the specific form of the utility function and the role we assigned to pollution. With a Stone–Geary function, adjustments to the subsistence parameters reduce discretionary income, due to the increased demands for health services. This follows because the subsistence parameter for this sector increases with the increase in the morbidity effects due to air pollution. At the benchmark solution in the U.K., these morbidity-induced service demands were about 5 per cent of the expenditures on services (and 1 percent of total expenditures).

TABLE 16.2
Comparison of Price and Trade Effects of Trade and Environmental Policy Changes

U.K.	−50% NTB for U.K. Durables			
	NE	BKV	ENDOW*	ES*
I. *Consumer Prices* (%Δ)				
Agriculture and food	0.8515	0.8503	0.8618	0.8591
Mining and energy	0.9018	0.9003	0.9119	0.9113
Non-durable manufacturing	0.8124	0.8110	0.8232	0.8213
Durable manufacturing	−1.1979	−1.1989	−1.1900	−1.1923
Construction and transportation	−1.0877	1.0859	1.1010	1.0996
Services	1.1669	1.1651	1.1815	1.1790
II. *Producer Prices* (%Δ)				
Agriculture and food	1.1074	1.1059	1.1210	1.1175
Mining and energy	1.1681	1.1663	1.1813	1.1805
Non-durable manufacturing	1.0695	1.0677	1.0840	1.0813
Durable manufacturing	0.9984	0.9967	1.0147	1.0098
Construction and transportation	1.0988	1.0970	1.1123	1.1108
Services	1.1872	1.1854	1.2021	1.1996
III. *Trade Durables*				
%Δ Exports				
To EC	16.195	16.199	16.159	16.169
To non-EC	−1.903	−1.886	−1.922	−1.899
%Δ Imports				
From EC	13.691	13.695	13.675	13.675
From non-EC	−4.139	−4.140	−4.168	−4.155

*When the 50 percent reciprocal reduction in non-tariff barriers for durables is implemented for these computations, the emission rates for all air pollutants generated by the U.K. durable sector are increased by 25 percent.

The endowment approach allows the labor endowment to adjust down with increased morbidity effects due to air pollution. Because this adjustment is endogenous, factor prices and final goods' prices also adjust, changing the product mix and, ultimately, the pollution levels and associated morbidity effects. Welfare effects take account of the change in relative prices between the two scenarios and the associated adjustment in demands.

When we consider the mortality effects entered as a separable effect on preferences, the distinctions between the NE and other scenarios become more pronounced. Mortality effects were 5 percent of U.K. total expenditures in the benchmark solution. Here we find that it is possible to *misinterpret* the net benefits from the change in trade policy. Considering trade liberalization alone, the reduction in non-tariff barriers enhances well-being. If, as we postulated, there is an accompanying increase in the emission rates for air pollution in this sector, then there is a net welfare *loss* considering both the separable and non-separable health-related effects. One reason for considering these elements in steps is to recognize that small omissions from the environmental consequences of trade policy may not be important to the measures of welfare consequences of the change. Of course, this is not guaranteed. The nature of the change, sector affected, and character of environmental impacts all matter. This is illustrated with our example where even this relatively small increase in total emissions (from 2.0 percent to 6.7 percent) was sufficient to change the verdict on the policy.

TABLE 16.3
Welfare Implications of Policy Evaluations

Country	−50% NTB for U.K. Durables			
	NE	BKV*	ENDOW*	ES*
I. *U.K.*				
A. *Welfare Change*				
1. BT/GDP (%)	0.1979	0.1653	0.1766	0.1767
(excluding mortality)	[0.12]	[0.06]	[0.00]	—
2. BT/GDP (%)	0.1979	−0.1773	−0.1660	−0.1662
(including mortality)	[2.19]	[0.07]	[0.00]	—
B. Environmental Effects (U.K.)				
% Emissions				
PM	0.161	6.714	6.687	6.693
NO$_X$	−0.111	2.087	2.061	2.064
SO$_X$	−0.098	2.427	2.402	2.406
% Health effects				
Morbidity	—	1.654	1.645	1.647
Mortality	—	6.714	6.687	6.693
II. *Germany*				
A. *Welfare Change*				
1. BT/GDP (%)	0.0134	0.0123	0.0120	0.0121
(excluding mortality)	[0.11]	[0.02]	[0.00]	—
2. BT/GDP (%)	0.0134	0.0070	0.0067	0.0070
(including mortality)	[0.91]	[0.00]	[0.04]	—

*When the 50 percent reciprocal reduction in non-tariff barriers for durables is implemented for these computations, the emission rates for all air pollutants generated by the U.K. durable sector are increased by 25 percent.

Of course, it is important to acknowledge that we postulated the 25 percent increase in emission rates from this sector. Comparison of the first column under category B with the other three columns illustrates how trade liberalization affected emissions in the absence of this exogenously specified increase in emission rates for this sector. Particulate matter would have increased by about 0.2 percent (in comparison to the 6.7 percent aggregate increase implied by the scenario).

While the primary focus of the analysis is on the U.K., the price effects and environmental effects have impacts beyond this region. Germany experiences welfare gains for the trade liberalization which would *also* be overstated if the environmental effects are not taken into account. This overstatement arises primarily from the effects of increased output in the durable manufacturing sector on emissions, and to a more limited degree from the transboundary effects of increased emissions from the U.K. for Germany.[8] The direction of these effects for both morbidity and mortality, and for methods of measurement, are comparable to what we found with the U.K. Here, we do not have the simultaneous increase in pollutants assumed to accompany trade liberalization in the U.K.'s durable manufacturing sector. Nonetheless, there is a large difference in the evaluation of the policy depending on whether (and how much of) the environmental effects are recognized. The differences among evaluation methods, BKV, endowment, and our approach are much smaller because the overall importance of the non-separable environmental effects is smaller as a fraction of GDP.

[8] Sulphur oxide is the only transboundary pollutant. Its effects are limited to one morbidity relationship (chest discomfort), not a mortality effect. The dispersion coefficient for SO$_X$ from the U.K. to Germany is derived as the product of the share of the U.K.'s SO$_X$ emissions received by Germany (0.015) and the SO$_X$ dispersion coefficient for Germany (0.01128 ppm/ton).

Overall, this comparison reinforces the issues identified in our review of the conceptual and CGE literature to date. To assume that environmental resources make separable contributions to preferences not only contradicts the logic used in revealed preference approaches to non-market valuation, it has an important effect on the relative price impacts and welfare measures derived for composite trade and environment policies. Equally important, there is a more subtle role played by non-separability. These specifications recognize the jointness in the general equilibrium solution between market *and* non-market interactions between economics agents. When environmental resources influence the marginal rates of substitution (in consumption) for market goods, we can expect that externalities will both *affect* and *be affected* by final good choices. The first of these influences is through the interaction outside markets, and the second is the result of market-based substitution on final production patterns and the level of emissions that result from them.

Preference-related impacts provide a potentially important feedback loop that has to date been ignored by conceptual and computational models. In these approaches, the exclusive message seems to be one where international markets magnify external effects. However, in a more general approach, these external effects can change the demands for marketed goods and services and, thereby, transmit impacts outside markets that nonetheless change domestic and international resource allocation decisions.

D. GEOGRAPHY, TRADE, AND THE ENVIRONMENT

Krugman's (1991) essays on the frontier issues in international economics argue that some of the most important research questions in trade theory, as well as in the analysis of trade policy, follow from greater recognition that geography matters to economic activities. In his models, the spatial or geographic concentration of economic activities arises from a collection of forces: increasing returns to a common location in some production activities that must be balanced against transport costs to the larger markets required to take advantage of these economies, as well as by the degree to which production can be "footloose."

He uses two primary approaches to argue for greater attention in international economic models to economic geography. The first is a bottom-up framework that looks at the micro forces leading to different regional patterns of activity that may transcend separate economies as they are traditionally defined. He concludes with a top-down perspective, that postulates countries as the primary source of restrictions to the movement of factors and goods. In his analysis, national economies are simply sources of inefficiency in resource allocations.

In both situations, space (or location) matters in providing external economies, but the factors giving rise to either the advantage or the costs must be postulated by the analyst. Increasing returns exist and their source is human capital or other forces often outside the model and thus *not* explained by it. An alternative explanation of the tendencies Krugman finds is the environmental resources that regionally and globally transmit external effect.[9]

Chichilnisky's analysis concludes that greater attention to trade and environmental interactions is warranted because international markets *magnify* externalities. Our discussion of how environmental externalities can influence general equilibrium evaluations of trade and environmental policies emphasizes the *feedback loop* or, more simply, the implications of interactions between agents that arise outside markets because of one or more environmental resources. Through non-separabilities these interactions can play a direct role in the substitutions that influence outcomes within markets. To the extent that non-marketed environmental resources make non-separable contributions to production and preferences, they influence the signals for marketed goods from both sides of market transactions. These issues are likely to be at least as important for developing economies, where the

---

[9] Copeland and Taylor (1995b) have formulated a two-country model with production externalities that encourage spatial separation of activities and trade that leads to environmental degradation. The logic of the analysis is comparable to Helpman and Krugman's (1985) analysis of the effects of increasing returns of trade.

environmental resources often serve a key role in agriculture (e.g., Lopez and Niklitschek's natural biomass) or as extractive outputs, but could also contribute (in a preserved status) to watershed protection, microclimatic functions, and amenities (e.g., Barbier and Rauscher, 1994).

What determines the importance of these interactions outside markets? One explanation is that environmental media define the relevant geographic extent of the positive and negative interactions outside markets. This was illustrated by the differences between local and transboundary pollutants. Differences in tariff, non-tariff barriers, and domestic taxes provide the basis for separating the regions within the HRW1 model. They provide the price wedges that define each economy. Without them some other exogenous constraint must be used to separate or distinguish activities within each country. Within CGE models, these restrictions provide the primary limit to footloose movement of production.

Once the restrictions are defined, however, it is possible to consider something completely different: the diffusion of air pollution through transfer coefficients that define, in a simple way, the basis for local and transboundary feedbacks in the international economy. But we could re-think matters and start with the environmental transfer function, and then consider how the price wedges enter.

Thus, a key lesson that has more general implications beyond discussion of coordination of trade and environmental policies is the basic premise of Krugman's essays. Regional analysis can inform discussions of the issues relevant to international economic modeling because it helps to focus on the forces that create positive and negative interactions. In Krugman's approach there are still omitted factors, entering as accidents of historical interactions that gave rise to the regions. Perhaps starting with the role of the environmental resources serving as one source of services with positive and negative effects would help take the mystery out of the origins of economically meaningful regions. Transport costs and footloose behavior are not the whole story.

The relevance of this perspective is, of course, not limited to the interactions of trade and environmental policy. Analyses of regional economic issues and national environmental policies are equally plausible candidates. Here institutional restrictions providing for differences in prices are less relevant. The properties of diffusion systems, whether linking or separating activities, now have the opportunity to influence economic behavior. Thus, we are left with another source of endorsement for Krugman's recommendation to look at regional patterns of economic activities for the role of constraints (and incentives), whether institutional, technological, *or environmental*, for the scale and dispersion of economic activities.

REFERENCES

Anderson, J. E. and Neary, J. P., A new approach to evaluating trade policy, working paper WPS1022, International Economics Department, The World Bank, Washington, D.C., 1992.

Anderson, K., The standard welfare economics of policies affecting trade and the environment, in *The Greening of World Trade Issues*, Anderson, K. and Blackhurst, R., Eds., University of Michigan Press, Ann Arbor, 1992.

Ballard, C. L. and Medema, S. G., The marginal efficiency effects of taxes and subsidies in the presence of externalities: a computational general equilibrium approach, *J. Public Econ.*, 52, 199, 1993.

Barbier, E. B. and Rauscher, M., Trade, tropical deforestation and policy interventions, *Environ. Resour. Econ.*, 4, 75, 1994.

Bergstrom, T., Blum, L., and Varian, H., On the private provision of public goods, *J. Public Econ.*, 29, 25, 1986.

Boyd, R., Krutilla, K., and Viscusi, W. K., Energy taxation as a policy instrument to reduce CO_2 emissions: a net benefit analysis, *J. Environmental Econ. Manage.*, 29, 1, 1995.

Chichilnisky, G., North–South trade and the global environment, *Am. Econ. Rev.*, 84, 851, 1994.

Copeland, B. R., International trade and the environment: policy reform in a polluted small open economy, *J. Environ. Econ. Manage.*, 26, 44, 1994.

Copeland, B. R. and Taylor, M. S., North–South trade and the environment, *Q. J. Econ.*, 109, 755, 1994.

Copeland, B. R. and Taylor, M. S., Trade and transboundary pollution, *Am. Econ. Rev.*, 85, 716–737, 1995.

Copeland, B. R. and Taylor, M. S., Trade, nonconvexities and the environment, UBC discussion papers No.95–19, Department of Economics, University of British Columbia, 1995b.

Desvousges, W. H., Banzhaf, H. S., Johnson, F. R., and Wilson, K. N., *Assessing Environmental Externality Costs for Electricity Generation in Wisconsin*, Vol. 1 and 3, Research Triangle Institute, Research Triangle Park, NC, 1994.

Espinosa, J. A. and Smith, V. K., Measuring the environmental consequences of trade policy: a non-market CGE analysis, *Am. J. Agric. Econ.*, 77, 772, 1995.

Gamaletsos, T., International Comparisons of Consumer Expenditure Patterns: An Econometric Analysis, Ph.D. thesis, University of Wisconsin, Madison, 1971.

Grossman, G. M. and Krueger, A. B., Environmental impacts of a North American Free Trade Agreement, in *The Mexico–U.S. Free Trade Agreement*, Garber, P., Ed., MIT Press, Cambridge, 1993.

Harrison, G. W., Rutherford, T. F., and Wooton, J., The economic impact of the European community, *Am. Econ. Rev. Pap. Proc.*, 79, 288, 1989.

Harrison, G. W., Rutherford, T. F., and Wooton, I., An empirical database for a general equilibrium model of the European community, *Empirical Econ.*, 16, 95, 1991.

Helpman, E. and Krugman, P. O., *Market Structure and Foreign Trade*, MIT Press, Cambridge, 1985.

Krugman, P., *Geography and Trade*, MIT Press, Cambridge, 1991.

Krutilla, K., Environmental regulation in an open economy, *J. Environ. Econ. Manage.*, 20, 127, 1991.

Lluch, C. and Powell, A. A., International comparisons of expenditure patterns, *Eur. Econ. Rev.*, 15, 786, 1975.

Lopez, R., The environment as a factor of production: the effects of economic growth and trade liberalization, *J. Environ. Econ. Manage.*, 27, 163, 1994.

Lopez, R., and Niklitschek, Dual economic growth in poor tropical areas, *J. Dev. Econ.*, 36, 189, 1991.

Markandya, A. and Rhodes, B., External costs of fuel cycles: an impact pathway approach — economic valuation, report for EC/US Fuel Cycle Study, Metroeconomica, Ltd., 1992.

Oates, W. E., Portney, P. R., and McGartland, A. M., The net benefits of incentive based regulation: a case study of environmental standard setting, *Am. Econ. Rev.*, 59, 1233, 1989.

Perroni, C. and Wigle, R. M., International trade and environmental quality: how important are the linkages?, *Can. J. Econ.*, in press.

Piggott, J., Whalley, J., and Wigle, R., International linkages and carbon reduction initiatives, in *The Greening of World Trade Issues*, Anderson, K. and Blackhurst, R., Eds., University of Michigan Press, Ann Arbor, 1992.

Pollak, R. A. and Wales, T. J., Demographic variable in demand analysis, *Econometrica*, 49, 1533, 1981.

Repetto, R., *Jobs, Competitiveness and Environmental Regulation: What Are the Real Issues?*, World Resources Institute, Washington, D.C., 1995.

Smith, V. K., Environmental costing for agriculture: will it be standard fare in the Farm Bill of 2000?, *Am. J. Agric. Econ.*, 74, 1076, 1992.

17 Carbon Taxes with Exemptions in an Open Economy: A General Equilibrium Analysis of the German Tax Initiative[1]

Christoph Böhringer and Thomas F. Rutherford

Sectoral exemptions from environmental regulation are applied in many countries to avoid adverse adjustment effects in specific industries. The problem with exemptions is that they can make environmental policy more costly. This chapter analyzes the welfare costs of exemptions in environmental policy together with the issue of unilateral carbon taxes in an open economy. Several countries within the EU have introduced or contemplate unilateral taxes to reduce anthropogenic CO_2 emissions. Taxes which are unilaterally imposed in an open economy can have significant impacts on production and employment of energy- and export-intensive industries. To save jobs by maintaining "international competitiveness," most CO_2 taxation schemes include exemptions for energy- and export-intensive industries as a compromise between environmental objectives and employment in these sectors. In this chapter, we show that such compromises are costly: as the tax base narrows, the dead-weight loss increases. Our calculations illustrate this point in the framework of a static general equilibrium model for West Germany calibrated to 1990 data. We evaluate the excess costs of exemptions as a means of saving jobs in specific industries relative to an alternative instrument, a uniform carbon tax cum wage subsidy which achieves an identical level of national emissions and employment at a fraction of the cost.

© 1997 Academic Press

A. INTRODUCTION

Unilateral environmental taxes can have significant implications for the international competitiveness of industries. As the relative production costs of environment-intensive goods rise in comparison to the relative costs of producing the same goods elsewhere, unilateral taxation implies a loss of comparative advantage. A loss in comparative advantage, i.e., international competitiveness, will be most severe for export-intensive sectors in which environment-intensive inputs represent a significant share of direct and indirect costs. To avoid adverse production and employment shocks in these industries, sectoral exemptions from environmental taxes are common in many countries. (For a recent survey on the structure and administration of environmental taxes in OECD countries see OECD 1995). The problem with exemptions is that they can induce significant excess costs for achieving environmental objectives.

[1] We are grateful to Uli Fahl and Tobias Schmidt for expert research assistance, and to James Markusen, Charles de Bartolome, and participants in seminars at Colorado, Stanford, Berne, and the European Commission. Three anonymous referees provided useful comments and suggestions. Research support from the Electric Power Research Institute and the German Federal Ministry of Education, Science, Research, and Technology (BMBF Grant 0329632A) is gratefully acknowledged. The ideas expressed here are those of the authors, who remain solely responsible for errors and omissions.

This chapter investigates the welfare implications of exemptions together with the issue of unilateral carbon taxes in an open economy. In our analysis we calculate the costs of environmental policies with respect to national CO_2 emission reduction, operating on the premise that trade-induced changes in CO_2 emissions of nonparticipating countries may be of second-order magnitude.[2] The empirical analysis in this paper focuses on three questions:

1. To what extent do exemptions magnify the costs of carbon taxes?
2. How much more costly are exemptions than direct wage subsidies as a means of saving jobs?
3. How do exemptions affect sectoral employment and exports?

We address these questions in the context of a static general equilibrium model for West Germany with 58 sectors, based on the most recently available consistent economic and energy data for 1990 (Statistisches Bundesamt, 1994).

The results presented here confirm standard intuition from the public finance and taxation literature: exemptions can significantly increase the welfare cost of taxes (see, for example, Ballard and Shoven, 1987; Wiegard, 1991). The welfare costs of exemptions increase with the target level of emission reductions and the share of the exempted sectors in economic activity and total emissions. Exemptions may retain jobs in subsidized sectors, but these jobs are expensive for society as a whole. Our calculations suggest that there may be far less costly methods of retaining employment in specific industries. Sector-specific wage subsidies applied in concert with a uniform environmental tax produce a Pareto-superior policy.

The economic effects of CO_2 reduction measures have been widely assessed over the past years. (For comprehensive surveys of studies see Grubb et al., 1993 and UNEP, 1995). Although several authors have addressed the issue of competitiveness (Pezzey, 1993, Proops et al., 1993), we are aware of no previous work which has focused on the efficiency implications of carbon taxes with exemptions. There are only a few "top-down" macroeconomic analyses of carbon taxation in Germany, and these assess either uniform taxation (Conrad and Schröder, 1991, Conrad and Wang, 1993) or carbon taxes with exemptions (Welsch, 1994). None of these studies analyzes the economic costs of exemptions subject to economy-wide reduction targets. In a footnote, Welsch states:

> The tax exemption for sectors is not relevant for explaining the reduced effectiveness of the national tax … {as compared with an EU-wide energy/carbon tax} … because these sectors only account for 10% of gross output value. (Welsch 1994, p. 23, footnote 5)

The results presented here dispute this conclusion. Welfare losses associated with exemptions are significant even though the share of exempted sectors in overall economic activity and carbon emission might be small.

The results presented in this chapter are subject to some important caveats. First of all, we recognize that our open economy model fails to account for potentially important trade-related impacts on global emissions. The German economy is closely linked to other European economies and therefore the *global* environmental impacts of unilateral policies cannot be assessed in isolation. On the other hand, if Germany contemplates unilateral cutbacks in order to signal a willingness to take action and encourage other EU member states to take similar steps, the key issue is the cost of reducing German emissions and not EU-wide emissions. A second limitation of our analysis is

---

[2] An increase in emissions by nonparticipating countries due to the relocation of energy-intensive production or reductions in the international oil price has been referred to as "carbon leakage" (Rutherford, 1993). If leakage rates are high, this can justify a tax rate differentiation across sectors to avoid the problem of carbon leakage and to increase efficiency of (global) CO_2 reduction (Hoel, 1994); however, this depends crucially on model structure (see Böhringer et al., in press). The magnitude of leakage rates is an open research topic with considerable disagreement on empirical values. For an overview and interpretation of different calculations, see Rutherford (1996).

the treatment of adjustment and intertemporal linkages. Adjustment costs relate to the speed with which policies are adopted and the ability of the human and physical capital stock to accommodate these policies. We believe that the economic links identified in our static framework can provide a useful starting point for subsequent analysis of the adjustment process.

The remainder of this chapter is organized as follows: Section B introduces the policy background of carbon taxation in Europe and lays out the specifics of the German carbon tax proposal which provides the policy framework for our empirical analysis. Section C gives an overview of the basic model structure and the model specification (benchmark data and elasticities). Section D describes the policy scenarios and reports the computational results. Section E provides a sensitivity analysis of our results with respect to changes in the underlying model structure and elasticities and Section F concludes.

B. POLICY BACKGROUND

To combat global warming, several international conferences (Toronto, 1988, Cairo, 1990, Rio, 1992, Berlin, 1995) have called for significant reductions in the combustion of fossil fuels. Due to the difficulties of identifying and implementing a "fair" world-wide CO_2-reduction schedule for all countries, no concerted policy action has yet been undertaken. The European Union which forcefully advocated taxes on carbon emissions as an effective instrument to reduce global carbon-dioxide emissions put EU-wide CO_2-reduction measures under the condition that other important developed economies (i.e., U.S. and Japan) would take similar steps (Commission of the European Communities, 1992a, b). The "conditionality" clause was introduced in the EU policy platform in part due to the political pressure of energy- and export-intensive industries which feared negative impacts on their international competitiveness. Given the lack of concrete EU-wide action, there is increasing political pressure for unilateral action in single EU member countries where domestic voters stress the need for taking a leading role. Five countries, Denmark, Finland, the Netherlands, Sweden, and Norway have already implemented some kind of carbon tax (OECD, 1995, pp. 32–36). Other EU member states such as Germany are contemplating a unilateral carbon tax (Enquete–Kommission, 1995). Policy makers at the national level must take account of impacts on energy- and export-intensive industries which face severe production and employment cutbacks as a consequence of the loss in international competitiveness. Moreover, many of the energy- and export-intensive industries (such as iron and steel or chemical products) have long been protected and treated as strategically important for economic development. Given the threat of high sectoral unemployment and facing the traditional strong lobby of these sectors, nearly every tax scheme which has been proposed involves tax exemptions to energy- and export-intensive industries in order to maintain comparative advantage and sectoral employment.[3]

As the main emitter of CO_2 within the EU,[4] Germany continues to affirm its objective of reducing CO_2 emissions by 25–30% by the year 2005, taking 1990 as the base year for the whole of Germany including the Five New Laender. In order to achieve these quantitative reduction targets a carbon tax is considered to be an appropriate instrument by policymakers. Exemptions are proposed for sectors whose energy cost share of gross production value is greater than 3.75% and whose export share of turnover is greater than 15%.[5] Table 17.1 specifies the seven industrial sectors which are potentially exempted from paying carbon taxes (Enquete-Kommission, 1994).

---

[3] In Sweden the total amount of CO_2 tax paid by energy-intensive industries is limited to 1.7% of the sales values of the goods produced (Commission of the European Communities, 1992c, p. 84). In other Nordic countries, (Denmark, Norway) the tax rates for industries are very low as compared to the final demand sector with some sectors (e.g., shipping, aviation) being fully exempted (TemaNord, 1994, p. 32–36).

[4] At 1990 levels, West Germany accounted for 25% of overall EU emissions (EU without East–Germany). East-Germany's emission amounted to an additional 11% of the EU emissions.

[5] Export-intensity of sectors according to the policy proposal does not imply that these sectors are net exporters. In the benchmark 10 table of 1990 some sectors are actually net importers (ceramic goods, nonferrous metals, paper).

TABLE 17.1
Sectors Qualifying for Exemptions Under Enquete–Kommission Plan

Sector (SIO)	Energy Share	Export Share	Output[a]
Chemical products (CHM, 9)	3.8%	40.5%	198.3
Geramic goods (CER, 14)	4.9%	30.7%	5.2
Glass (GLA, 15)	6.2%	27.5%	13.6
Iron and steel (ORE, 16)	10.1%	33.0%	52.1
Nonferrous metals (NFM, 17)	5.1%	29.8%	30.2
Casting (CAS, 18)	5.2%	15.9%	16.1
Paper products (PAP, 33)	8.6%	37.5%	20.3

Note: SIO, System of production sectors for Input-Output computations; (CHM, 9), (model-specific acronym, IO table index); Energy share, share of gross production value, 1990; Export share, export share of turnover, 1991.

[a] In billion DM_{1990}.

Source: Enquete-Kommission (1994).

C. MODEL STRUCTURE AND PARAMETRIC FRAMEWORK

Our model is a static open general equilibrium model designed to investigate the economic implications of CO_2 tax exemptions for energy- and export-intensive industries in the West German economy.[6] It has a disaggregate representation of 58 industries corresponding to the standard structure of German input-output tables (Statistisches Bundesamt, 1995). The energy goods identified in the model include coal and coal products (COA), crude oil (CRU), electric power, steam and warm water (ELE), natural gas (GAS), and refined oil products (OIL). This disaggregation is essential in order to distinguish energy goods by carbon intensity.

The costs of carbon taxes can significantly change in a second-best situation when the interaction of carbon taxes with existing taxes and the method of carbon tax recycling alter the marginal excess burden for public goods provision (see e.g., Goulder, 1994). To account for second-best effects, the model incorporates the main features of the German tax system: labor taxes (including social insurance contributions), capital taxes (corporate and trade taxes), other indirect taxes (e.g., mineral oil tax) and value-added taxes.

In international trade, Germany is treated as small relative to the world market. That is, we assume that changes in German import and export volumes have no effect on its terms of trade. Domestic and foreign energy goods are regarded as perfect substitutes. In other sectors, domestic and foreign products are distinguished by the Armington assumption.[7] We impose trade balance with respect to the rest of the world accounting for an exogenously specified net trade surplus and international capital flows.

1. FACTOR MARKETS

Primary factors of production are capital and labor, which are employed together with energy and materials inputs to produce the domestic output. The labor supply is responsive to the real wage through a labor leisure choice, with preferences calibrated to a target elasticity of labor supply. For

---

[6] An appendix containing the algebraic formulation of the model is available upon request from the authors.

[7] The Armington goods are aggregated with identical import shares for a given import good across all components of final and intermediate demand. On the export side, products of the Armington sectors destined for domestic and international markets are treated as imperfect substitutes, produced subject to a constant elasticity of transformation.

our baseline computations, we assume perfectly competitive factor markets in which the prices on factors adjust so that supply equals demand. Also, both factors are assumed to be homogeneous and perfectly mobile between sectors. Labor is only mobile domestically, whereas capital is mobile also at the international level; i.e., the return to capital is treated as exogenous in the model.

2. Production

Nested, separable constant elasticity of substitution production functions are employed to specify the substitution possibilities in domestic production between capital, labor, energy, and material inputs (KLEM).

The material aggregate in each sector is composed of nonenergy inputs between which there is a zero elasticity of substitution. Within the energy aggregate nonelectric and electric energy inputs trade off with elasticities of substitution characterized by a separable nested CES function.[8]

The above specification of technologies allows the comprehensive representation of the substitution possibilities in production, with interfuel substitution within the energy aggregate as well as substitution between energy and other production factors.

Intermediate energy demand is distinguished between feedstocks, i.e., intermediate energy goods which enter the material aggregate, and fuels, i.e., intermediate energy goods which enter the energy aggregate. This distinction between different types of energy inputs is important because only (fossil) fuel consumption leads to carbon emissions. We combine monetary flows of the German national input-output table with consistent physical flow data on the emission-relevant fuel use in production (and final demand) to produce sector- and energy-specific CO_2 emission coefficients. The CO_2 emission coefficients indicate tons of CO_2 per unit DM of fuel use in a specific sector and are the base for the application of carbon taxes.

3. Taxation and Public Expenditure

Government provides a public good which is produced with commodities purchased at market prices. These expenditures are financed with tax revenues. All of our simulations are based on revenue-neutral tax reforms. This is done by keeping the amount of the public good provision fixed and recycling any residual revenue lump-sum or through a reduction in labor or capital taxes.

4. Investment Demand

Investment is determined by the savings decisions of private households. This decision, in turn, depends on the expected rate of return. In our model with static expectations, the rate of return is represented by the ratio of the quasi-rent on capital to the cost of a unit of new capital, both evaluated at equilibrium prices. We specify preferences to match a target benchmark elasticity of savings with respect to the rate of return.

5. Final Demand

In order to focus on efficiency issues we use a representative consumer to model demand side. The consumer's welfare is determined by a nested separable CES function which describes the trade-offs between savings (future consumption), leisure, and consumption. At the top level savings enter with a composite of leisure and aggregate consumption at a constant elasticity of substitution, which is calibrated to produce an exogenous price elasticity of savings with respect to the rate of return. At the second level leisure and aggregate consumption combine with a constant elasticity of substitution, which is calibrated to be consistent to an exogenous elasticity of labor supply with respect

---

[8] Nonelectric fossil inputs (gas, refined oil and crude oil) excluding coal enter at the bottom nest with a constant substitution elasticity. At the next level nonelectric (noncoal) inputs and electricity combine with a constant elasticity of substitution. In the top nest, coal and the nonelectric-electric energy aggregate trade off with a constant elasticity of substitution.

TABLE 17.2
Key Elasticities Used in the Model

Index[a]	Description	Value
σ_{KLEM}	Elasticity of substitution between the material inputs and the composite of capital, labor, and energy inputs	0
σ_{KLE}	Elasticity of substitution between energy inputs and value added	0.5
σ_{KL}	Elasticity of substitution between labor and capital	1
σ_{COA}	Elasticity of substitution between coal and the aggregate of electricity and different fossil inputs (excluding coal) in the energy aggregate of sectoral production	0.25[b]
σ_{ELE}	Elasticity of substitution between electricity and the aggregate of different fossil inputs (excluding coal) in the (non-coal) energy input of sectoral production	1
σ_{FOS}	Elasticity of substitution between different fossil inputs (excluding coal) in the energy aggregate of sectoral production	2
σ_{NEC}	Elasticity of substitution between the energy aggregate and the non-energy aggregate of household demand	0.3
σ_{NC}	Elasticity of substitution between different non-energy inputs into the non-energy bundle of household demand	1
σ_{EC}	Elasticity of substitution between different energy inputs into the energy aggregate of household demand	2
σ_{DM}	Elasticity of substitution between domestic goods and imports (Armington)[c]	4
σ_{DX}	Elasticity of substitution between domestic goods and exports (Armington)[c]	4
σ_{LS}	Uncompensated labor supply elasticity	0.1
σ_{S}	Elasticity of savings with respect to the rate of return	0.4

[a] See appendix for related notations.
[b] Except for electricity where the value is set equal to 3.
[c] The elasticity of substitution is set to infinite for Heckscher–Ohlin goods.

to the real wage. Expenditures on the aggregate consumption bundle are collected between an aggregate energy commodity and an aggregate nonenergy commodity according to a CES function. At the bottom level the aggregate energy commodity is composed of different energy goods which trade off with a constant elasticity of substitution, whereas the aggregate nonenergy commodity consists of nonenergy goods whose substitution patterns are described by a Cobb–Douglas function.

6. BENCHMARK EQUILIBRIUM

As is customary in applied general equilibrium analysis, the model is based on economic transactions in a particular benchmark year (1990 in this case). Benchmark data determine parameters of the functional forms from a given set of benchmark quantities, prices, and elasticities. Table 17.2 shows the key elasticity values employed in the model. These values are similar to empirical estimates used in comparable studies.[9]

Table 17.3 presents total tax payments as well as the marginal excess burden (MEB) of Germany's major taxes in the benchmark equilibrium. The MEB represents the excess cost to society of raising an additional DM of government revenue using a particular tax. In our model, the cost of raising one additional DM of public funds using the capital tax thus costs 1.73DM in terms of forgone consumer welfare. Our numbers for the marginal excess burden of taxation are in line with recent estimates in the literature (Ewringmann and Hansmeyer, 1994). Due to assumed international capital mobility within the EU, the excess burden of capital taxation is highest, followed by labor taxation and VAT. This suggests that the costs of carbon taxes might

---

[9] For the nesting structure and the substitution elasticities in production see Manne and Richels (1992). The substitution patterns in final demand between savings, leisure, and consumption are taken from Ballard et al. (1985).

TABLE 17.3
Tax Revenues and Marginal Excess Burden in the Benchmark

	Tax Revenue[a] (billion DM_{1990})	Marginal Excess[b] Burden (%)
Capital tax[c]	129	72.8
Labor tax[d]	585	36.4
VAT	158	23.4

[a] Statistisches Bundesamt (1995).

[b] Authors' calculation.

[c] Capital taxes in the model represent all corporate and trade taxes.

[d] Labor taxes in the model represent income taxes on labor as well as total payments of employers and employees to social insurances.

TABLE 17.4
Benchmark Data for 1990

	XEM	ES	ROI	FD	Total
	Economic Data in Billion DM_{1990}				
Output	386	188	4078		4652
	(8%)	(4%)	(88%)		(100%)
Wages	79	28	1199		1306
	(6%)	(2%)	(92%)		(100%)
Exports	133	8	567		708
	(19%)	(1%)	(80%)		(100%)
	Carbon Emissions in Million Tons CO_2 (Total: 699 million tons)				
Coal	50	228	13	5	297
	(17%)	(77%)	(4%)	(2%)	(100%)
Oil	13	23	119	138	293
	(4%)	(8%)	(41%)	(47%)	(100%)
Gas	23	24	31	32	110
	(21%)	(22%)	(29%)	(29%)	(100%)
Total	86	275	163	175	699
	(12%)	(40%)	(23%)	(25%)	(100%)

Note: XEM, potentially exempt energy-intensive; ES, energy producing and processing sectors; ROI, other industrial and service sectors; FD, final demand.

Source: Statistisches Bundesamt (1994).

be significantly lowered by revenue-neutral swaps of carbon taxes against distortionary capital and labor taxes compared to a revenue-neutral lump-sum rebate of taxes or an increase in government expenditures.

Table 17.4 provides an overview of the structure of economic activity and the pattern of emissions by energy good and sector in the benchmark year. We have grouped sectors from the model into energy sectors (EG), exempted sectors (XEM), rest of industry (ROI), and final demand (FD).

Potentially exempt sectors amount to roughly 8% of gross output, 6% of employment, 19% of exports, and 12% of carbon emissions. The small share of exempt sectors in gross output and carbon emissions suggests that overall welfare losses due to exemptions will be moderate. The ratio between emissions and gross output is significantly higher for exempted sectors than for the rest of industry, which indicates a potentially higher scope of CO_2-substitution possibilities in exempted sectors. Given exogenous reduction targets, exemptions for those sectors with high carbon intensity can significantly increase the burden of cutbacks to nonexempted sectors with low carbon mitigation possibilities.

D. SCENARIOS AND RESULTS

1. POLICY SCENARIOS

In our simulations we consider three tax policies. The first of these ("uniform") considers an economy-wide tax on carbon dioxide emissions stemming from fossil fuel combustion in domestic production and consumption. Carbon taxes do not apply to embodied carbon in imported goods. Similarly, carbon taxes do not apply to exports of any kind, and there are no rebates of carbon taxes paid on inputs to exported goods. Given technical and political constraints for an increase of inter-European electricity trade, we fix trade in electricity at the benchmark level. This setting avoids an extreme increase in electricity imports which are not subject to the carbon tax.

The second policy scenario ("exempt") involves a carbon tax with exemptions for selected industries. The basis for qualification as an exempted sector is taken as given, i.e., we do not model the endogenous response of firms which could conceivably modify their cost or sales shares in order to qualify for exemption.

The third policy scenario ("subsidies") involves a uniform carbon tax with the addition of a wage subsidy for the industries exempted in the previous scenario. The subsidy rate is set so that there is no decrease in the aggregate employment of potentially exempted sectors relative to the benchmark.

We generate results for each of these three policy scenarios using three different assumptions regarding the compensating adjustments in other taxes. The government is assumed to keep tax changes revenue-neutral by recycling carbon tax income either lump-sum or through a reduction in labor or capital taxes.

In our counterfactuals we compute equilibria with and without exemptions for target reductions of 10, 20, and 30% from the benchmark carbon emission level. Exemptions imply that carbon taxes for the remaining sectors must increase.

Tables 17.5 and 17.6 summarize the results for the three policy scenarios (uniform taxes, tax exemptions, and uniform taxes with wage subsidies) and the different assumptions about revenue recycling (lump-sum, labor taxes, and capital taxes).

2. EMPLOYMENT EFFECTS AND THE WELFARE COSTS OF EXEMPTIONS

Table 17.5A reports Hicksian-equivalent variations in income as a measure of the welfare effects resulting from the imposition of different policy instruments to meet the given emission targets. Our most significant and robust finding is that exemptions significantly magnify the costs of emission reduction when compared to uniform taxes. The excess burden of exemptions increases with the target level of emission reduction. For a 30% reduction in carbon emissions, exemptions raise the total costs of emission abatement by roughly one-fifth. Tax exemptions for energy- and export-intensive sectors responsible for 12% of the benchmark emissions cause fossil energy consumption of these sectors to decline less and shifts the burden of emission reduction to other sectors with lower carbon substitution possibilities. As a consequence marginal tax rates and induced welfare losses are higher to achieve a given economy-wide target of emission reduction.

The excess burden of exemptions becomes more evident when we measure the specific welfare costs related to the use of exemptions for retaining jobs in energy-intensive sectors. In order to provide a meaningful measure for the specific costs of employment policy we compute an effective general equilibrium wage premium in DM per worker by dividing the Hicksian equivalent variation by the number of workers whose jobs are "saved". For the 20% reduction target, these wage premia range between annual 13,400 and 46,200 DM per worker depending on the recycling strategy of carbon

TABLE 17.5
Model Results of the Effects on Welfare, Employment, and Exports

	Uniform			Exemption			Wage Subsidy		
	10	20	30	10	20	30	10	20	30
A. Welfare in money-metric benchmark utility (billion DM$_{90}$)									
Lump-sum	−6.1	−16.0	−31.7	−6.5	−18.3	−38.7	−5.6	−15.0	−30.5
L Tax	−4.4	−12.2	−25.5	−4.5	−13.4	−30.1	−3.8	−11.0	−23.8
K Tax	−2.9	−9.4	−21.5	−2.8	−10.2	−26.7	−2.4	−8.2	−20.0
B. Wage premium (1000 DM$_{90}$ per job saved)									
Lump-sum	0.0	0.0	0.0	19.1	46.2	86.9	−18.3	−14.5	−10.0
L Tax	0.0	0.0	0.0	1.9	21.4	52.6	−22.8	−18.9	−13.9
K Tax	0.0	0.0	0.0	−6.2	13.4	45.6	−26.9	−22.3	−15.7
C. Employment in energy- and export-intensive sectors (% change from benchmark)									
Lump-sum	−4.2	−10.0	−17.3	−2.0	−5.2	−9.8	0.0	0.0	0.0
L Tax	−3.9	−9.5	−16.6	−1.6	−4.4	−8.4	0.0	0.0	0.0
K Tax	−3.2	−8.0	−14.4	−0.5	−1.8	−3.6	0.0	0.0	0.0
D. Exports of energy- and export-intensive sectors (% change from benchmark)									
Lump-sum	−8.4	−19.1	−31.8	−3.9	−10.0	−18.2	−2.6	−5.8	−9.7
L Tax	−8.2	−18.8	−31.4	−3.6	−9.2	−17.0	−2.8	−6.2	−10.3
K Tax	−6.7	−15.7	−27.2	−1.3	−3.9	−7.5	−2.2	−4.9	−8.2

TABLE 17.6
Model Results of Carbon Tax Rates and Replacement Tax Rates

	Uniform			Exemption			Wage Subsidy		
	10	20	30	10	20	30	10	20	30
A. Carbon tax rates (DM$_{90}$ per tonne of carbon)									
Lump-sum	23.0	59.0	119.0	29.0	82.0	178.0	24.0	63.0	129.0
L Tax	23.0	61.0	124.0	30.0	86.0	190.0	24.0	65.0	134.0
K Tax	25.0	65.0	134.0	33.0	96.0	223.0	26.0	69.0	144.0
B. Carbon tax revenue (% of benchmark labor and capital tax revenue)									
Lump-sum	2.0	4.6	8.2	2.2	5.5	10.2	2.1	5.0	9.0
L Tax	2.0	4.8	8.6	2.3	5.8	10.9	2.1	5.1	9.3
K Tax	2.2	5.2	9.3	2.5	6.4	12.5	2.2	5.4	9.8
C. Replacement tax rates (change relative to benchmark index (=100))									
Lump-sum	−0.7	−1.6	−2.7	−0.9	−2.1	−3.9	−0.7	−1.7	−2.9
L Tax	−1.8	−4.1	−7.1	−2.2	−5.6	−10.5	−1.9	−4.4	−7.7
K Tax	−8.9	−20.2	−34.3	−11.1	−27.5	−52.6	−9.3	−21.5	−36.9

taxes. For a 30% reduction target, the premia increase to values between 45,600 and 86,900 DM per worker and year (see Table 17.5B).

If jobs are the sole justification for exemptions, Tables 17.5A–5C show that a uniform carbon tax combined with wage subsidies to potentially exempted sectors retains more jobs than an exemption policy and costs far less. As pointed out by Dixit (1985), we find *that targeted instruments* (carbon taxes together with wage subsidies) are clearly superior to *blunt instruments* (a CO$_2$ tax with exemptions).

In a model with a variety of benchmark taxes, uniform carbon taxes are not always optimal. Given our representation of the German economy and tax system, we find that a policy which combines uniform carbon taxes with wage subsidies to potentially exempted sectors is Pareto-superior due to second-best effects.[10] First, direct wage subsidies lower the distortionary effects of labor taxes.[11] Second, wage subsidies for potentially exempted sectors which are labor-intensive relative to the rest of the economy cause an increase in capital use in the rest of the economy, thereby indirectly offsetting some of the excess burden of capital taxes.[12]

The comparison of carbon tax recycling across different instruments indicates that the costs of emission reduction are significantly reduced by a revenue-neutral cut of existing distortionary taxes. Given the higher marginal excess burden on (internationally mobile) capital in this model, a compensating reduction in capital taxes is superior to reductions in labor taxes from an efficiency standpoint.

3. Export Performance

Next, we turn to the analysis of changes in exports for energy- and export-intensive sectors as given in Table 17.5D. As with sectoral employment, exports decrease in spite of the exemptions, although less than with a uniform tax. Comparing with the third policy package, however, we see that wage subsidies in fact cause exports to fall less precipitously than uniform taxes or exemptions. Tables 17.5C and 17.5D suggest that energy- and export-intensive sectors will face serious adjustment problems with increasing targets of emission reduction.

4. Marginal Costs of Abatement

Table 17.6A shows the marginal costs of carbon emission reductions. The marginal costs of emission reduction, i.e., the implicit carbon tax rates, range between 119 DM and 184 DM per tonne of carbon for a 30% reduction ($74–$140 per tonne at DM 1.6 per US$). This is roughly consistent with values obtained in other studies. The high inframarginal welfare cost of exemptions are reflected in higher tax rates for the exemption scenario.

5. Carbon Tax Revenue and Replace Tax Rates

Revenues from carbon taxation reflect carbon tax rates at different reduction targets. Tables 17.6B–6C show that high carbon tax revenues in the exemption case allow for high cut-backs of existing distortionary taxes. As evident from Table 17.5A this policy is less effective because the strong negative tax interaction effect (due to the high marginal tax rates) dominates the positive revenue recycling effect. The differences of welfare costs across alternative carbon tax recycling strategies (see Table 17.5A) are reflected in the magnitudes of cutbacks in distortionary taxes: high cutbacks in relatively distortionary taxes are most beneficial.

E. SENSITIVITY ANALYSIS

We have done a number of additional calculations to better understand how different economic assumptions affect our conclusions. This section summarizes our findings.

---

[10] As reported in an earlier version of this chapter, the ranking between uniform taxation and uniform taxation cum wage subsidies is reversed for the case of a first-best economy without prior tax distortions while the exemption policy clearly remains most inefficient.

[11] The endogenous wage subsidy rates are small relative to the tax rates imposed on labor. The cut-back in producers' labor costs due to subsidies are on the order of magnitude of 10% of the labor taxes.

[12] Similar second-best mechanisms apply for the exemption policy. Carbon tax exemptions work, however, only as an indirect wage subsidy and generally do not offset the distortions induced by shifting away the burden of carbon reduction mainly on labor.

1. CAPITAL MOBILITY AND LABOR SUPPLY ELASTICITY

We have repeated our basic calculations in a model with restricted international capital mobility. This framework greatly reduces the marginal excess burden of capital taxes in the benchmark and reverses the ranking of tax recycling strategies. Labor taxes are relatively more distortionary with an excess burden of 16%, and carbon tax recycling through cuts in labor taxes are superior to capital tax cuts. This model illustrates the theoretical point often made in public finance literature (see e.g., Goulder, 1994), namely, the cost of achieving a given level of reduction depends on both the revenue recycling and the tax interaction effect. The welfare costs are reduced for all policies because the benchmark tax system is less distorted and thus the negative tax interaction effects with additional carbon taxes are significantly lower than in the baseline model. The excess burden of exemptions over uniform carbon taxation stays in the same order of magnitude with a wage premia in the range of 69 to 92 thousand DM per worker at a 30% reduction target for labor and capital tax recycling.

If we make capital sector-specific in order to account for short-run adjustment costs, the marginal excess burden of the benchmark tax system further decreases. Decreased intersectoral capital mobility, however, increases the cost of achieving any reduction target by reducing the size of the production possibility frontier. As the overall cost increases, so does the excess cost of exemptions relative to uniform taxation. The wage premium of exemptions increases to between 91 and 118 thousand DM per job.

Augmenting the elasticity of labor supply with respect to real wages (from 0.1 to 0.15) increases the benchmark excess burden of taxes and leads to slightly higher welfare costs in all scenarios compared to the baseline results. The quantitative and qualitative ranking of different tax policies and recycling options remains unchanged.

2. NUMBER OF EXEMPTED SECTORS

The baseline computations suggest that the exemption of energy- and export-intensive sectors creates a large excess of costs even though the exempted sectors have only a small share of overall economic activity and carbon emissions. In order to see how the excess costs change as the number of exempted sectors increases, we vary the qualification requirements for exemptions. Not surprisingly, we find that the welfare costs of tax exemptions rise with the number of exempted industries.

3. BASE YEAR

A well-recognized short-coming of calibrated equilibrium modeling is that temporary fluctuations over the business cycle are always present in input-output tables. When these data are subsequently used to calibrate utility and cost functions, temporary disequilibria may distort the behavioral models for consumers and producers. In order to establish that our results were not unduly affected by extraordinary features of the 1990 input-output table, we have repeated our calculations using two alternative base years for which comparable input-output data are available, 1986 and 1988. We find that changing the base year does not alter our qualitative results.

F. SUMMARY AND CONCLUSION

In this study, we have considered the economic implications of sectoral exemptions from environmental regulation with specific reference to carbon taxation with exemptions for energy- and export-intensive industries. We have analyzed this issue in the context of a static general equilibrium model calibrated to 1990 data for West Germany. The insights which emerge from the simulations are as follows:

1. Welfare losses associated with exemptions can be substantial even when the share of exempted sectors in overall economic activity and carbon emission is small. Holding emissions constant, exemptions for some sectors imply increased tax rates for others and higher costs for the economy as a whole.
2. Results from our simulation of a uniform carbon tax suggest that political pressure from labor unions and industries is likely because many of these sectors decline in employment, output, and exports as a result of carbon taxes. Our model gives insights into trade-offs between sector-specific concessions for energy- and export-intensive industries and the impacts of exemptions on economy-wide efficiency.
3. Our comparison of a "tax expenditure" (i.e., a carbon tax exemption) with a direct wage subsidy revealed that the first of these is extraordinarily costly if it is viewed solely as a means of maintaining jobs in the affected industries.

Carbon emissions are a global externality, so the results and conclusions presented here are predicated on the assumption that unilateral carbon reduction in a single country does not induce significant changes of carbon emissions elsewhere. If leakage rates are low, tax exemptions of energy- and export-intensive industries do not provide an efficient policy for unilateral action. In the case of high leakage rates partial tax exemptions of industries might increase efficiency of (global) CO_2 reduction (Hoel, 1994; Böhringer and Rutherford, 1995). As leakage rates are crucial for the efficient design of a unilateral abatement policy, future research should be directed toward the measurement of these values in a multiregional framework.

REFERENCES

Ballard, C. et al., A general equilibrium model for tax policy evaluation, NBER Monogr., University of Chicago Press, 1985.

Ballard, C. and Shoven, J. B., The efficiency cost of achieving progressivity by using exemptions, in *Essays in Honor of Arnold Harberger*, Blackwell, New York, 1987, 109.

Böhringer, C., Ferris, M., and Rutherford, T. F., Alternative CO_2 abatement strategies for the European Union, in *Economic Aspects of Environmental Policy Making in a Federal State*, Braden, J. and Proost, S., Eds., Edgar Elgar, in press.

Commission of the European Communities, COM(92) 226 final, Brussels, 1992a.

Commission of the European Communities, The climate challenge, *European Economy*, 51, 1992b.

Commission of the European Communities, The economics of limiting CO_2 emissions, *European Economy*, Special Edition No.1, Brussels, 1992c.

Conrad, K. and Schröder, M., The control of CO_2 emissions and its economic impact, *Environ. and Resour. Econ.*, 1, 289, 1991.

Conrad, K. and Wang, J., Gesamtwirtschaftliche Auswirkungen einer CO_2–Besteuerung in Deutschland (West), Diskussionspapier, Universität Mannheim, 1993.

Dirkse, S. and Ferris, M., *The PATH Solver: A Non-Monotone Stabilization Scheme for Mixed Complementarity Problems, Optimization Methods & Software*, 1995.

Dixit, A., Tax policy in open economies, in *Handbook of Public Economics*, Vol. 1, North–Holland, Amsterdam, 1985.

Enquete–Kommission, Vorsorge zum Schutz der Erdatmosphäre des Deutschen Bundestages, Arbeitspapier: Michaelis–Vorschlag–Steuerbefreiung von Export- und Energieintensiven Industrien, *Economica Verlag*, Bonn, 1994.

Enquete–Kommission, Vorsorge zum Schutz der Erdatmosphäre des Deutschen Bundestages, Band 3, *Economica Verlag*, Bonn, 1995.

Ewringmann, D. and Hansmeyer, K. H., Beurteilung ökologischer Steuerreformvorschläge vor dem Hintergrund des Bestehenden Steuersystems, Studie im Auftrag des BDI, Universität Köln, 1994.

Goulder, L. H., Environmental Taxation and the Double Dividend: A Reader's Guide, mimeo, Stanford, 1994.

Grubb, M. et al., The costs of limiting fossil-fuel CO_2 emissions: a survey and analysis, *Annu. Rev. of Energy and the Environ.*, Vol. 18, 1993.

Hoel, M., Should a Carbon Tax be Differentiated Across Sectors?, mimeo, Department of Economics, University of Oslo, 1994.

Manne, A. S. and Richels, R. G., The EC proposal for combining carbon and energy taxes — the implications for future CO_2 emissions, *Energy Policy*, June 1992.

OECD, *Environmental Taxes in OECD Countries*, Paris, 1995.

Pezzey, J., Impacts of greenhouse gas control strategies on U.K. a competitiveness, Department of Economics, University of Bristol, 1993.

Proops, J., Faber, M., and Wagenhals, G., *Reducing CO_2-Emissions — A Comparative Input-Output Study for Germany and the UK*, Springer–Verlag, Heidelberg, 1993.

Rutherford, T., Welfare effects of fossil carbon restrictions, in *The Costs of Cutting Carbon Emissions: Results from Global Models*, Organization for Economic Cooperation and Development, Paris, 1993, 95.

Rutherford, T., General equilibrium modelling with MPSGE as a GAMS subsystem, mimeo, Department of Economics, University of Colorado, 1994.

Rutherford, Thomas F., Carbon dioxide emission restrictions in the global economy: Leakage, competitiveness, and the implications for policy design, in *An Economic Perspective on Climate Change Policies*, American Council for Capital Formation, Center for Policy Research, 1996, 203.

Statistisches Bundesamt, Volkswirtschaftliche Gesamtrechnungen — Input-Output-Tabellen, *Fachserie*, 18, Reihe 2, Wiesbaden, 1994.

Statistisches Bundesamt, Volkswirtschaftliche Gesamtrechnungen — Hauptbericht, *Fachserie*, 18, Reihe 1.3, Wiesbaden, 1995.

TemaNord, *The Use of Economic Instruments in Nordic Environmental Policy*, Copenhagen, 1994.

UNEP, Climate change 1995: Economic and social dimensions of climate change, *Inergovernmental Panel on Climate Change*, Cambridge University Press, 1995.

Welsch, H*., Gesamtwirtschaftliche Auswirkungen von CO_2–Minderungsstrategien, Teilstudie C2*, Enquete–Kommission zum Schutz der Erdatmosphäre, Bonn, 1994.

Wiegard, W., Exemption versus zero rating: a hidden problem of VAT, *J. of Public Econ.*, 46, 307, 1991.

shifts toward toxic-intensive structures in the 1970s and 1980s. This is because import-substituting industrialization protected mainly capital- and pollution-intensive sectors. Using partial equilibrium analysis, Anderson (1992) shows that even if a country has comparative advantages in the production of pollution-intensive goods, free trade would still raise welfare unambiguously, so long as an optimal pollution tax is introduced. Thus, previous studies suggest that trade does not necessarily lead to degradation of national environment, trade policy is never the first-best policy to remedy environmental problems, and an efficient environmental policy is one that would equalize marginal social costs and benefits of production.[2]

The objective of this chapter is to show empirically that a combination of trade liberalization and a cost-effective tax policy would not only raise the country's welfare, but it can also improve the environmental quality. We use applied general equilibrium analysis to examine the environmental implications of trade and tax policies in Indonesia. This is a country well suited to our analysis because it has comparative advantages in dirty industries and its trade has historically conferred asymmetric environmental effects, inducing a net transfer of environmental costs from its trading partners, particularly Japan.

In Section B, we present some statistics on the embodied pollution service trade between Indonesia and Japan during 1965–1990. Section C describes the two-country calibrated general equilibrium (CGE) model used in this study, followed by the appraisal of the environmental implications of Indonesia's trade liberalization in Section D and the evaluation of the welfare effect of pollution abatement by alternative instruments in Section E. Conclusions are summarized in Section F.

B. INTERNATIONAL TRADE AND PATTERNS OF EFFLUENT TRANSFER

This section offers some historical evidence on how international trade influences the transfer of environmental effects. We introduce the concept of embodied effluent trade (EET) to capture the idea that traded commodities embody an environmental service, i.e., the amount of pollution emitted when goods are produced domestically. If countries impose different environmental costs on pollution, then the ability to pollute becomes a source of comparative advantage. One would thus expect to see a pattern with relatively high EET in exports from countries with low environmental standards and relatively low EET in their imports, while the opposite would prevail in countries with higher environmental standards. This is indeed the case for trade between Indonesia and Japan, which exhibits a striking imbalance in EET.

The database on the Industrial Pollution Projection System of the World Bank is used to measure domestic effluent intensities in production. It provides emission levels per unit of output for a variety of pollutants at a four-digit ISIC level of sectoral detail for U.S. manufacturing.[3] The data are then mapped to four-digit output share data for Indonesia and Japan, computed from a 128-sector bilateral input-output table constructed by the Institute of Developing Economies (1991), to obtain weighted emission intensities for the 19 sectors of the model.[4] Table 18.1 presents the results of this conversion for Indonesia.[5]

---

[2] See Dean (1992) for a survey of literature on trade and the environment. O'Connor (1994) reviews the recent Asian Pacific experience. A survey of Indonesia's trade and adjustment policies can be found in Roland–Holst (1992).

[3] See Martin et al. (1991) and Wheeler (1992).

[4] Such detailed emission intensities are at the moment only available for U.S. manufacturing sectors, obliging us to apply them to both Indonesia and Japan. The 19 sectors are: (1) agriculture, forestry and fishing, (2) petroleum, (3) mining, (4) processed food, (5) textiles, (6) lumber and wood products, (7) pulp and paper products, (8) industrial chemicals, (9) other chemicals, (10) plastics, (11) non-metallic mineral products, (12) steel, (13) nonferrous metals, (14) metal products, (15) machinery and precision instruments, (16) electrical machinery, (17) transport equipment, (18) other manufactures, and (19) services.

[5] The results for Japan are similar to those for Indonesia although there are some differences in the emission intensities at the 19-sector level of disaggregation between the two countries because of different output shares at the four-digit ISIC level.

18 The Environment and Welfare Implications of Trade and Tax Policy

Hiro Lee and David Roland–Holst

Developing countries with comparative advantage in dirty industries face the risk of environ
degradation unless appropriate policies are implemented. Using applied general equilibrium an
we examine how trade influences the environment and assess the welfare and environmental appli
of alternative pollution abatement policies for Indonesia. Our results indicate that unilatera
liberalization by Indonesia would increase the ratio of emission levels to real output for alm
major pollution categories. More importantly, when tariff removal is combined with a cost-ef
tax policy, the twin objectives of welfare enhancement and environmental quality improvement
to be feasible. This sheds new and positive light on the role of trade in sustainable developmen

© 1997 Elsevier Scien

Key words: trade and environment, pollution, Indonesia, applied general equilibrium model

(*JEL* classifications: F13; O53; Q28.)

A. INTRODUCTION

International trade can exert an important influence on the environment via its effec
composition of domestic production activities. Countries with less stringent environment;
tions may have comparative advantage in dirty industries. This leads to the export of
services embodied in goods made with technologies that do not meet the environmental
of the importing countries. In the course of trade, one observes pollution being transferr
a life cycle from more to less advanced nations. A number of empirical studies (e.g., (
and Krueger, 1992; Hettige et al., 1992; Lucas et al., 1992) have shown an inverse
relationship between GDP per capita and industrial pollution intensity.[1] The normal good (
istic of environmental quality, relatively high costs of monitoring and enforcing pollution s
and an increase in output shares of manufactures during industrialization are some of
factors leading to relatively high pollution levels per unit in developing countries (Bir
Wheeler, 1992).

Although there has been intense pressure from environmentalists and some policy i
include environmental standards in trade agreements, economists have long argued that tr
the root cause of environmental damage. Low and Safadi (1992) suggest that freer trade ma
benefits to the environment through its effects on resource allocation and income levels. Lt
(1992) find that among developing countries, the more closed economies experienced v

---

[1] Hettige et al. (1992) and Lucas et al. (1992) suggest that the declining portion in the inverse U-shaped relatio
solely to a shift from industry to services and not a result of a shift toward a less toxic mix of manufacturing

TABLE 18.1
Sectoral Effluent Intensities in Indonesia (tons/year/$ million unless indicated otherwise)

		PARTIC[a]	SO2	NO2	LEAD	VOC	CO	BOD[b]	SS	TOX[c]	METAL
1	Agriculture	n.a.	n.a.	n.a.	n.a.	n.a.	n.a.	n.a.	n.a.	n.a.	n.a.
2	Petroleum	5.36	16.28	3.47	6.54	1.60	0.65	0.48	0.58	1.15	0.06
3	Mining	4.19	20.12	1.65	0.40	2.05	19.76	8.17	117.64	4.05	2.79
4	Processed food	0.53	0.63	1.84	0.02	0.37	0.51	7.55	1.95	0.29	0
5	Textiles	0.40	2.83	4.91	0.00	1.20	0.93	0.02	0.03	1.49	0.07
6	Lumber and wood products	4.15	1.42	3.10	0.00	4.19	5.33	0	0	2.05	0.02
7	Pulp and paper	1.35	13.37	5.41	0.69	2.85	7.72	9.9	39.91	3.1	0.02
8	Industrial chemicals	0.61	4.09	3.95	0.03	4.15	5.60	9.41	21.09	13.57	0.08
9	Other chemicals	0.50	5.50	2.69	0.10	2.58	3.30	2.54	0.72	1.7	0.02
10	Plastics	0.14	1.36	0.40	0.00	4.52	0.12	0	0	3.39	0.09
11	Non-metallic mineral products	4.71	6.90	7.62	0.29	0.54	1.38	0	0	1.54	0.38
12	Steel	1.61	4.42	1.99	6.11	1.05	15.27	0.02	12.41	3.47	1.86
13	Non-ferrous metals	4.12	20.36	1.54	0.00	2.08	21.00	8.67	125.23	4.23	2.97
14	Metal products	0.25	0.17	0.82	0.22	4.23	0.09	0.55	11.71	2.08	0.3
15	Machinery and precision instruments	0.68	0.37	0.28	0.20	0.83	0.10	0	0	0.71	0.11
16	Electrical machinery	0.05	0.17	0.10	0.15	2.04	0.12	0	0.02	0.82	0.14
17	Transport equipment	0.15	0.09	0.06	0.00	0.85	0.02	0	0.01	0.5	0.02
18	Other manufactures	0.26	0.22	0.07	0.00	4.12	0.03	0	0	1.23	0.27
19	Services	n.a.	n.a.	n.a.	n.a.	n.a.	n.a.	n.a.	n.a.	n.a.	n.a.

[a] Air pollutants: particulates (PARTIC), sulfur dioxide (SO2), nitrogen dioxide (NO2), lead (tons/year/$ billion), volatile organic compounds (VOC), carbon monoxide (CO).
[b] Water pollutants: biochemical oxygen demand (BOD), suspended solids (SS).
[c] Toxic pollutants/all media: total toxic release (TOX), bioaccumulative metals (METAL).

Sources: Martin et al. (1991), Wheeler (1992), and authors' calculations.

As the database is limited to industrial pollution, agriculture and services are omitted from the effluent database. Since environmental damages related to agriculture and forestry, such as soil erosion and loss of soil fertility from deforestation are particularly important for Indonesia, their omission would understate the pollution content of domestic production. Of the remaining 17 sectors, petroleum, mining, lumber and wood, pulp and paper, industrial chemicals, non-metallic minerals (consisting of cement and stone products), steel, and nonferrous metals may be regarded as pollution-intensive sectors. For example, petroleum has high effluent intensities of particulates, SO_2, NO_2, and lead, while mining and nonferrous metals have high emission coefficients on particulates, SO_2, carbon monoxide, the two water pollutants (biochemical oxygen demand and suspended solids), and the two toxic pollutants (total toxic release and bioaccumulative metals).

Let ε_{ih} denote sectoral effluent intensities of pollutant h. To measure average effluent levels embodied in tradeable commodities, the acute human toxic linear (AHTL) index developed by Wheeler (1992) is used. The AHTL index is a weighted average of various effluents with weights representing their human health risk. The index of sectoral effluent output is defined as

$$e_i = \frac{\varepsilon_{i,A}}{\sum_i \varepsilon_{i,A} q_i}, \tag{18.1}$$

where $\varepsilon_{i,A}$ is the sectoral AHTL emission rate per unit of output in U.S. manufacturing and q_i is sector i's share in total domestic output. If these indices are multiplied by 1985 U.S. sectoral output shares, they sum to unity. For any other country, such a sum measures the effluent potential of domestic output in units relative to the United States. In 1985, for example, Japanese output shares give a value of $E_q = \Sigma_i e_i q_i = 0.86$, indicating that, under the same technologies, the effluent intensity of Japanese domestic production would be 14% below that of the U.S. by this index. The comparable figure for Indonesia is 2.45. Thus E_q serves as an index of aggregate effluent levels for a given composition of domestic production. As the structure of the economy shifts toward relatively cleaner activities, such as services, this index will decline. It is unaffected by the absolute level of output, but simply allows comparison across countries of one representative unit of domestic product.

In light of differing environmental standards in the two countries, the disparity in E_q is likely to be greater than the indices would indicate. Japan's effluent controls are more stringent than those of the U.S., and thus the compositional index for the former is likely to overstate Japanese effluent levels. Likewise, Indonesia's environmental controls are weaker than the reference country, so its actual effluent levels are underestimated by E_q.

This measure can also be used to evaluate the implicit effluent content of trade. The indices

$$E_x^f = \sum_i e_i x_i^f \tag{18.2}$$

and

$$E_m^f = \sum_i e_i m_i^f \tag{18.3}$$

measure the embodied effluent content of exports and imports, respectively; x_i^f and m_i^f are the sectoral shares of exports to destination f (f = bilateral partner, rest of world (ROW)) and the sectoral shares of imports from origin f. If E_x^f exceeds unity, for example, the composition of the country's existing exports represents (in their production) a higher level of pollution per unit than that of the representative output in the United States. Values less than unity mean that the country's overall exports are cleaner than overall U.S. domestic output.

The indices E_x and E_m thus measure the embodied effluent trade for a given composition of exports and imports per unit of trade. E_x and E_m for Indonesian-Japanese bilateral trade and their trade with the rest of the world are presented in Table 18.2. These estimates were constructed for the 1965–90 period at five-year intervals, with detailed trade data from the United Nation's COMTRADE tables. The ratio of E_x and E_m are also given in the table.[6]

The most arresting feature of Table 18.2 is the imbalance in direct EET between the two trading partners. Over the last two and a half decades, Indonesia's production for export to Japan has been about six times more effluent intensive than have Japanese exports to Indonesia. In a long-term situation of relatively balanced bilateral trade, this implies a sustained and significant transfer of environmental costs from Japan to Indonesia. Although the trend in recent years has reduced this disparity, it is still quite significant.

These results are even more striking when compared to each country's trade with the rest of the world. Indonesia's imports from Japan are about half as effluent intensive as what it buys from other countries and its exports to Japan are about 30% as effluent intensive as other countries' exports to Japan. Trade between countries at different stages of modernization has long exhibited hierarchical properties that are correlated with technology levels and environmental effects.

---

[6] The pollution of U.S. exports and imports has been estimated by Walter (1973). He estimated environmental-control loadings entering U.S. trade flows and found that the pollution content of U.S. exports exceeded that of imports in 31 out of 78 sectors in 1971.

TABLE 18.2
Trends in Embodied Effluent Content of Exports and Imports[a]

	1965	1970	1975	1980	1985	1990	Average
Indonesia							
Exports to							
Japan	11.32	11.45	15.34	13.43	11.77	10.41	12.29
Rest of world	7.28	6.49	14.14	12.20	10.59	7.23	9.66
Imports from							
Japan	2.10	2.17	2.03	1.80	1.99	1.67	1.96
Rest of world	2.29	2.73	2.79	4.44	4.16	3.34	3.29
Effluent trade ratio (E_x/E_m)							
Japan	5.38	5.29	7.57	7.47	5.93	6.24	6.31
Rest of world	3.18	2.38	5.06	2.75	2.54	2.17	3.01
Japan							
Exports to							
Indonesia	2.10	2.17	2.03	1.80	1.99	1.67	1.96
Rest of world	1.75	1.62	1.69	1.60	1.52	1.54	1.62
Imports from							
Indonesia	11.32	11.45	15.34	13.43	11.77	10.41	12.29
Rest of world	4.09	3.87	7.63	8.86	7.39	4.78	6.10
Effluent trade ratio (E_x/E_m)							
Indonesia	0.19	0.19	0.13	0.13	0.17	0.16	0.16
Rest of world	0.43	0.42	0.22	0.18	0.21	0.32	0.30

[a] These indices measure embodied effluent content of exports and imports relative to the emission intensity of overall U.S. domestic output. Values greater than unity imply that the country's overall exports by destination or imports by origin are more pollution-intensive than the U.S. output.

Sources: Wheeler (1992), United Nation's COMTRADE database, and authors' calculations based on Equation 18.2 and Equation 18.3 in the text.

Given that the effluent indices are derived assuming U.S. technology and environmental standards, economic structure alone could explain Indonesia's higher pollution intensities in production, both for domestic and foreign consumption. There are significant differences in sectoral and trade structure between the two countries. Indonesia's heavy export dependence on petroleum (64% of total exports in 1985) is the dominating factor for its high effluents embodied in overall exports. On average, the petroleum sector has been responsible for more than 50% of Indonesia's industrial emissions and about 90% of the EET in exports. Lumber and wood and nonferrous metals have each accounted for between 3% and 4% of the effluents embodied in exports.[7] By contrast, except for steel, Japan's exports are concentrated in sectors with low pollution intensities, resulting in low effluents embodied in its exports. Thus, the composition of output in Indonesia is substantially more pollution intensive than that in Japan.[8]

---

[7] Because log exports were banned in 1985, the production and exports of wood products have increased sharply. In our industrial classification, logging is included in agriculture where no emission data are incorporated. Thus, a shift in exports from logs to lumber and wood products would also raise the value of embodied effluent content of Indonesian exports although the export ban could slow deforestation.

[8] Since the U.S. effluent coefficients are applied to Indonesia and Japan, differing levels of technology and environmental regulations between the two countries do not affect our results. If country-specific data were available, the results would have yielded even larger asymmetries. There are significant technological disparities between the two countries in a variety of industrial activities, with Japan's environmental regulation being more stringent than Indonesia's.

C. TWO-COUNTRY CGE MODEL FOR INDONESIA AND JAPAN

Calibrated general equilibrium models have been increasingly used as tools for detailed empirical analysis of the long-run implications of economic policy. They have been used extensively in general equilibrium assessments of CO_2 abatement policies (e.g., Burniaux el at., 1992; Jorgenson et al., 1992; Perroni and Rutherford, 1993; OECD, 1994). Because a CGE model can distinguish dirty industries and capture a variety of indirect effects, such as interindustry and trade linkages, it is well-suited to analyzing the impact of trade and tax policy on the environment and economic welfare.

The Indonesia–Japan CGE model is calibrated to the 1985 social accounting matrix (SAM) of the two countries.[9] An important feature of this model is its endogenous specification of domestic supply, demand, and bilateral trade for the two countries at the sectoral level. This is particularly important for Indonesia as its bilateral trade with Japan as a percentage of the total trade was 47% for exports and 21% for imports in the base year. While trade between the two countries is modeled endogenously, we assumed that their individual trade flows with the rest of the world (ROW) are each governed by the small country assumption.[10] The resulting six sets of sectoral trade flows are then directed by two endogenous price systems (Indonesia–Japan imports and exports), and four exogenous price systems (Indonesia–ROW and Japan–ROW imports and exports).

As has been employed in many other CGE models, a differentiated product specification is used for the demand and supply for tradeable commodities. Domestic demand is a CES composite of goods differentiated by origin. For each product category,

$$D_i = \overline{A}_{D_i} \left[\sum_k \beta_i^k \left(D_i^k \right)^{(\sigma_i - 1)/\sigma_i} \right]^{\sigma_i/(\sigma_i - 1)} \tag{18.4}$$

where k = {Indonesia, Japan, ROW}. D_i^k consists of domestic goods, imports from the bilateral trading partner, and imports from ROW, σ_i is elasticities of substitution among D_i^k, and $\overline{A}_{D_i}$ and β_i^k are intercept and share parameters. Similarly, domestic production is supplied to differentiated destinations (domestic market, exports to bilateral partner, and exports to ROW), which is specified as CET composite:

$$S_i = \overline{A}_{S_i} \left[\sum_k \delta_i^k \left(S_i^k \right)^{(\lambda_i + 1)/\lambda_i} \right]^{\lambda_i/(\lambda_i + 1)} \tag{18.5}$$

where λ_i is elasticities of transformation among S_i^k, and $\overline{A}_{S_i}$ and δ_i^k are intercept and share parameters.

Every sector is characterized by constant returns to scale and perfect competition.[11] The production function is given by

$$S_i = \min \left\{ CES \left(L_{D_i}, K_{D_i}; \phi_i \right), V_{ji}/a_{ji}, \dots, V_{ni}/a_{ni} \right\}, \tag{18.6}$$

---

[9] See Lee and Roland–Holst (1993b) for a complete set of equations describing the model.

[10] Lee and Roland–Holst (1993a) treat Japan as a large country so as to affect prices in the ROW market. For the moderate trade flow adjustments for Japan described in this study, however, the small country assumption makes almost no change in the results of simulation experiments.

[11] While varying returns to scale does affect the magnitude of impact of trade and tax policies, the key results of this paper are robust and not affected by different specifications on market conduct (e.g., oligopolistic behavior) and returns to scale.

where L_{D_i} and K_{D_i} are labor and capital demands, ϕ_i is the elasticities of substitution between labor and capital, a_{ji} is input-output coefficients, and $V_{ji} = a_{ji}S_i$ is demand for intermediate good i in sector j. The zero-profit condition implies

$$\left(1 - t_{S_i}\right)P_{S_i} = AC_i,$$ (18.7)

where P_{S_i} and AC_i are prices and average costs of composite supply.[12]

Ad valorem tax rates on supply, t_{S_i}, are the sum of ad valorem indirect taxes, t_{X_i}, and ad valorem effluent taxes:

$$t_{S_i} = t_{X_i} + \sum_n \tau_{ih}\varepsilon_{ih},$$ (18.8)

where τ_{ih} are excise taxes on emissions (\$/ton of pollutant h).

We assume both countries have a fixed aggregate stock of domestic productive capital which is mobile between sectors, while the economy-wide average rental rate adjusts to equate aggregate capital demand to the fixed total supply. We also assume that labor in both countries is mobile between sectors, but the total labor supply is specified as function of the wage rate and household income. In the product markets, prices are normalized by a fixed numeraire chosen to be the GDP price deflator. Finally, we assume that the real exchange rate is flexible while the current account balances for the two countries are fixed at the baseline values.[13]

Sectoral emission levels by pollutant and destination of supply are computed as

$$EMI_{ih}^k = \varepsilon_{ih}P_{S_i}^k S_i^k.$$ (18.9)

The matrix of effluent intensities by sector and type of pollutant, $\{\varepsilon_{ih}\}$, forms the basis for calculating environmental effects resulting from policy changes, such as tariff liberalization and effluent taxes. A limitation of this approach at the moment is that there is no scope for technical substitution within sectors, and thus emissions are proportional to output regardless of relative prices and differential effluent taxes. The main advantage of this approach over previous modeling with these coefficients is the general equilibrium nature of the simulations, which allow for changing composition of domestic output, a large medium-term source of pollution mitigation.[14]

D. TRADE AND DOMESTIC POLLUTION IN INDONESIA

The two-country CGE model is used to assess the linkage between trade and the environment by removing Indonesia's nominal tariffs on all imports. Table 18.3 summarizes the aggregate results. The tariff removal leads to an increase in Indonesia's real GDP by 0.87% and economy-wide employment by 1.87%. Equivalent variation (EV) income, which measures the change in real consumer purchasing power, rises by less than the increase in real GDP because of a fall in the bilateral terms of trade with Japan.[15] The wage rate and the rental rate on capital both increase, but the latter increases more because an increase in labor supply in response to higher wages raises the natural productivity of capital. Indonesia's tariff removal induces real rupiah depreciation and

---

[12] The composite supply price is given by $P_{S_i}S_i = \Sigma_k P_{S_i}^k S_i^k$.

[13] Since there are no assets in the model, the real exchange rate is the relative price of tradeables to nontradeables.

[14] Compare to e.g., Anderson (1992) and ten Kate (1993).

[15] The terms of trade with the rest of the world are unaffected because of the small country assumption.

TABLE 18.3
Aggregate Results of Indonesia's Tariff Liberalization (percentage changes)

	Indonesia	Japan
Real GDP	0.87	0.00
EV income	0.53	0.03
Wage rate	1.10	0.05
Employment	1.87	0.00
Rental rate on capital	3.31	0.04
Real exchange rate	5.26	–0.07
Total imports	5.81	0.14
Total exports	5.71	–0.07

the subsequent increase in its exports. It has a negligible effect on the Japanese economy, with all aggregate measures changing by a small fraction of one percent.

Trade liberalization leads to dramatic shifts in the composition of Indonesia's sectoral trade and output that are driven by sharp changes in relative prices. It induces an expansion of output in petroleum and mining, lumber and wood, nonferrous metals, and services.[16] The real rupiah depreciation leads to increased demand for Indonesian exports in most of the major sectors.[17] Results for Japan are again small, except for adjustments in bilateral trade with Indonesia. There is some shift of resources toward Japanese export sectors and a slight diversion of import demand in response to the rupiah depreciation.

While removing tariff protection leads to expanded trade and greater economy-wide efficiency, in the absence of new technologies it entails an increase the total emission levels. Tables 18.4 and 18.5 report the effects on emission levels of each pollutant embodied in domestic output supplied to different destination (domestic market, bilateral partner, and ROW). Given the extensive compositional shifts in production that occur in response to trade liberalization in Indonesia, emissions from the production of goods supplied domestically increase for three pollutants (particulates, SO_2, and lead) while those decrease for the other pollutants (column (1) of Table 18.4). In percentage terms, these changes are relatively small (column (5)). Trade expansion would, however, increase emissions quite substantially from the production of goods that are exported to both Japan and the rest of the world for all pollution categories included in the study (columns (2)–(3), (6)–(7)). The net effect is an increase in the emission level of all the pollutants generated from total output (columns (4), (8)). Table 18.5 reports the effects on emissions from output produced in Japan resulting from Indonesia's tariff removal, which are significantly smaller than those in Indonesia in percentage terms.

The result that trade liberalization leads to higher pollution levels is not surprising because it leads to an increase in real output. A more interesting result is that it leads to an increase in the relative output shares of dirty industries, causing higher average pollution intensities for almost all major pollution categories. The only exception is biochemical oxygen demand (water pollution) whose emission level rises by a smaller percentage (0.51%) than the increase in real output. For all other pollutants, the percentage change in emission levels (1.43–3.73%) exceeds the percentage change in real output, resulting in higher emission intensities.

---

[16] The sectoral results are available upon request from the authors.

[17] The sectors that experience a fall in exports (pulp and paper, non-mineral metallic products, steel, and metal products) have small export shares.

TABLE 18.4
Changes in Emission Levels by Destination of Supply: Indonesia[a]

	Absolute changes[b]				Percentage changes			
	(1)	(2)	(3)	(4)	(5)	(6)	(7)	(8)
	Domestic	Japan	ROW	Total	Domestic	Japan	ROW	Total
PARTIC	773	2298	2297	5369	0.89	4.94	7.40	3.27
SO_2	2461	7146	6338	15945	1.04	5.01	7.23	3.41
NO_2	−209	1443	1619	2853	−0.24	4.89	6.96	2.04
LEAD	1.17	2.54	2.23	5.94	1.57	4.79	7.02	3.73
VOC	−599	747	1029	1178	−1.15	5.07	7.80	1.47
CO	−778	1112	1307	1641	−1.20	6.71	9.49	1.73
BOD	−657	536	608	487	−0.83	6.43	8.66	0.51
SS	−1497	5094	3973	7571	−0.76	7.66	11.71	2.55
TOX	−368	649	951	1232	−0.89	5.46	9.38	1.95
METAL	−62	137	106	181	−0.91	6.99	9.80	1.83
AHTL index	−0.10	1.17	1.56	2.64	−0.16	5.16	8.76	2.64

[a] Changes in emission levels of pollutant embodied in output supplied to different destination resulting from unilateral tariff liberalization by Indonesia.

[b] Absolute changes in thousand tons of pollutant except the AHTL index.

E. RELATIVE COST OF ALTERNATIVE TRADE AND TAX POLICIES IN CURTAILING POLLUTION

For Indonesia, the emission results of Section D amplify the policy challenge of addressing the environmental consequences of trade-based economic growth. If the marginal social damage caused by an increase in emissions of a particular pollutant is known, then in the absence of other distortions in the economy an optimal policy would be the imposition of an effluent tax which would internalize the social damage (Pigou, 1920). However, true marginal damage is unknown, and reliable estimates on marginal damage functions associated with externalities are unavailable. This is an important direction for future research as economic accounting of environmental costs and benefits would be essential for comprehensive integration of economic and environmental policies.[18]

Since the uncertainties on marginal benefits of pollution abatement would make the calculation of an optimal tax rate impossible, our approach is to set a particular level of emission target and assess empirically the relative cost of alternative instruments that achieve the target.[19] In the first three experiments, we evaluate the cost of mitigating emissions of various pollutants by 5% using three policy instruments: an export tax, sector-specific effluent taxes, and a uniform effluent tax.[20] Different abatement targets have also been tried, but the relative efficiency of these instruments were not affected by the choice of abatement targets. No taxes are levied on the agricultural or service sector because no emission data are incorporated for these sectors. An AHTL (human health risk index) tax is equivalent to a set of taxes on the major air pollutants (particulates, SO_2, NO_2,

[18] For a survey of this kind of environmental valuation, see O'Connor (1992).

[19] An optimal rate of effluent tax would equalize marginal damage and marginal abatement cost of 1 ton of pollutant. On the cost side, Hartman et al. (1994) provide comprehensive estimates on abatement of 7 air pollutants for 37 U.S. manufacturing industries.

[20] Since our primary objective is the evaluation of the combined effects of trade liberalization and cost-effective tax policy, we limited the number of tax instruments to three. For evaluation of alternative policy instruments, such as pollution abatement and control expenditure (PACE) equalization tax and the polluter-pay principle, see Low (1992) and Low and Safadi (1992).

TABLE 18.5
Changes in Emission Levels by Destination of Supply: Japan[a]

	Absolute changes[b]				Percentage changes			
	(1) Domestic	(2) Indonesia	(3) ROW	(4) Total	(5) Domestic	(6) Indonesia	(7) ROW	(8) Total
PARTIC	−1058	51	−49	−1057	−0.08	2.82	−0.07	−0.08
SO$_2$	−3619	83	−147	−3683	−0.10	1.58	−0.07	−0.09
NO$_2$	−520	84	−38	−473	−0.03	2.97	−0.04	−0.03
LEAD	−0.96	0.01	−0.05	−1.00	−0.09	0.70	−0.06	−0.09
VOC	−417	83	−72	−406	−0.03	2.77	−0.06	−0.03
CO	−2111	50	−221	−2282	−0.06	0.67	−0.08	−0.06
BOD	−495	25	−46	−517	−0.02	1.05	−0.07	−0.02
SS	−9796	126	−638	−10309	−0.09	0.75	−0.12	−0.09
TOX	−550	56	−72	−565	−0.04	1.45	−0.06	−0.03
METAL	−276	9	−30	−297	−0.06	1.00	−0.08	−0.06
AHTL index	−0.64	0.08	−0.12	−0.69	−0.03	1.11	−0.05	−0.03

[a] Changes in emission levels of pollutant embodied in output supplied to different destination resulting from unilateral tariff liberalization by Indonesia.

[b] Absolute changes in thousand tons of pollutant except the AHTL index.

lead, volatile organic compounds, and carbon monoxide) which act to reduce the AHTL index by 5%. In the fourth experiment, the combination of a uniform effluent tax and tariff removal is simulated to evaluate whether it is possible to increase real output and reduce emissions at the same time. While these experiments were conducted for each pollutant, for simplicity we only report the aggregate results for SO$_2$ and the AHTL index in Table 18.6.[21]

In the first experiment, an export tax is chosen as a policy instrument because in Indonesia exports are on average considerably more pollution intensive than goods supplied domestically (Section B). Since the root cause of the pollution problem is production (regardless of destination), however, the imposition of a tax only on exports would be less efficient than a tax on output supplied domestically and exported. Columns (1a) and (1b) of Table 18.6 indicate that the cost of achieving the emission target with an export tax, in terms of lost real GDP or EV income, is highest among the three policy instruments. The reduction in total exports is partly offset by a large real depreciation of the rupiah but is still over 10% of the baseline quantity. The sharp contraction of trade causes additional reduction in real GDP and EV.

In experiment 2, sector-specific effluent taxes are levied to lower SO$_2$ emissions or the AHTL index by 5% in every sector.[22] While effluent taxes are imposed on all output regardless of destination, an enforcement of the same abatement target in all industrial sectors imposes an extremely high cost on some. This is clearly illustrated in Table 18.7, which summarizes the sector-specific SO$_2$ taxes required to achieve alternative emission targets within each sector. These taxes approximate the marginal cost of mitigating SO$_2$ emissions compared with the baseline levels for each sector.[23] SO$_2$ abatement costs in metal products, transport equipment, and other manufactures

[21] SO$_2$ is chosen because it is a pollutant that is known to adversely affect local environmental conditions, including acidification of soils and water and corrosion of materials.

[22] Sector-specific taxes required to mitigate emissions by 5% will lead to the same results regardless of the pollutant chosen except for the effluent tax results, which depend upon the effluent intensities of the pollutant in different sectors.

[23] These are not directly comparable with the econometric estimates on U.S. sectoral marginal abatement costs by Hartman et al. (1994). As in other CGE models, our estimates take into account input-output linkages but are based on the database of the benchmark year (1985) and the embedded model structure.

TABLE 18.6
Aggregate Results for Alternative Trade and Tax Policies (percentage changes)

	Export tax[a]		Sector-specific effluent taxes[b]	Uniform effluent tax[c]		Uniform tax and liberalization[d]	
	(1a)	(1b)	(2)	(3a)	(3b)	(4a)	(4b)
	SO_2	AHTL		SO_2	AHTL	SO_2	AHTL
Real GDP	−1.65	−2.26	−1.22	−0.56	−0.54	0.30	0.33
EV income	−1.18	−1.67	−1.14	−0.34	−0.45	0.25	0.14
Employment	−3.07	−3.99	−2.46	−0.15	−0.97	1.82	0.92
Wage rate	−2.53	−3.57	−2.16	−1.15	−1.09	0.03	0.14
Rental rate on capital	−5.75	−7.68	−5.46	−2.49	−2.78	0.80	0.56
Real exchange rate	17.67	24.70	4.81	3.61	3.54	9.02	9.03
Total imports	−11.87	−15.58	−1.94	−1.45	−1.47	4.20	4.09
Total exports	−10.87	−14.25	−2.26	−2.15	−1.64	3.31	3.81
SO_2 emissions	−5.00	−6.42	−5.00	−5.00	−3.74	−2.03	−0.61
AHTL index	−3.98	−5.00	−5.00	−3.44	−5.00	−1.10	−3.07

[a] (1a) export tax to cut SO_2 emissions by 5%; (1b) export tax to lower the AHTL index by 5%.
[b] Sector-specific effluent taxes to lower SO_2 emissions or the AHTL index by 5% in every sector.
[c] (3a) uniform effluent tax to cut SO_2 emissions by 5%; (3b) uniform effluent tax to lower the AHTL index by 5%.
[d] (4a) combination of (3a) and tariff removal; (4b) combination of (3b) and tariff removal.

TABLE 18.7
Summary of SO_2 Tax Results Under Alternative Emission Targets ($/ton of SO_2 at 1985 prices and 1985 exchange rate)

	Sector-specific taxes	Emission reduction compared with the baseline				
		1%	3%	5%	7%	10%
1	Agriculture	n.a.	n.a.	n.a.	n.a.	n.a.
2	Petroleum	1.05	3.07	5.01	6.87	9.50
3	Mining	1.16	3.39	5.51	7.52	10.34
4	Processed food	30.20	89.87	148.59	206.40	291.49
5	Textiles	3.65	10.83	17.85	24.71	34.70
6	Lumber and wood	14.21	42.00	68.96	95.09	132.75
7	Pulp and paper	0.70	2.05	3.36	4.61	6.39
8	Industrial chemicals	0.22	0.60	0.93	1.20	1.48
9	Other chemicals	4.27	12.64	20.78	28.69	40.14
10	Plastics	11.63	34.85	58.05	81.25	116.12
11	Non-metallic mineral products	5.89	16.85	26.79	35.78	47.65
12	Steel	3.55	10.40	16.90	23.07	31.70
13	Nonferrous metals	0.34	0.99	1.61	2.19	3.02
14	Metal products	89.25	262.54	429.05	588.94	816.70
15	Machinery and precision instruments	17.64	52.66	87.35	121.71	173.03
16	Electrical machinery	47.47	139.98	229.23	315.23	438.08
17	Transport equipment	87.66	259.92	428.18	592.51	831.85
18	Other manufacturing	90.97	270.54	446.95	620.20	874.09
19	Services	n.a.	n.a.	n.a.	n.a.	n.a.
20	Weighted average[a]	1.79	5.27	8.62	11.85	16.46
21	Uniform tax	0.74	2.19	3.61	4.99	7.02

[a] The sectoral SO_2 emission shares are used for the weights.

are found to be more than 400 times those in industrial chemicals. Thus, regulation, which would require every sector to cut emissions by the same proportion, would be highly inefficient.

Given the large disparity in marginal abatement costs, a uniform effluent tax would significantly reduce the costs of achieving a given emission curtailment target (experiment 3). The cost of cutting SO_2 emissions by 5% in terms of a loss in real GDP under a uniform tax is less than half (0.56 vs. 1.22%) compared with sector-specific taxes (Table 18.6, columns 2 and 3a). Under this scheme each sector will abate SO_2 until the marginal abatement cost is equal to the uniform tax rate. Thus, the industrial chemical and nonferrous metal sectors will abate more than 10% of SO_2, pulp and paper over 5%, and petroleum and mining between 3 and 5% (Table 18.7). Many high abatement cost sectors will not abate any SO_2 emissions at all. The uniform tax rate required to achieve a given target is also substantially lower than the emission weighted average of the sector-specific taxes (Table 18.7, rows 20 and 21).

While a uniform effluent tax will tend to minimize the cost of a given mitigation target, those sectors with low marginal abatement cost would bear much of the cost in terms of loss in real output. A system of tradeable emission permits is an alternative cost-effective instrument to a uniform tax, but which can be more supportive to equity issues. Under this system a fixed number of permits to emit a specific quantity of the pollutant is issued to emitters. Those firms or sectors with low abatement cost can sell permits to those with high abatement cost at a market-clearing permit price, thereby receiving compensation for further abatement in emissions.

The equilibrium permit price is determined by demand and supply of permits, which should equal the uniform tax rate required to achieve the same emission curtailment target. In the absence of transaction costs and regulatory distortions, a uniform tax and tradeable emission permits would both achieve a given level of environmental quality at minimum cost.[24]

In the final experiment, the same uniform tax scheme implemented in the third experiment is combined with removal of all tariffs. This experiment is conducted to illustrate a critical point that the combination of trade liberalization and a cost-effective emission abatement instrument can lead to both an improvement in welfare (in terms of real GDP or EV) and a reduction in pollution (Table 18.6, columns 4a and 4b). This is possible because the benefits of tariff removal are greater than the cost of cutting pollution by the magnitude which more than offsets pollution induced by trade liberalization. The twin objectives of a welfare improvement and an emission curtailment can be achieved under a range of ex ante abatement targets for each pollutant. For SO_2, for example, the implementation of an ex ante abatement target of between 3.2 and 7.4% by a uniform tax, combined with complete tariff removal, will lead to realization of both objectives. It should be recalled that no pollution externalities have been introduced in our model because of the uncertainties regarding marginal damage. In the presence of externalities, therefore, the net social benefits of the combined policy would be even greater than our estimates would suggest.

F. CONCLUSIONS

Indonesia's historical trade orientation has been environmentally asymmetric, effecting significant transfers of pollution services from its trading partners, particularly Japan, to the domestic economy. While trade liberalization would improve Indonesian real income, it would also raise the emission level of major industrial pollutants. In light of this tradeoff between outward-oriented industrialization and the environment, we have assessed the relative cost of curtailing pollution with a variety of instruments, including export taxes, sector-specific effluent taxes, and uniform effluent taxes. In

---

[24] Hahn and Stavins (1992) point out that these incentive-based approaches are not well suited when there are political and technological constraints. For example, source-specific standards may be more appropriate for highly localized pollution problems with nonlinear damage functions. It is beyond the scope of this study, however, to incorporate such additional features as emission source type and political constraints.

addition, a combination of uniform tax and tariff removal is simulated to examine the possibility of lowering domestic emissions and raising material welfare simultaneously.

Our simulation results indicate that a uniform effluent tax is the most cost-effective instrument in abating SO_2 emissions. This result holds for abatement of other industrial pollutants and for different abatement targets. Neither the imposition of an export tax, nor uniform emissions reduction with sector-specific taxes are recommended as an alternative policy. Pollution abatement using these instruments would result in a loss of real GDP that is significantly greater than achieving the same target using a uniform effluent tax.

The most important result of this chapter is that it is possible to abate industrial pollution while maintaining or even increasing real output when uniform taxation is combined with trade liberalization. In other words, trade liberalization should not be discouraged because of its environmental effects, and environmental taxation need not be contractional if distortions can be removed elsewhere. While the present model does not incorporate the benefits of reduced pollution in the utility function or EV calculation, their inclusion would only strengthen our conclusion.

ACKNOWLEDGMENTS

Research was conducted under the auspices of the OECD Development Center's 1993–1995 research program on Sustainable Development, Environment, Resource Use, Technology and Trade and CEPR's research project on Market Integration, Regionalism, and the Global Economy. Funding from the Japan Foundation Center for Global Partnership (CGP) and the Permanent Delegation of the Netherlands to the OECD is gratefully acknowledged. We thank Sebastian Dessus, David O'Connor, David Turnham, Dominique van der Mensbrugghe, and anonymous referees for their helpful comments. The opinions expressed here are those of the authors and should not be attributed to their affiliated institutions.

REFERENCES

Anderson, K., Effects of trade and environmental policies on the environment and welfare, in *The Greening of World Trade Issues*, Anderson, K., and Blackhust, R., Eds., University of Michigan Press, Ann Arbor, 1992.

Birdsall, N. and Wheeler, D., Trade policy and industrial pollution in Latin America: where are the pollution havens?, in *International Trade and the Environment*, Discussion Paper 159, Low, P., Ed., World Bank, Washington, D.C., 1992.

Burniaux, J. M., Nicoletti, G., and Martins, J. O., GREEN: A global model for quantifying the costs of policies to curb CO_2 emissions, *OECD Econ. Stud.*, 19, 49, 1992.

Dean, J. M., Trade and the environment: a survey of the literature, in *International Trade and the Environment*, Discussion Paper 159, Low, P., Ed., World Bank, Washington, D.C., 1992.

Grossman, G. M. and Krueger, A. B., Environmental aspect of a North American Free Trade Agreement, Working Paper 644, Center for Economic Policy Research, London, 1992.

Hahn, R. W. and Stavins, R. N., Economic incentives for environmental protection: integrating theory and practice, *Am. Econ. Rev., Pap. and Proc.*, 82, 464, 1992.

Hartman, R. S., Wheeler, D., and Singh, M., The cost of air pollution abatement, mimeo, Policy Research Department, World Bank, Washington, D.C., 1994.

Hettige, H., Lucas, R. E. B., and Wheeler, D., The toxic intensity of industrial production: global patterns, trends, and trade policy, *Am. Econ. Rev., Pap. and Proc.*, 82, 478, 1992.

Institute of Developing Economies, International Input-Output Table: Indonesia–Japan, 1985, IDE, Tokyo, 1991.

Jorgenson, D. W., Slesnick, D. T., and Wilcoxen, P. J., Carbon taxes and economic welfare, *Brookings Papers on Economic Activity: Microeconomics*, 393, 1992.

Lee, H. and Roland–Holst, D., Cooperation or confrontation in US–Japan trade? Some general equilibrium estimates, *Irvine Economics Paper 92-93-08*, University of California, Irvine, CA, 1993a.

Lee, H., and Roland–Holst, D., International trade and the transfer of environmental costs and benefits, *Technical Paper 91*, OECD Development Center, Paris, 1993b.

Low, P., Trade measures and environmental quality: the implications for Mexico's exports, in *International Trade and the Environment*, Discussion Paper 159, Low, P., Ed., World Bank, Washington, D.C., 1992.

Low, P. and Safadi, R., Trade policy and pollution, in *International Trade and the Environment*, Discussion Paper 159, Low, P., Ed., World Bank, Washington, D.C., 1992.

Lucas, R. E. B., Wheeler, D., and Hettige, H., Economic development, environmental regulation and the international migration of toxic industrial pollution: 1960–1988, in *International Trade and the Environment*, Discussion Paper 159, Low, P., Ed., World Bank, Washington, D.C., 1992.

Martin, P., Wheeler, D., Hettige, M., and Stengren, R., The industrial pollution projection system: concept, initial development, and critical assessment, mimeo, World Bank, Washington, D.C., 1991.

O'Connor, D., Measuring the costs of environmental damage: a review of methodological approaches with an application to Bangkok, Thailand, mimeo, OECD Development Center, Paris, 1992.

P'Connor, D., Managing the environment with rapid industrialization: lessons from the East Asian experience, *Developing Center Studies*, OECD, Paris, 1994.

OECD, Policy response to the threat of global warming, Working Party No. 1 *Report of the Economic Policy Committee*, ECO/CPE/WP1(94)5, OECD, Paris, 1994.

Perroni, C. and Rutherford, T., International trade in carbon emission rights and basic materials: general equilibrium calculations for 2020, *Scand. J. of Econ.*, 95, 257, 1993.

Pigou, A. C., *The Economics of Welfare*, Macmillan, London, 1920.

Roland–Holst, D., Stabilization and structural adjustment in Indonesia: an intertemporal general equilibrium analysis, *Technical Paper 83*, OECD Development Center, Paris, 1992.

ten Kate, A., Industrial development and the environment in Mexico, Working Paper 1125, Policy Research Department, World Bank, Washington, D.C., 1993.

Walter, I., The pollution content of American trade, *West. Econ. J.*, 11, 61, 1973.

Wheeler, D., The economics of industrial pollution control, Industry Series Paper 60, Industry and Energy Department, World Bank, Washington, D.C., 1992.

19 The Impact of NAFTA on Mexico's Environmental Policy

Bryan W. Husted and Jeanne M. Logsdon

A major controversy during the debate over the North American Free Trade Agreement focused on the impact of NAFTA on Mexico's environment. This chapter examines the evidence of impact specifically on Mexico's environmental policy. Criteria of impact are developed, and comparisons made for three periods: before 1990 as the baseline period; 1990–1993 when NAFTA was being negotiated; and beginning in 1994 when NAFTA came into effect. Much evidence indicates that Mexico's environmental policymaking and enforcement did improve in the early 1990s while NAFTA was being debated. Some evidence also suggests that the NAFTA-influenced environmental commitment was sustained during the 1995 financial crisis. Thus, it is concluded that NAFTA has contributed significantly to Mexico's environmental policy.

A. INTRODUCTION

The North American Free Trade Agreement (NAFTA) brought to public attention the question of the impact of trade on environmental protection in countries with different levels of economic development. In 1990, the United States was one of the wealthiest nations in the world, with a substantial regulatory and industrial infrastructure dedicated to environmental protection. On the other hand, Mexico was a middle-income country that was just emerging from a severe economic crisis during the 1980s. During that decade its environmental conditions had been deteriorating in terms of industrial pollution and population-related environmental degradation.

NAFTA's supporters tended to separate trade issues from environmental issues and advocated that NAFTA be evaluated only as a trade agreement. When pressed about NAFTA's environmental impacts on Mexico, they responded with the argument that greater economic growth from free trade would bring greater resources that could be devoted to environmental protection. Critics feared that NAFTA would allow "dirty" industries to migrate more easily to Mexico where they could pollute with impunity. Many doubts were expressed about the will of Mexico's government to rigorously enforce environmental regulations and its private sector to comply with them.

This chapter addresses the issue of environmental impacts of NAFTA, specifically on environmental policy in Mexico, by first examining the arguments about trade and environment in order to develop criteria by which the Mexican experience can be evaluated. Then we analyze Mexico's expressed and operative environmental policy before 1990 and after 1990 when the drive to create NAFTA began in earnest. This comparison of pre-NAFTA and NAFTA-induced policies and practices will lead to an assessment of the impact of NAFTA on Mexico's environmental policies and practices to date in urban manufacturing areas and in border cities.

B. DEBATE ON TRADE, ENVIRONMENT, AND ENVIRONMENTAL POLICY

Two lines of argument dominate the debate about the relationship between free trade, environmental quality, and environmental regulation in developing countries. The conventional economic position advocates that developing nations should incorporate themselves into the world trading system as a necessary step for economic growth (e.g., Srinivasan, 1982). With economic growth and higher per capita incomes will come the resources to invest in pollution control and the ability of consumers to select less "environmentally intensive" products (e.g., Globerman, 1993). Free trade also permits the importation of pollution control technologies that have been developed elsewhere (Bhagwati, 1993). A variation of this argument focuses on the role of knowledge about the consequences of environmental degradation as citizens of developing nations are exposed to ideas and values from industrialized nations (Schmidheiny, 1992). As living standards improve because of increased trade, there will be greater concern with the quality of life and more information available about environmental hazards. This, in turn, will lead to stricter environmental regulations and ultimately greater protection of the environment (Bhagwati, 1993).

The opposing position to the conventional economic argument views trade liberalization as a potentially serious threat to environmental quality for developing countries. One argument holds that First World countries use free trade to export the ecological costs of capitalist production to the Third World. According to this reasoning, high-polluting industries will tend to locate away from affluent nations where their costs are higher and instead choose to locate in poorer countries where local governments are more desperate for economic development and will tolerate little or no pollution control (Faber, 1992; Daly, 1993). A variation of this argument is that the local business in poor countries will be forced to compete with more technologically sophisticated firms from richer countries and, therefore, limit investments in pollution control equipment in order to keep their costs low. Under each of these scenarios, whether the polluters are local or from more industrialized nations, environmental policies will be set at low levels or not be adequately enforced. A final argument that free trade leads to environmental deterioration in developing nations is that with economic growth come higher incomes and ultimately higher consumption, which lead to more environmental degradation. According to steady-state environmental thinking (e.g., Daly, 1993), the only way to control environmental degradation is to stimulate development without economic growth per se.

In order to analyze whether the conventional economic or the critics' arguments are the more accurate, researchers will examine the state of Mexico's environment and environmental policy after NAFTA is fully implemented in the year 2009. It is not now empirically possible to evaluate which of these two positions will more accurately explain the long-term impacts. However, neither of these lines of argument addresses the hypothesis that Mexico's policies and practices toward the environment might have been strengthened in order to generate sufficient support for the passage of NAFTA itself (e.g., Pastor, 1993). Some analysts have speculated on a related but distinct hypothesis that Mexican regulatory enforcement might be at best cosmetic and serve only to fulfill the minimum expectations required in order to gain support at home and abroad for its economic reforms (e.g., Mumme, 1992b). In order to analyze these two hypotheses, evidence about the following questions before 1990, between 1990 and 1993, after 1993[1] would be relevant:

- What trends in environmental laws, standards, and regulations were evident in the three time periods? Specifically, were environmental rules being strengthened?
- How effectively were these laws and regulations enforced in each period?
- Did the behavior of polluting businesses change after 1990, and could any behavioral changes be attributed to NAFTA? Were behavioral changes sustained once NAFTA came into effect?
- To the extent that can be determined at this time, has Mexico's environment been improved or degraded as a result of NAFTA?

---

[1] On June 10, 1990, U.S. President George Bush and Mexico's President Carlos Salinas de Gortari met in Washington, D.C., and announced that a comprehensive free trade agreement would be in the best interests of both countries. Canada soon after expressed an interest in participating in the negotiation process. NAFTA negotiations were concluded and accepted by the three signatory governments in 1993, and the trade agreement went into effect on January 1, 1994. The three periods in this study reflect this history.

The first hypothesis, that the government's environmental policy was used to generate support or at least reduce criticism of the agreement, would be supported by evidence that (1) Mexico's environmental laws and regulatory standards were strengthened after 1990; (2) regulatory enforcement was strengthened after 1990; (3) many companies did change their environmental practices to reduce environmental damage and to comply with regulations; and (4) some significant reductions in environmental degradation are occurring. The second hypothesis, that the foregoing positive impacts of NAFTA were merely cosmetic and would not be sustained after NAFTA was approved, focuses on the period after January 1994 to examine whether any advances occurring between 1990 and 1993 are continuing.

C. MEXICO'S ENVIRONMENTAL POLICY AND PERFORMANCE BEFORE 1990

Before the NAFTA debate emerged in 1990, Mexico's environment was deteriorating (e.g., Weitzenfeld, 1992). Increased urbanization due to structural adjustment[2] in the economy added greater demographic stress on an already inadequate infrastructure in both the cities and the border region (e.g., Sanchez, 1991). Poverty and industrialization contributed to ever greater levels of environmental degradation. The environment was a low priority for Mexico's government until the administration of President Miguel de la Madrid (1982–88). Although Mexico's first environmental agency, the Subsecretaria de Mejoramiento del Ambiente (SMA), had been formed in the Ministry of Health in 1972, the agency had little impact on halting environmental deterioration, even in Mexico City where most of its efforts were directed (e.g., Mumme, 1988). For economic reasons, Mexico continued to produce and sell gasoline with a very high lead content at lower prices than unleaded gas. Another obvious problem involved industrial water use. Industrial pre-treatment in Mexico was virtually nonexistent. The SMA was not oriented toward inspection or enforcement (Bath, 1982).

Despite De la Madrid's support for higher environmental quality, his 6-year term of office coincided with the debt crisis that was precipitated by falling oil prices. In December 1982, De la Madrid's government created a new federal environmental policy agency, the Secretaria de Desarrollo Urbano y Ecologia (SEDUE), with more authority than the SMA, but only very modest resources were allocated to it. SEDUE suffered from very high turnover in its top leadership, having four ministers during De la Madrid's term of office. However, it was during this period that a latent environmental movement began to surface among the intellectual elite and within the middle class (Nuccio, 1991; Mumme and Sanchez, 1990).

Environmental concerns began to play a role in presidential politics during the 1988 elections. Even the PRI's candidate, Salinas de Gortari, voiced strong support for a clean environment along with economic growth. Although the impact of public concerns on the electoral outcome is unclear, the poor showing of the PRI in the Federal District (Mexico City), where voters had placed a high priority on environmental quality, suggests that the public was not convinced by Salinas' expressions of support for environmental improvement (Mumme, 1992a).

De la Madrid, in coordination with the incoming Salinas administration, championed comprehensive changes to environmental legislation by sponsoring the new General Law of Ecological Equilibrium and Environmental Protection in 1988. In many ways this new law was patterned after U.S. statutes. For example, the law established some specific environmental standards for the first time and gave SEDUE the authority to develop other regulatory standards. The law also established a "police" approach to regulatory enforcement for the first time, with strict fines and even jail terms

---

[2] The term "structural adjustment" encompasses a series of policies promoted by the International Monetary Fund to balance aggregate supply and demand and to promote the production of exportable goods in order to ease balance of payments problems. These policies usually include lower government expenditures, increases in taxes, higher domestic interest rates and higher producer prices, policies to increase capital investment, trade liberalization, and wage restraints (Khan and Knight, 1985). Unsurprisingly, such policies are not politically popular (Haggard, 1985).

TABLE 19.1
Environmental Conditions in Mexico City, Guadalajara, and Monterrey

	Mexico City	Guadalajara	Monterrey
Population	19,000,000	4,000,000	3,000,000
Manufacturing firms	35,000	6,695[1]	5,000
Motor vehicles	2,500,000	445,612[2]	500,000
Solid waste (daily)	19,000 tons	3,295 tons[3]	2,500 tons
Waste collected (%)	80	70–75	60
Air pollution due to:			
Motor vehicles (%)	83	38[4]	40
Industry (%)	13	41[4]	50
Natural sources (%)	5	21[4]	10

Sources: Mexico City information is based on Gamboa de Buen (1991).
Monterrey information is based on García Ortega (1991).
The Guadalajara information is based on the following different sources:
1. Mercamétrica (1994). Data are for 1988.
2. Mercamétrica (1994). Data are for 1990.
3. *El Universal* (1994).
4. *La Jornada* (1993). Data are for 1992.

for violations. In a few instances the 1988 General Law exceeded U.S. requirements (Gresham and Bloomfield, 1993).[3] However, SEDUE's budget was very low, only $4.3 million in 1989 (U.S. General Accounting Office, 1992b). Environmental rhetoric and rules were developing but without resources to implement them.

1. URBAN MANUFACTURING SECTOR

The three largest metropolitan areas in Mexico — Mexico City, Guadalajara, and Monterrey — accounted for 33% of the entire population and almost 59% of the nation's manufacturing GNP in 1988 (INEGI, 1994). While some manufacturing takes place in other smaller cities, very little systematic information is publicly available about the environmental problems of these cities (*La Jornada*, 1993). Thus, we will focus on the three largest industrial cities. A summary comparison of these three cities is found in Table 19.1.

The Mexico City metropolitan area constitutes the world's second most populated metropolis after the Tokyo–Yokohama area. It was widely believed to have the most polluted air in the world (*New York Times*, 1991). From 1986 to 1991, air quality in Mexico City deteriorated steadily (Quadri and Sanchez, 1992), from 102 days in 1986 when its air was reported as "satisfactory" to 33 days in 1989[4] (INEGI 1995). Probably the most serious environmental hazard was caused by the dust from unpaved streets that mixed with untreated human fecal matter to produce a very unhealthy exposure to bacteria (*La Jornada*, 1994).

Guadalajara is Mexico's second most populated city. For the most part, industry in Guadalajara is dominated by small firms and has excelled in such areas as food, commerce, and electronic

---

[3] For example, environmental impact statements would now have to be prepared for any new construction that might adversely affect the environment, not just for federal government projects as in the United States (Gresham and Bloomfield, 1993). However, this new requirement was routinely violated (U.S. General Accounting Office, 1992b).

[4] Mexico City's air quality is measured with an indicator called the IMECA (indice metropolitano de la calidad del aire). Readings up through 100 points indicate "satisfactory" air quality. The lowest number of days of satisfactory air quality occurred in 1991 for 26 days (INEGI, 1995).

assembly. Little data have been published about Guadalajara's environmental conditions before 1990. Regular air quality monitoring was not initiated until December 1993. Most of the municipal dumps in Guadalajara were not covered, and untreated waste contaminated the soil (*El Universal*, 1994). During the 1980s, efforts to control pollutants were essentially nonexistent.

Monterrey is Mexico's third largest metropolitan area and its second most industrial city. Monterrey is a major center for the production of cement, glass, steel, and other heavy industry. As a result of its large manufacturing base, it was tied with Guadalajara in terms of its levels of pollution despite its smaller size (*El Sol de Mexico*, 1994). However, more programs to control Monterrey's pollution were begun earlier than in Guadalajara. Wastewater treatment and solid waste collection systems were in place by the end of the 1980s (Garcia Ortega, 1991).

Overall, the environmental conditions in Mexico's urban manufacturing sector were deteriorating before the NAFTA debate began in 1990. Government policy consistently favored economic development over the environment (Mumme, 1991). Economic policy focused on industrialization, in both its import substitution and export promotion varieties. For the greater part of the post-war period until the early 1970s, Mexico pursued a policy of import substitution while almost entirely ignoring its environment. During the 1970s, Mexico began to follow an export-driven model of development, but its attention to environmental quality remained largely symbolic through the 1980s. Both sets of policies tended to increase environmental degradation in the urban manufacturing sector unless they provided resources for basic public services. Only in Mexico City were serious efforts begun to reduce air pollution. It is in this context that the environmental debate surrounding the NAFTA must be placed.

2. The Border Zone

Mexico created its Border Industrialization Plan in 1965 to provide incentives for direct foreign investment, especially by U.S. companies. Under the 1965 border plan, foreign firms could bring raw materials, components, and capital equipment into Mexico duty-free to plants within a prescribed distance of the border, initially 12 miles, as long as the final products were exported from Mexico. The key attraction of the plan for foreign companies was the low labor cost for simple assembly work. By most economic indicators, the Border Industrialization Plan worked very well, as indicated in Table 19.2a.

While initially the maquiladora plants had been restricted to a short distance from the border, after 1972 the only geographical restriction prohibited these plants from the heavily industrialized cities of Mexico City, Guadalajara, and Monterrey (George and Tollen, 1985). However, the plants remained in the border area for the most part.[5] Jobs brought population from inland Mexico to increasingly crowded border cities (see Table 19.2b). Border workers earned some of the highest per capita incomes in Mexico. However, border infrastructure for the rapidly growing cities was inadequate, and this affected the environment in a number of ways (e.g., Kelly et al., 1991; U.S. General Accounting Office, 1991; Logsdon, 1993).

The most critical environmental issue was inadequate-to-nonexistent sewage treatment in most border cities. For example, in the Tijuana–San Diego area, a public health quarantine was declared in 1980 for a 4-kilometer beach north of the border. Data from Calderon (1990) confirmed the very high measurements of fecal coliform bacteria in the Tijuana coastal region. A bilateral agreement to reduce untreated wastewater dumped into the Tijuana River was negotiated in 1985, but this solution was not adequate. By 1990, 12 million gallons of raw sewage flowed into the Tijuana River daily. Sewage treatment was a problem in all border cities (Kelly et al., 1991; U.S. Environmental Protection Agency, 1992).

---

[5] In 1981, 81% of the maquiladoras were still located along the border, and the majority of jobs (56%) were in just two cities, Ciudad Juarez and Tijuana. By 1985, these plants provided Mexico's second highest amount of export revenue after petroleum (Stoddard, 1987). A 15% growth rate was predicted for the early 1990s (Mirowski and Helper, 1989).

TABLE 19.2A
Data on Border Zone Development

Year	No. of Plants	Average No. Employees	Foreign Exchange Earnings ($ mil.)
1970	120	20,327	81
1975	454	67,214	454
1980	620	119,546	773
1985	760	211,968	1,267
1986	890	249,833	1,295
1987	1,125	305,253	1,598
1988	1,396	369,489	2,338
1989	1,467	418,533	3,001
1990	1,707	447,606	3,552

Sources: For 1970–1980, Stoddard (1987); for 1985–1990, U.S. General Accounting Office (1992a).

TABLE 19.2B
Population in the Border Zone, 1990

Largest Sister City Pairs	Metropolitan Population	Total Population
Tijuana, Baja California	742,686	
San Diego, California	2,498,016	3,240,702
Ciudad Juarez, Chihuahua	797,679	
El Paso, Texas	591,610	1,389,289
Reynosa, Tamaulipas	376,676	
McAllen, Texas	383,545	760,221
Mexicali, Baja California	602,390	
Calexico, California	109,303	711,693
Matamoros, Tamaulipas	303,392	
Brownsville, Texas	260,120	563,512
Total Mexican border population	3,500,038	
Total U.S. border population	5,722,694	9,222,732

Source: U.S. Environmental Protection Agency (1992).

In terms of air pollution, unpaved streets and roadways contributed to very high levels of particulates. This problem was especially noticeable in Ciudad Juarez where particulates were a major source of haze when temperature inversions reduced air circulation. Juarez had over 3,000 km of unpaved roads (Gray et al., 1989). Air emissions from leaded gasoline remained very high because over 80% of Juarez's sampled vehicles that were intended to operated on unleaded gasoline had the fuel-filter restrictor removed so that the cheaper leaded gas could be used (Gray et al., 1989).

Solid waste generation and disposal were increasingly becoming a problem during the 1980s. Of the 3,286 metric tons of garbage generated in Mexico's border cities every day, only 46% was collected by public units, and then 65% of the collected garbage was dumped into open pits because conventional landfills did not exist in many communities (Herrera, 1992).

Industrial and commercial sources of pollution did contribute significantly to certain environmental problems, and they were widely suspected in others, even when documentation was absent. Data on actual pollutant creation and control practices were scarce, but evidence indicated that compliance with environmental regulations was extremely low. For example, in a 1990 study by

SEDUE, only 6% of the maquiladora facilities complied with operating license requirements (Gresham and Bloomfield, 1993). The industry-generated environmental problem that created most concern involved the use of hazardous materials. At least part of the fear came from the lack of information about the quantities used and disposal practices (Applegate and Bath, 1990). Furthermore, the available data were said to be "uniformly unreliable because of underreporting, insufficient monitoring, underfunded public agencies, and fragmentation of governmental authority" (Guidotti, 1990). Studies stimulated by the NAFTA debate implicated maquiladora plants as sources of hazardous effluents (Lewis et al., 1991; Taylor, 1992).

When discussion of the free trade agreement arose in 1990, attention began to focus on the state of Mexico's urban and border environments. Critics found much to be concerned about and expressed fears that Mexico would become a "pollution haven" for companies that did not want to comply with U.S. and Canadian environmental regulations. The 25-year experience of freer trade along the border was hardly the model that NAFTA's supporters could hold up for close scrutiny. Nor could the condition of Mexico's three major industrial cities, especially Mexico City itself, give much confidence that the environment would be protected if manufacturing grew with free trade.

D. NAFTA'S IMPACT ON MEXICO'S ENVIRONMENTAL POLICY AND PERFORMANCE (1990–93)

Once the possibility of NAFTA was raised with the meeting of Presidents Salinas and Bush in June 1990, Mexico's environment, particularly along the Mexico–U.S. border, came under public scrutiny. Critics of NAFTA began to link free trade and environmental damage in a very vocal attack against any loosening of trade restrictions (e.g., Vogel, 1995). NAFTA's supporters did not agree that free trade would have negative impacts on environmental conditions, but they nonetheless were forced to address the deteriorating environmental conditions in Mexico, if only to demonstrate goodwill about the highly politicized issue.

A first step toward more effective environmental regulation of industry's pollution involved the development of ecologically based technical norms, which were recommended but not required standards for pollutants. This effort began early in the Salinas administration (Mumme, 1991).[6] Another related indication of this shift was evident in SEDUE's budgets.[7] Its allocation increased dramatically from $4.3 million in 1989 to $66.8 million in 1992. About $30.6 million of the 1992 budget was designed specifically for border infrastructure improvements (U.S. General Accounting Office, 1992b). A major shift to more enforcement began to occur. The number of personnel in inspection and enforcement increased from 81 in 1989 to 250 in 1992 (U.S. General Accounting Office, 1992b), and the number of inspections increased, as indicated in Table 19.3.

In May 1992, a new agency at the cabinet level, the Secretariat of Social Development (SEDESOL), was created. SEDESOL was charged with urban development issues as well as ecological concerns. The structural change was intended to increase the visibility and authority of environmental regulators. Rulemaking and permitting would now be separated from enforcement (Gresham and Bloomfield, 1993). The National Institute of Ecology (INE) was given the responsibility for creating regulations and approving permits. Enforcement would now be undertaken by a new Federal Attorney's Office for Environmental Protection (PROFEPA). PROFEPA was given the authority to inspect, fine, and even close plants that were found in violation of the regulations developed by the INE. PROFEPA was also obligated to investigate citizen's complaints, called "denuncia popular," about polluting facilities.

---

[6] In 1989, seven normas tecnicas ecologicas or NTEs were designed; in 1990, three were added; and in 1991, 20 new NTEs were developed (U.S. General Accounting Office, 1992b). In 1992, the NTEs were replaced by normas oficiales mexicanas or NOMs, which are standards enforceable by regulators and not merely recommended standards.

[7] A substantial portion of SEDUE's budget increase came from a World Bank loan of $50 million and grant of $30 million, negotiated in April 1992 for a 4-year period. Mexico agreed to provide matching funds of $46.6 million over the period for a total increase in funds of $126.6 million for 1992-96 (U.S. General Accounting Office, 1992b).

TABLE 19.3
Environmental Enforcement Activities (1992–94)

State	Population (1990)	Contribution to manufacturing GNP (1988) (%)	Inspections (1992)	Inspections (1/1/93–6/30/94)
Aguascalientes	719,650	0.80	95	310
Baja California	1,657,927	1.81	202	887
Baja Cal. Sur	317,326	0.13	196	227
Campeche	528,824	0.15	75	572
Chiapas	3,203,915	0.74	0	319
Chihuahua	2,439,954	2.65	113	238
Colima	424,656	0.14	32	46
Coahuila	1,971,344	3.87	139	1,015
Distrito Federal*	8,236,960	23.38	1,329	11,738
Durango	1,352,156	1.27	15	60
Guanajuato	3,980,204	3.20	0	270
Guerrero	2,622,067	0.36	4	95
Hidalgo	1,880,632	1.86	30	0
Jalisco*	5,278,987	7.10	31	237
México, Edo de*	9,815,901	18.43	60	1,042
Michoacan	3,534,042	1.41	142	186
Morelos	1,195,381	1.47	48	174
Navarit	816,112	0.56	99	261
Nuevo León*	3,086,466	10.07	117	382
Oaxaca	3,021,514	0.98	0	0
Puebla	4,118,059	3.08	0	1,000
Querétaro	1,044,227	2.23	199	57
Quintana Roo	493,605	0.13	0	0
San Luis Potosi	2,001,966	2.11	26	107
Sinaloa	2,210,766	0.90	0	0
Sonora	1,822,247	1.72	268	756
Tabasco	1,501,183	0.59	59	362
Tamaulipas	2,244,208	1.72	206	680
Tlaxcala	763,683	0.76	159	307
Veracruz	6,215,142	5.28	0	0
Yucatan	1,363,540	0.95	48	157
Zecatecas	1,278,279	0.18	21	200
Total	81,140,923	100.00	3,713	21,685

*Most manufacturing occurred in these states.

Sources: Informe de Actividades de la Secretaría de Desarrollo Social, México, D. F., (SEDESOL), 1993, 1994. Sistema de Cuentas Naciones de México. Instituto Nacional de Estadística, Geografía, e Informática, México, D. F., (INEGI), 1994.

With respect to the impact of NAFTA on regulatory enforcement, we can observe in Table 19.4 that from 1987 through 1990, SEDUE conducted 6,882 plant inspections throughout Mexico (SEDESOL, 1993).[8] A doubling of the number of inspections occurred in 1991 over 1990, and the

---

[8] A word of caution should be mentioned. The government figures shown in Table 19.4 are not the result of a rigorous reporting process. The biannual reports of SEDESOL are compiled based on informal reports of how many inspections or plant closings occurred during a certain period. They are not the result of an actual count of inspections and closings, but rather are good estimates. As a result, some of the figures do not coincide. It is impossible to reconcile the figures.

TABLE 19.4
Regulatory Inspections and Plant Closings, 1982–95*

| | Inspections | | Plant Closings | |
Period	Total	Average Monthly Rate	Total	Average Monthly Rate
1982–84	1,209	33.5	N.A.	N.A.
1985–86	3,525	146.9	N.A.	N.A.
1987	1,034	86.4	N.A.	N.A.
1988	2,501	208.4		
1989	1,922	160.2	983[1]	27.3[1]
1990	1,425	118.8		
1991	**3,119**	**259.9**	**N.A.**	**N.A.**
1992	**3,713**	**309.4**	**653**	**54.4**
1993	**14,387**	**1,198.9**	**632**	**52.7**
1994	8,187	682.3	207	34.5
1995	13,993	1,166.1	280	23.3

[1] These figures relate to the period 1988–90.

Sources: Zagaris (1992); SEDESOL (1993, 1994); INEGI (1995, 1996); SEMARNAP (1996); Diario Official (1996). Note that data comparability between 1982–86 (Zagaris, 1992) and 1987–95 (Mexican government sources) is uncertain. Boldface indicates NAFTA-induced shifts in enforcement.

1992 data indicate another 19% increase. With the creation of PROFEPA in 1992, inspectors began to close polluting plants more frequently.

In terms of overall trends, one can discern that the organization of PROFEPA as the enforcement arm of SEDESOL in July 1992 had a major impact on the regulatory activity taking place within Mexico. Interestingly, the number of plant closings due to violations of environmental standards did not continue to increase annually. A change in PROFEPA's policy occurred to focus more on prevention rather than police-style enforcement. The environmental audit became an important instrument to correct environmental problems. The audit allowed government inspectors to evaluate the environmental compliance of a given company and to develop with company officials a plan of action to bring the company into compliance. During the period in which the plan is in effect, no fines can be imposed or plants closed, thus creating an incentive for executives to cooperate and even volunteer for an audit. Since PROFEPA's creation in 1992, 541 environmental audits have been initiated, of which 425 were concluded and 116 were still in process in early 1996 (Diario Official, 1996).

1. Urban Manufacturing Sector

Looking at regulatory activity by geographical area in Table 19.3, we see that in Mexico City, Guadalajara, and Monterrey, which account for 59% of the country's manufacturing GNP, the number of inspections increased from 1,509 in 1992 to 13,399 for the 1/1/93–6/30/94 period. In terms of proportion of total inspections, these industrialized areas received 40.6% of the inspections in 1992, and then 61.8% in the latter period. It is apparent that enforcement activity was not only increasing dramatically, but also concentrating more and more on the industrialized areas of the nation.

The foregoing data describe an overall strengthening of environmental law with an increase in enforcement activity beginning in 1991. What has been the response of business to these changes in environmental regulation and enforcement? Measuring response is somewhat problematic because firms are usually not willing to disclose data concerning compliance with emission require-ments or figures dealing with environmental budgets. One approximation of corporate response is the establishment of an environmental department as a way to focus regulatory compliance efforts. In 1993, a telephone survey of 100 large businesses in Monterrey was conducted to determine the extent to which large businesses had established an environmental department or related function (Husted, 1994). It was found that 45% of large Monterrey businesses had an environmental function of some sort, at least a manager, if not an entire department. Of those, 60% were created after January 1, 1990.

Clearly, many large Monterrey businesses have responded organizationally since 1990. This response seems to be due to the combination of governmental actions, including the passage of the General Ecology Law in 1988, the debate surrounding NAFTA which began in 1990, and the organization of PROFEPA in 1992. The survey does not capture the activity of other businesses that may have responded without necessarily organizing a department for that purpose. It should be remembered that large businesses accounted for about 56.5% of manufacturing GNP in 1980 (Ruiz Duran and Zubiran Schadtler, 1992). Interviews reveal that many small and medium-sized firms, which produced the remaining 43.5% of manufacturing GNP, had not yet begun to comply with environmental regulations (Husted, 1994).[9]

What does this mean in terms of environmental quality? Since more hard data are available about air pollution than other forms of pollution, most of the evaluation is framed in terms of air pollution. In Mexico City, a number of programs were implemented to reduce air pollution in the early 1990s (Simonian, 1995). These programs began to show significant results in 1994. While from January to November 1993 there were only 31 days in which the air quality met minimum international standards (El Financiero, 1993), by August 1994 there was a 90% reduction in lead in the atmosphere, a 70% reduction in sulfur dioxide, and a 38% reduction in carbon monoxide from June 1992 levels (Reforma, 1994b). Suspended particles were reduced by 75% in the period from June 1993 to May 1994 (La Jornada, 1994), and nitrogen dioxide levels remained stable (Reforma, 1994b). Ozone remained the biggest problem, showing only a 2% reduction in levels during the same period (Reforma, 1994b). Ozone levels exceeded international standards at least once a day more than 200 days in 1993 (Reforma, 1994a).

In Guadalajara, it appears that air quality has continued to deteriorate. In the period following the initiation of the NAFTA debate, ozone emissions tripled from 1990 to 1994 (El Sol de Mexico, 1994). Lead concentrations in the air also increased during that period. Most of this deterioration appears to be due to population growth and the consequent explosion in the number of motor vehicles on the city's roads. The motor vehicle population increased from about 445,612 units in 1990 (Mercamétrica, 1994) to over 1,000,000 vehicles by November 1993 (La Jornada, 1993). The city of Guadalajara had not implemented a vehicle inspection program as Mexico City and Monterrey had done.

In Monterrey as in Mexico City, it appears that some reductions in pollution have been made. According to state government sources, 70% of motor vehicles have participated in the vehicle verification program, and large business firms have invested at least 400 million pesos in pollution control equipment (Silverstein, 1992). PEMEX and the Comision Federal de Electricidad (the electric power company) have switched from fuel oil to natural gas as a source of energy. As a result of fuel oil use declining from 44% in 1992 to 17% in 1994, sulfur oxide emissions have been reduced by 16,000 tons, and carbon monoxide emissions have decreased by 30,278 tons (Casas Sauceda, 1994).

In summary, progress in environmental protection occurred in the early 1990s in the major urban manufacturing areas but was uneven. Governmental enforcement increased substantially, particularly in Mexico City. Monterrey's big business sector became more actively involved in trying to pre-empt regulation by making environmental protection its own concern. Only in Guadalajara was

---

[9] The earliest reports of assignment of the environmental function in the Husted (1994) survey were in 1954 in the first company and 1969 in the second company. Eight more companies made such assignments in the 1970s, and another eight companies in the 1980s.

environmental deterioration unmitigated. In terms of the state of Mexico's environmental quality, the evidence shows less pollution in some urban areas and further deterioration in others.

2. IN THE BORDER ZONE

During the NAFTA debate, SEDUE stepped up enforcement of environmental standards in the border area, and companies began to comply more frequently. For example, in a 1990 study by SEDUE, only a 6% of the maquiladora facilities complied with operating license requirements, but in 1991 compliance improved to 54.6% (cited in Gresham and Bloomfield 1993). Compliance with SEDUE's rules to describe waste handling in border plants and to return hazardous waste to the country of origin also improved. The number of companies declaring production of hazardous waste increased from 30% in 1990 to 55% in 1991. The number of plants returning waste to the country of origin grew from 14.5% in 1990 to 31% in 1991 (Gresham and Bloomfield, 1993).

A report on the state of the maquiladoras' environmental conditions in June 1992 indicated that 73% of 1,502 plants had a license to operate from SEDESOL, 25% had some type of air pollution control system, and 11% had a wastewater treatment system (INEGI 1995). In terms of enforcement, only 135 inspections had taken place. This lack of regulatory oversight changed dramatically in 1993 when 1,117 inspections of maquiladora plants occurred. Sixty-four inspections (5.7%) resulted in partial or total closure of facilities; 884 inspections (79.1%) found slight irregularities; and 169 (15.1%) found no irregularities (SEDESOL, 1994). This evidence suggests higher compliance rates in plants along the border than in industrialized cities. Environmental scrutiny of maquiladora plants also stimulated actions by some U.S. parent corporations (e.g., *Business Week*, 1992). According to a survey by the American Chamber of Commerce of Mexico, companies reported an average 85% increase in environmental expenditures over the previous five years (Askari, 1993).

To deal with criticisms about border pollution that threatened to derail the NAFTA agreement itself, the Bush and Salinas administrations devised the first phase of the Integrated Border Environmental Plan in 1991 and early 1992. Its major focus was on infrastructurally related environmental problems, especially sewage treatment plants (U.S. Environmental Protection Agency, 1992). The plan promised unprecedented resources in the amount of $589 million over two years. Of this amount, $460 million would be provided by Mexico. While the plan demonstrated a higher commitment on both sides of the border to upgrade and coordinate policies and resources than in the past, it did not satisfy critics (Mumme, 1992a). They said that the plan did not go much beyond funding commitments already made for sewage treatment plants and called for more investigation of hazardous waste problems (U.S. House of Representatives, 1991).

With estimates of $6.5 to $8 billion needed for border infrastructure over a 10-year period, clearly more permanent solutions were needed beyond the interim approach of the Integrated Plan. The incoming Clinton administration was more sympathetic to environmental concerns surrounding NAFTA and supported more long-lasting institutional innovations to handle border pollution. Two new bilateral organizations were announced just before the NAFTA debate in U.S. Congress in fall 1993; the Border Environment Cooperation Commision (BECC) and the North American Development Bank (NADBank). The three initial priorities for BECC attention were established in the initial announcement as wastewater treatment, drinking water supply, and municipal solid waste projects (U.S. General Accounting Office, 1994). While local governments were responsible for seeking the best financing they could, the BECC was expected to assist in locating financial support for approved projects, often through the newly established NADBank and also through other government programs and private sources (Taylor, 1995).[10]

---

[10] The NADBank was established to provide financing for BECC-approved projects from a fund of $450 million, half from Mexico and half from the U.S., over a 4-year period. These funds could be leveraged up to the authorized limit of $2.55 billion. A study by the U.S. Corps. of Engineers estimated the costs of needed sewage treatment plants for 11 Mexican border cities at $2.5 billion, at least some of which would be provided by NADBank funding (Noah, 1994).

TABLE 19.5
Assessing the Evidence

Criteria	Pre-1990 (baseline)	1990–93 (NAFTA-induced)	1994–95
Environmental policy	Weak	Upgraded	Maintained
Environmental enforcement	Poor	Significantly improved	Maintained
Business behavior	Poor	Improving, esp. large firms	Improving
Environmental quality	Deteriorating	Slower rate of deterioration	?

E. ASSESSING THE EVIDENCE FOR THE FIRST HYPOTHESIS

The evidence about the initial impact of NAFTA on Mexico's environment leads to the following assessment. Mexico began to move toward a greater awareness of environmental issues during the administration of De la Madrid (1982–88), primarily because of increasing environmental degradation from population growth, increasing urbanization with inadequate infrastructure, and higher industry-generated pollution from increased production levels. However, resources were not provided for effective environmental standards and enforcement until the 1990s when NAFTA brought greater scrutiny of Mexico's deteriorating environmental conditions.

NAFTA brought the promise of higher GDP with which to provide resources for environmental protection and clean-up. Without NAFTA on the drawing board, it is unlikely that the Salinas administration would have been able to negotiate for World Bank funds in order to increase enforcement. It is also unlikely that the U.S. and Mexican governments would have negotiated the Integrated Border Environmental Plan to provide $589 million for border infrastructure and other environmental projects. The Salinas Administration did implement a number of measures to strengthen standards and enforcement. Mexico and the U.S. also designed several new bilateral programs and organizations to clean up border pollution and improve border infrastructure. Action at the local level also gained support because of linkage with NAFTA-triggered environmental scrutiny (e.g., Blackman and Bannister, 1998).

We conclude that it is unlikely that the multi-faceted attention to environmental issues in Mexico in the early 1990s would have occurred in the absence of the NAFTA debate. It appears that strengthening Mexico's environmental policy was perceived as essential in order to generate support for freer access to U.S. and Canadian markets (Vogel, 1995). Thus, as indicated in Table 19.5, the first hypothesis is supported.

F. HAS IMPROVEMENT BEEN SUSTAINED?
EVIDENCE ABOUT THE SECOND HYPOTHESIS

Now the question shifts to whether the early-1990s commitment to strengthening environmental policies would be sustained. Once NAFTA became effective in January 1994 and the presidential elections were held late in 1994, how would the newly elected Mexican president, Ernesto Zedillo, deal with environmental policies and problems? Critics had argued that increased Mexican environmental enforcement was at best a response to U.S. pressure for environmental responsibility and was a kind of "preemptive reform" to reduce political pressure by responding to critics without making fundamental changes (Mumme, 1992b). According to this view, well publicized plant closings in Mexico City were nothing more than cleverly designed photo opportunities to gain the support of members of the U.S. Congress for the Free Trade Agreement. Once the agreement was signed, critics expected a reduction in commitment to protect the environment while NAFTA supporters expected a continuation of gradual improvement as Mexico became more productive and wealthy.

President Zedillo revealed his environmental concerns early in his term with a structural change to create a new cabinet-level environmental department, the Secretariat of Environment, Natural Resources, and Fishing (SEMARNAP). SEMARNAP includes the National Institute of Ecology (INE) for standard-setting and PROFEPA for enforcement as well as several other agencies focused specifically on natural resources, water, and fishing. This higher status was intended to demonstrate Zedillo's understanding of the increasing importance of the environment, both domestically and internationally.

A significant challenge to Zedillo's environmental commitment was raised by the December 1994 devaluation of the peso and the resulting economic crisis. Given the austerity program that the government was undergoing and the refusal of many firms and individuals to pay taxes, many questioned whether the government's environmental efforts, especially enforcement, would continue. One area to investigate is the budget for SEMARNAP. Since it was estimated that the Treasury collected about 20% less in taxes during the first three months of 1995 (*El Financiero*, 1995c), it was feared that the economic crisis could provoke an environmental crisis if the government appeared to be decreasing SEMARNAP's funding. In fact, the 1995 budget was $4.07 billion (almost $600 million in U.S. dollars). This represented about a 48% increase over 1994 funding in real dollars (Carabias Lillo, 1995). About 70% of the 1995 budget was committed to building wastewater treatment plants and other projects of the Commision Nacional de Agua. Although no environmental personnel were added, no personnel cuts were made among the approximately 4,000 staff directly involved in environmental management at SEMARNAP (Diario Official, 1996). Therefore, the standard-setting process could continue within INE. The number of official technical norms increased from 62 in 1994 (INEGI, 1995) to 83 in 1995 (SEMARNAP, 1996b).

The most critical test would be the amount of enforcement during the economic crisis. As shown in Table 19.4, during 1995 there were 13,993 inspections carried out by PROFEPA. The average monthly rate of 1,166.1 inspections represents a decrease of only 2.7% in enforcement activity from the high reached in 1993 during the NAFTA debate. In the border area, inspections numbered 993 in 1995, with 14 (1.4%) resulting in partial or total closure of facilities (SEMARNAP, 1996a). No irregularities were found in 196 inspections (29.8%). This compares favorably with the 26.1% of inspections with no violations throughout the nation (SEMARNAP, 1996a). Border plants appear to be maintaining their lead in compliance over plants located elsewhere in Mexico.

With regard to the two new border environmental institutions, they were slow to be organized and funded (U.S. General Accounting Office, 1994; Taylor, 1995). It was not until February 1995 that an American general manager and a Mexican deputy manager were named to head the BECC. Guidelines for BECC projects were not finalized until late 1995. The U.S. share of the NADBank authorization for fiscal 1996, $56 million, was threatened during the fall 1995 budget debate, but funding was restored in later negotiations and matched by Mexico's equal contribution. Whether the remaining three years of NADBank funding will be provided is not certain, but there continues to be strong U.S. presidential and legislative support for border environmental projects (Richardson, personal communication, 1996).

This evidence leads to rejection of the second hypothesis. The peso devaluation brought to the forefront many of the critics' concerns about this new model of development for Mexico. The fear was that interest in the environment would dissipate with serious domestic economic pressures on government and businesses and without much foreign pressure. However, the economic crisis presented Mexico with the opportunity to test its commitment. Despite extraordinary pressure on policymakers and budgets because of reduced resources, Mexico's commitment to environment protection was maintained.

G. SUMMARY AND CONCLUSION

In summary, a number of signs indicate that Mexico's environmental policymaking and enforcement did improve in the early 1990s while NAFTA was being debated. There is also evidence to infer that the NAFTA-influenced environmental commitment has been maintained. Despite the 1995 financial crisis, environmental policymaking and enforcement have not been subjected to budget cuts that were applied to other government programs. Also the technical norms for establishing regulatory standards have continued to be developed. All indications suggest that the structural changes are in place upon which to base higher levels of environmental protection. NAFTA has left an indelible mark on environmental policy in Mexico.

The impacts of the new bilateral environmental institutions that have been created by NAFTA are not yet fully known. They are still evolving, and progress in setting them up has been slow. It is clear that Mexico's environmental performance will be much more visible to outside observers in the future, and critics will have more access to information about environmental quality and problems than in the past (Mumme, 1994). Such information may allay the concerns of those who fear a return to the past approach of valuing economic development over a healthy environment.

Unfortunately, the NAFTA-induced leap in environmental regulation and enforcement has not yet been sufficient to create significant changes in overall environmental quality. While the rate of deterioration has slowed and, in some cases, the pollutant concentrations have been substantially reduced, much remains to be done. Compliance among business firms remains a serious problem. Most firms are not being inspected, which is true in every nation. However, only 26.1 percent of those inspected are in full compliance, and some skeptics might doubt even this low figure. The respect for law and rules is not deeply entrenched in Mexico's cultural value system, yet abiding by environmental rules is necessary to avoid sanctions under NAFTA (Quinones, 1995). It is likely that pressure on companies to improve compliance and begin to incorporate pollution prevention into planning and operations will increase. In addition to external scrutiny, the growing domestic concern about the state of their environment by middle- and upper-class Mexicans will provide support for continued government and business attention to pollution control and enforcement (North American Institute, 1994).

With regard to the fundamental theoretical question about the relationship between free trade and the environment, as stated early in the chapter, the evidence presented here is not able to address the long-term question of whether the degradation of Mexico's environment will be halted once the full impacts of NAFTA are felt after the year 2009. However, it is clear that environmental concerns have become legitimated in terms of future trade negotiations (Vogel, 1995). With the NAFTA negotiations as the precedent, environmental impacts and interest groups will be integrated into the previously separated world of trade. The NAFTA experience suggests that this linkage will strengthen environmental policy and lay the foundation for higher environmental quality in the future.

ACKNOWLEDGMENTS

The authors acknowledge the research assistance of Elvira Gonzalez and Roberto Perez at the Instituto Tecnologico y de Estudios Superiores de Monterrey (Mexico) and Robert Bitto and Tim Carroll at the University of New Mexico. They also appreciate the comments of three anonymous reviewers.

REFERENCES

Applegate, H. G. and Bath, C. R., Hazardous and toxic substances along the U.S.–Mexico border, in *Environmental Hazards and Bioresource Management in the United States–Mexico Borderlands*, Ganster, P. and Walter, H., Eds., UCLA Latin American Center Publications, Los Angeles, 1990, 119–129.
Askari, E., The greening of Mexico, *Albuquerque J.*, May 9: B1, B4, 1993.

Bath, C. R., U.S.–Mexico experience in managing transboundary air resources: problem, prospects, and recommendations for the future, *Natural Resour. J.*, 22, 1147, 1982.

Bhagwati, J., The case for free trade, *Sci. Am.*, 269, 42, 1993.

Blackman, A. and Bannister, G., Cross-border environmental management and the informal sector: The Ciudad Juarez Brickmakers' Project, in *Cooperation and Conflict: Environmental Management on North America's Borders*, Wirth, J., and Kiy, R., Eds., Island Press, Washington, D.C., in press.

Business Week, How do you clean up a 2,000-mile garbage dump?, July 6, 31, 1992.

Calderon, J. L., Policies and strategies for the control of contamination of water on the northern Mexican border, in *Environmental Hazards and Bioresource Management in the United States-Mexico Borderlands*, Ganster, P. and Walter, H., Eds., UCLA Latin American Center Publications, Los Angeles, 1990, 31–47.

Carabias Lillo, J., Orientaciones generales de politica de medio ambiente, recursos naturales y pesca, speech delivered on February 15, 1995, accessed on May 13, 1996 at internet site, Http://semarnap.conabio.gov.mx/orienta/orienta.htm, 1995.

Casas Sauceda, D., Logran empresas nuevoleonesas reducir emisiones por el combio de combustible, *El Universal*, April 1:1, 1994.

Daly, H. E., The perils of free trade, *Sci. Am.*, 269, 50, 1993.

Diario Official, Programa de medio ambiente: 1995–2000, Diario Official de la Federación, April 3, First Section, pgs. 73–84, Second Section, pgs.1–128, 1996.

El Financiero, Respirables sólo 31 dias de este año, según las normas internacionales, Nov. 26, 88, 1993.

El Financiero, La industria, marginada del plan emergente, Apr. 18, 26, 1995a.

El Financiero, Sin responder la industria maquiladora a la devaluación, Apr. 18, 5B, 1995b.

El Financiero, Cae 20% la captación tributaria en terminos reales dice Hacienda, Mar. 9, 1, 1995c.

El Financiero, Aplicó profepa multas por más de 4 mdnp a empresas, May 22, 58, 1995d.

El Sol de México, Guadalajara y Monterrey siguen al DF en contaminación, Feb. 9, 57, 1994.

El Universal, Nulo tratamiento a la basura en Guadalajara, Feb. 11, 70, 1994.

Faber, D., The ecological crisis of Latin America: a theoretical introduction, *Lat. Am. Perspect.*, 19, 3, 1992.

Gamboa de Buen, J. and Revah Locoutre, J. A., Servicios urbanos y medio ambiente: El caso de la Ciudad de México, in *Servicios Urbanos, Gestión Local y Medio Ambiente,* Schteingart, M., and d'Andrea, L., Eds., El Colegio de México and Centro di Ricerca e Documentazione Febbraio '74, D. F., México, 1991, 375–388.

García Ortega, R., Area metropolitana de Monterrey: problemática ecológica, servicios y medio ambiente (antecedentes y situación actual), in *Servicios Urbanos, Gestión Local y Medio Ambiente*, Schteingart, M., and d'Andrea, L., Eds., El Colegio de México and Centro di Ricerca e Documentazione Febbraio '74, D. F., México, 1991, 399-413.

George, E. Y. and Tollen, R. D., The economic impact of the Mexican border industrialization program, Center for Inter-American and Border Studies, El Paso, TX, 1985.

Globerman, S., The environmental impacts of trade liberalization, in *NAFTA and the Environment*, Anderson, T., Ed., Pacific Research Institute, San Francisco, CA, 1993, 27.

Gray, R., Reynoso, J., Diaz, C., and Applegate, H., *Vehicular Traffic and Air Pollution*, Texas Western Press, El Paso, TX, 1989.

Greshan, Z. O. and Bloomfield, T. A., The North American Free Trade Agreement and Mexico: environment requirements and trade opportunities, *Environ. Manage. Rev.*, 29, 116, 1993.

Guidotti, T. L., Hazardous substances in the San Diego–Tijuana conurbation, in *Environmental Hazards and Bioresource Management in the United States–Mexico Borderlands*, Ganster, P. and Walter, H., Eds., UCLA Latin American Center Publications, Los Angeles, 1990, 131.

Haggard, S., The politics of adjustment: lessons from the IMF's extended fund facility, *Intern. Organ.*, 39, 505, 1985.

Herrera, S., The ecological factor in the NAFTA, *Bus. Mex.*, 2, Apr., 28, 1992.

Husted, B., Environmental regulation and big business in Mexico: evidence from Monterrey, N.L., unpublished working paper, Instituto Tecnológico y de Estudios Superiores de Monterrey, 1994.

INEGI, *Sistema de Cuentas Nacionales de México*, Instituto Nacional de Estadística, Geografía, e Informática, D. F., México, 1994.

INEGI, *Estadísticas del Medio Ambiente: México 1994*, Instituto Nacional de Estadística, Geografía, e Informática, D. F., México,1995.

INEGI, *Anuario Estadístico de los Estados Unidos Mexicanos 95*, Instituto Nacional de Estadística, Geografía, e Informática Aguas Calientes, Ags., 1996.

Kelly, M. E., Kamp, D., Gregory, M., and Rich, J., U.S.–Mexico free trade negotiations and the environment, *Columbia J. World Bus.*, 26, 42, 1991.

Khan, M. S. and Knight, M. D., Fund-supported adjustment programs and economic growth, Occasional Paper No. 41, International Monetary Fund, Washington, D.C., 1985.

La Jornada, Critica, la contaminación en seis cuidades, en algunas duplica la norma internacional, Nov. 18, 70, 1993.

La Jornada, Se redujo 75% la concentración de partículas suspendidas totales, Sept. 12, 39, 1994.

Lewis, S. J., Kaltofen, M., and Ormsby, G., *Border Trouble: Rivers in Peril*, National Toxic Campaign Fund, Boston, 1991.

Logsdon, J. M., Environmental performance in the U.S.–Mexico border area: what is known and unknown, in *Proceedings of the Fourth Annual Meeting of the International Association for Business and Society*, Pasquero, J. and Collins, D., Eds., 1993, 512.

Mercamétrica, *Mercamétrica de 80 cuidades Mexicanas*, Vol. 1, Mercamétrica Ediciones, D. F., México, 1994.

Mirowski, P. and Helper, S., Maquiladoras: Mexico's tiger by the tail? *Challenge*, 32, 24, 1989.

Mumme, S. P., Complex interdependence and hazardous waste management along the U.S.–Mexico border, in *Dimensions of Hazardous Waste Politics and Policy*, Davis, C.E. and Lester, J.P., Eds., Greenwood Press, New York, 1988, 223.

Mumme, S. P., Clearing the air: environmental reform in Mexico, *Environment*, 33, 26, 1991.

Mumme, S. P., New directions in United States–Mexican transboundary environmental management: a critique of current proposals, *Nat. Resour. J.*, 32, 539, 1992a.

Mumme, S. P., System maintenance and environmental reform in Mexico: Salinas' preemptive strategy, *Lat. Am. Perspect.*, 19, 123, 1992b.

Mumme, S. P., Mexican environmental reform and NAFTA, *North Am. Outlook*, 4, 87, 1994.

Mumme, S. P. and Sanchez, R., Mexico's environment under Salinas: institutionalizing reform, *Rev. of Lat. Am. Stud.*, 3, 44, 1990.

New York Times, Mexico City emits more heat over pollution, Nov. 25, A5, 1991.

Noah, T., Cleaning up, *Wall Street Journal*, Oct. 28, R8, 1994.

North American Institute, The North American Commission for Environmental Cooperation: Early implementation. Report and recommendations from a workshop sponsored by the North American Institute, Vancouver, B.C., March, 1994.

Nuccio, R. A., The possibilities and limits of environmental protection in Mexico, in *Economic Development and Environmental Protection in Latin America*, Tulchin, J.S., Ed., Lynne Rienner Publishers, Boulder, CO, 1991, 109.

Pastor, R. A., NAFTA's green opportunity, in *Assessments of the North American Free Trade Agreement*, Moss, A. H. Jr., Ed., Transaction Publishers, New Brunswick, NJ, 1993, 19.

Quadri de la Torre, G., and Sánchez Cataño, L. R., *La ciudad de México y la contaminación atmosférica*, Limusa/Noriego Editores, D. F., México, 1992.

Quinones, S., Mexico takes new approach to environmental policy, *Albuquerque J.*, Oct. 22, B5, 1995.

Reforma, Aún no se controla el ozono en DF, Feb. 12, 76, 1994a.

Reforma, Disminuye contaminación 90%, Sept. 2, 7, 1994b.

Richardson, L., Personal communication with legislative aide to Sen. Pete Domenici, May 29, 1996.

Ruiz Duran, C. and Zubiran Schadtler, C., *Cambios en la Estructura Industrial y el Papel de las Micro, Pequenas y Medianas Empreses en México*, Nacional Financiera, D. F., México, 1992.

Sánchez, R. A., El tratado de libre comercio en América del norte y el medio ambiente de la frontera norte, *Frontera Norte*, 3, 5, 1991.

Schmidheiny, S., *Changing Course: A Global Business Perspective on Development and the Environment*, MIT Press, Cambridge, 1992.

SEDESOL, *Informe de la situación general en materia de equilibrio ecológico y protección al ambiente, 1991–1992*, Secretaría de Desarrollo Socia, Instituto Nacional de Ecología, D. F., México, 1993.

SEDESOL, *Informe de la situación general en materia de equilibrio ecológico y protección al ambiente, 1993–1994*, Secretaría de Desarrollo Socia, Instituto Nacional de Ecología, D. F., México, 1994.

SEMARNAP, Internet Home Page, Secretaríade Medio Ambiente, Recursos Naturales y Pesca, accessed on May 10, 1996 at internet site, http://semarnap.conabio.gob.mx/opinion/profepa.htm, 1996a.

SEMARNAP, La simplificación administrativa en materia de impacto ambiental no significa un relajamiento de la normatividad, Document of the Secretaría de Medio Ambiente, Recursos Naturales y Pesca, Dec. 5, accessed on May 13 at internet site, http://semarnap.conabio.gob.mx/simplifi.htm, 1996b.

Silverstein, J., Monterrey clean up: corporations take the lead, *Bus. Mex.*, 2, 28, 1992.

Simonian, L., *Defending the Land of the Jaguar: A History of Conservation in Mexico*, University of Texas Press, Austin, 1995.

Srinivasan, T. N., Why developing countries should participate in the GATT system, *World Economy*, 5, 85, 1982.

Stoddard, E. R., *Maquila: Assembly Plants in Northern Mexico*, Texas Western Press, El Paso, TX, 1987.

Taylor, L., The fast track trade agreement — help or hurt for the U.S.–Mexico border environment? *The Workbook*, 17, 50, 1992.

Taylor, L., Cleaning up the border: will sustainability be a priority? *The Workbook*, 20, 50, 1995.

U.S. Environmental Protection Agency, Integrated environmental plan for the Mexican–U.S. border area (first stage, 1992–1994), Environmental Protection Agency, Washington, D.C., 1992.

U.S. General Accounting Office, U.S.–Mexico trade: information on environmental regulations and enforcement, Report to the Chairman, Committee on Commerce, Science and Transportation, U.S. Senate, U.S. General Accounting Office, Washington, D.C., May 1991.

U.S. General Accounting Office, North American Free Trade Agreement: U.S.–Mexican trade and investment data, U.S. General Accounting Office, Washington, D.C., Sept. 1992a.

U.S. General Accounting Office, U.S.–Mexico trade: assessment of Mexico's environmental controls for new companies, U.S. Government Accounting Office, Washington, D.C., Aug. 1992b.

U.S. General Accounting Office, North American Free Trade Agreement: Structure and status of implementing organizations, U.S. General Accounting Office, Washington, D.C., Oct. 1994.

U.S. House of Representatives, Committee on Small Business, Protecting the environment in North American Free Trade Agreement negotiations, Hearing before the Subcommittee on Regulation, Business Opportunities, and Energy, Serial No. 102–144, U.S. Government Printing Office, Sept. 30, 1991.

Vogel, D., *Trading Up: Consumer and Environmental Regulation in a Global Economy*, Harvard University Press, Cambridge, MA, 1995.

Weitzenfeld, H., Contaminación atmosférica y salud en América Latina, *Infectología*, 12(5), 403, 1992.

Zagaris, B., The transformation of environmental enforcement cooperation between Mexico and the U.S. in the wake of NAFTA, *N. C. J. of Law and Commer. Regul.*, 18, 61, 1992.

20 The Empirical Relationship between Trade, Growth, and the Environment

Lewis R. Gale and Jose A. Mendez

This note reestimates Grossman and Krueger's (1993) SO_2 emissions regression, including regressors to capture the effects of scale, trade, and trade policy. Several new results are obtained. Increases in economic activity have a negative effect on the environment separate from changes in per capita income, whose relation to the environment is now positive and linear, not inverted U-shaped. The trade policy measure is not significant, but its effect is ambiguous *a priori*. Finally, in line with specialization patterns based on traditional sources of comparative advantage, pollution rises with the capital abundance of a country (since this favors capital-intensive and generally dirtier industries) and falls with increases in labor and land abundance.

(*JEL* classifications: Q2, O1, F1)

A. INTRODUCTION

The debate over the North American Free Trade Agreement (NAFTA) led to numerous studies of trade's likely impact on the environment (e.g., see the references cited in Pastor, 1992 and Cline, 1994). By far, the most often cited and influential is that of Grossman and Krueger (1993) (GK), who analyzed the relationship between trade, growth, and the environment by regressing measures of environmental quality for cities in various countries on per capita income, trade intensity, and other site characteristics.[1] This approach yielded two novel results. First, GK discovered a particular relationship between environmental quality and economic growth: at low per capita income levels, environmental quality falls with increases in per capita; however, beyond a per capita income somewhere between $4,000 to $5,000, increases in income per capita lead to improved environmental quality. Second, GK found that a country's trade intensity (the sum of imports and exports as a share of GDP) is positively related to environmental quality, that is, the more open the economy, the less it pollutes.

In this chapter we reexamine these two findings. GK contend that the inverted U-relationship between pollution and per capita income may reflect the changing strength of two influences on the environment, the **scale** and **technique** effects: initially increases in economic activity generate more pollution, but as incomes and living standards continue to rise, citizens' increased desire for a cleaner environment leads to more stringent pollution standards and the subsequent replacement of older technologies by newer, cleaner ones.[2] However, GK use only one proxy, income per capita, to capture both influences and, as a result, it is unclear that the inverted U-relationship reflects these

---

[1] Several empirical papers have looked at aspects of GK's work not considered here: they are referenced in Grossman and Krueger (1995) and Grossman (1995).

[2] Crossman (1995) also contains a mathematical decomposition intended to highlight the separate influence of these effects on the environment.

two influences or whether they operate in the manner hypothesized. To better differentiate between these two effects, we develop a city-specific measure of scale while retaining GDP per capita at the national level to reflect the influence of more stringent environmental standards on production techniques. Our results are supportive of GK's contention of separate influences. We find that increases in economic activity do degrade the environment (although this result is weak in a statistical sense), while higher per capita income is associated with improved environmental quality. However, once scale is controlled for, the relation between per capita income and pollution emissions is no longer inverted U-shaped but linear.

In contrast to their first finding, GK found the second puzzling.[3] The trade intensity variable was used in part to capture **composition** effects, the environmental effects arising from trade policy induced changes in an economy's output mix. While GK recognized that these effects could be positive if traditional sources of comparative advantage determine specialization patterns, they were puzzled by their finding since their intent was to explore whether cross-country differences in environmental regulations determine competitive advantage.[4] Since their finding suggests that specialization may be linked to traditional sources of comparative advantage, we tested this hypothesis by substituting determinants of trade intensity consistent with this possibility: country-specific measures of factor abundance and a trade policy measure. Our results are striking. We find that cross-country differences in endowments do exert an influence on the environment and in a direction that accords well with intuition.[5] Pollution emissions rise with the capital abundance of a country, which we know favors production of more capital-intensive and generally more polluting sectors, while increases in labor and land abundance are associated with lower pollution levels, which would favor relatively cleaner sectors and industries. Greater land abundance may also entail greater absorptive capacity and, consequently, lower overall pollution levels. The coefficient on the trade policy measure is not statistically significant, but *a priori* reasoning suggests that trade policy's impact on pollution is likely to be ambiguous.

The following section briefly outlines GK's methodology and our revisions of their specification. The empirical results are presented in Section C and our summary in Section D.

B. GROSSMAN AND KRUEGER'S METHODOLOGY

To better understand the changes we make to GK's analysis, first consider their methodology. Data on three measures of air quality from various urban areas in different countries and across time were regressed on national per capita income and a number of site- and city-specific variables. The three measures are ambient levels of sulphur dioxide (SO_2), dark matter suspended in the air, and the mass of suspended particles found in a given volume of air.[6] Site- and city-specific independent variables were included to control for factors likely to affect the pollution reading at a site independent of the level and nature of economic activity.[7] A trade intensity variable (the sum of imports and exports divided by GDP) was included for the country in which the site is located. Finally, since data for three years (1977, 1982, and 1988) are pooled, a time trend is included to control for the possibility that air quality may have been improving or worsening secularly.

---

[3] For instance, GK (1993, p.17) wrote "we have no good economic explanation for this finding."

[4] Composition effects are damaging if trade openness shifts production to countries with lax environmental standards.

[5] Although their model is highly stylized, Copeland and Taylor (1994) show that composition effects are likely to be very important, and one can argue that our endowment measures are picking up this effect. They also show consistent with that of our empirical results, that increased capital abundance raises pollution, while increased labor abundance lowers pollution by encouraging specialization in the labor-intensive, nonpollution-intensive good. We thank a referee for these insights and bringing Copeland and Taylor (1994) to our attention.

[6] These data were obtained from the Global Environmental Monitoring System, a joint project of the World Health Organization and the United Nations Environment Programme.

[7] They are the last seven variables defined and listed in Table 20.1.

TABLE 20.1
Variable Definitions, Means, and Standard Deviations

Variable	Definition (Mean, Standard Deviation)
SO_2	Annual median concentration of SO_2 in micrograms per cubic meter at site i, in city j, in country k (46.3, 38.3)
SCALE	GDP (millions of U.S. $) of country k scaled by the ratio of city j population to country k population (14776.8, 18668.7)
y	Per capita GDP (1,000s of U.S. $) of country k (5908.2, 4097.0)
CAPITAL	Country k share of world capital stock divided by country k share of world GDP (0.896, 0.210)
LABOR	Country k share of world nonprofessional, literate workers divided by country k share of world GDP (2.367, 3.005)
LAND	Country k share of world humid, mesothermal climate divided by country k share of world GDP (2.489, 4.153)
POLICY	Index of overall domestic price distortions for country k (1.58, 0.28)
T	Trade intensity: the sum of exports and imports divided by GDP in country k (0.412, 0.318)
COAST	Coastal city dummy: 1 when city j in country k is located along a coastline and 0 when not (0.52, 0.50)
CENTR	City center dummy: 1 when site i located in city j, country k, is located in the city center, 0 when not (0.51, 0.50)
INDSTRL	Industrial use dummy: 1 when site i located in industrial area, 0 when not (0.315, 0.467)
RESD	Residential use dummy: 1 when site i located in residential area, 0 when not (0.348, 0.479)
COMMU	Communist dummy: 1 when site i located in country governed by communist government, 0 when not (0.0033, 0.179)
BUBB	Gas bubbler dummy: 1 when measurement device at site i is a gas bubbler, 0 otherwise (0.717, 0.453
POPDN	Population density (10,000 people per square mile) of the city in which the measurement device is located (30769.8, 47369.8)

GK found that the following specification fit the data for concentrations of SO_2 emissions and dark matter:[8]

$$
\begin{aligned}
\left(SO_2\right)_{ijk} = &\ \beta_0 + \beta_1 y_k + \beta_2\left(y_k^2\right) + \beta_3\left(y_k^3\right) + \beta_4 POPDN_{jk} + \beta_5 T_k + \beta_6 COAST_{jk} \\
&+ \beta_7 CENTR_{ijk} + \beta_8 INDSTRL_{ijk} + \beta_9 RESD_{ijk} + \beta_{10} COMMU_k + \beta_{11} BUBB_{ijk} + \omega_i
\end{aligned}
\tag{20.1}
$$

where subscripts i, j, and k refer to site i, located within city j, in country k. Variables with only a k subscript refer to the country where the measurement device is located, while those with a jk subscript refer to the city where the site is located. All variables are defined in Table 20.1, but note that y is the per capita GDP and T the trade intensity of the country where the measurement device is located. ω is a random error term consisting of a common-to-city component μ and an idiosyncratic component ε, or $\omega = \mu + \varepsilon$.

Note that SO_2 emissions appear as a cubic function of y. It is this result that gives rise to GK's inverted U relationship between pollution and per capita income, which the authors contend reflects the influences of both the scale and technique effects. However, note that an increase in GDP (the scale measure) is never independent of an increase in GDP per capita, and even though GK do introduce population density (POPDN) to control for scale, it is ineffective. T was found to be negatively related to SO_2, but the authors confessed that they were puzzled by the result, expecting trade openness to be associated with increased pollution.

We modify GK's specification as follows. First, we introduce a separate city-specific scale measure of economic activity that is more independent of y. The measure, SCALE, is obtained by weighing a country's GDP by the city's share of the country's population.[9] As hypothesized by

---

[8] The function for suspended particles was decreasing in per capita income.
[9] City and country population data are from the *World Almanac and Book of Facts* (1982).

GK, we expect SCALE to be positively related to pollution concentration. We retain y to capture the technique effect, which is very appropriate since environmental standards are generally set nationally, and expect it to be negatively related to pollution concentration.[10]

Second, we replace T by variables that would be its determinants if specialization were based on traditional sources of comparative advantage: country specific measures of factor abundance and a measure of the restrictiveness of a country's trade policy. The endowment measures capture the effects of the extent and pattern of specialization on the environment. Trade theory has demonstrated that the more dissimilar a country's endowments are to the rest of the world, the more specialized its production and as a consequence the greater its trade intensity.[11] In addition, the pattern of a country's factor abundance or scarcity affects the composition of commodities produced and this in turn influences the amount of pollution generated in a location. For instance, countries that are capital abundant may have higher pollution levels since this favors capital-intensive and generally dirtier industries. On the other hand, labor and land abundant countries will tend to specialize in industries or sectors such as agriculture, light assembly and services that are often less polluting or dirty.[12] As suggested by the Heckscher–Ohlin–Vanek theory, we measure a country's relative factor abundance by dividing the country's share of the world supply of a factor by its share of total world production. From Leamer (1984) we obtained endowment data for physical capital (CAPITAL), nonprofessional and literate workers (LABOR), and land with a humid and mesothermal climate (LAND).

Finally, we use a measure of the restrictiveness of a country's trade policy since trade intensity should be a function of trade policy; the more restrictive the trade policy, the lower the trade volume and hence trade intensity. But, as noted by GK, the effect on pollution is ambiguous since it is not possible to predict how a trade policy induced shift in the composition of production may affect pollution. For instance, in the case of developing countries, restrictive trade policies such as those followed under an import-substitution industrialization strategy, may lead to greater pollution by encouraging capital-intensive and dirty industries like autos, chemicals, and steel.[13] The reverse may occur in industrialized countries since protectionist policy favors more labor-intensive and cleaner industries. To measure the likely restrictiveness of trade policy, we expand to industrial countries an index developed in World Bank (1983) to gauge developing country openness during the 1970s. The index uses practical approximations commonly used in policy to measure distortions in domestic prices, the price of foreign exchange, capital, labor, and infrastructure services (utilities). These distortions are classified as low (a value of 1), medium (a value of 2), and high (a value of 3) for seven categories and we use the average of these to obtain our index of openness (POLICY).[14]

With the above in mind, our specification is

$$
\left(SO_2\right)_{ijk} = \alpha_0 + \alpha_1 SCALE_{jk} + \alpha_2 y_k + \alpha_3 CAPITAL_k + \alpha_4 LABOR_k + \alpha_5 LAND_k
$$

$$
+ \alpha_6 POLICY_k + \alpha_7 COAST_{jk} + \alpha_8 CENTR_{ijk} + \alpha_9 INDSTRL_{ijk}
$$

$$
+ \alpha_{10} RESD_{ijk} + \alpha_{11} COMMU_k + \alpha_{12} BUBB_{ijk} + \omega_i \tag{20.1$'$}
$$

Note that since y now only controls for the technique effect, this eliminates the need for its nonlinear treatment.

---

[10] Data for per capita GDP were obtained from World Bank (1981).

[11] Helpman and Krugman (1985) show formally that the more dissimilar a country's factor composition to that of its trading partners, the more specialized the country in products in which it possesses a comparative advantage and the greater the volume of trade.

[12] One reason why agriculture is less polluting than manufacturing is that fossil fuel use is lower. For evidence of this, see Gale (1994).

[13] Gale (1994) finds that tariff elimination under NAFTA results in a shift away from the most polluting sectors in Mexico.

[14] Sources used to construct the variable POLICY were Krugman and Obstfeld (1991), Thomas and Nash (1991) and World Bank (1981, 1983, and 1991).

TABLE 20.2
Least-Squares Regression Results for 1979 Median Sulfur Dioxide Concentration (Standard Errors in Parentheses)

Variable	Specification	
	(1)	**(1′)**
y	0.018*	−0.0054**
	(0.011)	(0.0019)
y–squared	−3.02E–4	....
	(2.11E–4)	
y–cubed	1.23E–8	....
	(1.09E–8)	
POPDN	4.54E–3	....
	(9.40E–3)	
T	−2.07	....
	(11.94)	
SCALE	....	2.49E–2#
		(1.71E–2)
CAPITAL	....	35.33**
		(20.32)
LABOR	....	−4.16**
		(2.01)
LAND	....	−1.84**
		(0.74)
POLICY	....	3.40
		(15.18)
Adj. R^2	0.246	0.259
F	3.701	3.657
Sample size	92	92

Notes: Equations also include an intercept, and dummy variables for coastal location, city center, industrial location, residential location, communist government rule, and type of measurement device. White's variance-covariance method was used to correct for heteroscedasticity in both specifications.

**(*) Statistically significant at 0.05 (0.10) level for a two-tailed t-test.
Statistically significant at 0.10 level for a one-tailed t-test.

C. ESTIMATION AND FINDINGS

We employed the same environmental data used by GK, with two differences. First, our sample was restricted to data for 1979 since Leamer's endowment data were unavailable for the other two years. Our dependent variable is the observed annual median concentration of SO_2 in 1979 across 34 cities in 25 different countries. Second, again because of the unavailability of endowment data, we omitted three countries under communist rule. The cities and countries included in the sample are listed in the Appendix.

Prior to applying our specification, we first attempted to replicate GK's results in order to gauge whether our smaller sample conveyed similar information. The OLS estimates of equation (1) are listed in the first column of Table 20.2. The results are similar to GK's. (To save space, we have not listed the estimated coefficients for the site and city-specific dummy variables.) The two differences that arise are that the coefficients for T and y-squared and cubed are not significant, although of the right sign. The lack of significance may be due to the loss of degrees of freedom, the strong possibility of multicollinearity particularly for the y terms, and the fact that reducing the sample across time reduces the variability of the y variables. In any case, an F-test of the joint hypotheses that the coefficients of the squared and cubed y terms are zero was rejected.

The OLS estimates obtained by applying Equation (20.1′) are listed in the second column of Table 20.2.[15] Our results are striking and supportive of our specification. First, note that by controlling for the influence of scale, the coefficient on y is now unambiguously negative and significant. Moreover, not only were the coefficients on y squared and cubed statistically insignificant, but we could not reject the joint hypothesis that they were zero. Second, the coefficient on SCALE is statistically significant and positive (although the result is weak statistically). Thus, as hypothesized by GK and claimed by opponents of economic growth, an increase in the scale of economic activity does have a detrimental effect on environmental quality. Third, the endowment variables are all significant and contribute to explaining the cross-country variation in SO_2 concentrations. Cities located in capital abundant countries have higher SO_2 pollution, whereas those located in labor and land abundant countries have lower concentrations. These results accord well with intuition since one would expect greater capital abundance to favor production of more capital-intensive and generally more polluting sectors, whereas greater labor and land abundance should favor relatively cleaner sectors and industries. Greater land abundance may also entail greater absorptive capacity and, consequently, lower overall pollution levels. Finally, we find that the estimated coefficient for trade policy is not statistically significant which in a sense is in line with *a priori* reasoning that it would be ambiguous.

D. SUMMARY

In their highly influential empirical analysis, Grossman and Krueger (1993) discovered an inverted-U shaped relationship between per capita income and pollution. The authors conjectured that the relationship partly reflected the changing strength of **scale** and **technique** effects: initially increases in economic activity generate more pollution, but as living standards rise, citizens' pressure for more stringent pollution standards results in newer, cleaner production technologies. However, GK provided no evidence to support these conjectures since only one variable (income per capita) was used to capture both influences. To better differentiate between these two effects, we developed a city-specific measure of scale while retaining GDP per capita at the national level. Our results are supportive of GK's contention of separate influences. Increases in economic activity to degrade the environment and higher per capita income is associated with improved environmental quality. However, the relation between income per capita and pollution is no longer cubic but linear.

GK also used a trade openness regressor to capture **composition** effects, the environmental effects of trade-induced changes in the mix of production. Their intent was to examine whether international competitiveness was determined by cross-country differences in environmental standards, but their result suggested that specialization patterns are more likely due to traditional sources of comparative advantage. We tested this hypothesis by substituting determinants of trade openness consistent with such a possibility: country-specific measures of factor abundance and a measure of the restrictiveness of a country's trade policy. Our results are again very plausible and theoretically appealing. The trade policy measure was not statistically significant, but *a priori* reasoning suggests that it would be ambiguous. Even more interesting, our results indicate that cross-country differences in endowments do exert an influence on the environment and in a direction that accords well with intuition. Pollution rises with the capital abundance of a country (since this favors capital-intensive and generally dirtier industries) and falls with increases in labor and land abundance. To our knowledge, this is the first empirical study to link pollution emissions to cross-country differences in endowments.

---

[15] The estimated coefficients for the site- and city-specific dummy variables are similar to those obtained using Equation (20.1) and are again not listed to save space.

ACKNOWLEDGMENTS

We wish to thank Josef Brada, Seung Ahn, and two anonymous referees for particularly valuable suggestions. We also thank Alan B. Krueger for kindly making the data used in his paper available to us.

APPENDIX

Cities Included in Sample

City	Country	City	Country	City	Country
Amsterdam	Netherlands	Dublin	Ireland	Milan	Italy
Athens	Greece	Frankfurt	Germany	New York	United States
Auckland	New Zealand	Glasgow	United Kingdom	Osaka	Japan
Bangkok	Thailand	Gourdon	France	Santiago	Chile
Bogota	Colombia	Hong Kong	Hong Kong	Sao Paulo	Brazil
Bombay	India	Houston	United States	St. Louis	United States
Brussels	Belgium	Kla.Lpur	Malaysia	Stockholm	Sweden
Calcutta	India	London	United Kingdom	Sydney	Australia
Cali	Colombia	Madrid	Spain	Tel Aviv	Israel
Chrst.ch	N. Zealand	Manilla	Philippines	Tokyo	Japan
Copenhagen	Denmark	Medellin	Colombia	Zagreb	Yugoslavia
Delhi	India				

REFERENCES

Cline, W. R., *International Economic Policy in the 1990s*, MIT Press, Cambridge, 1994.

Copeland, B. R., and Taylor, M. S., North–South trade and the environment, *Q. J. of Econ.*, (August), 755, 1994.

Gale, L. R., Essays on the environmental impact of the North American free trade agreement in Mexico, Ph.D. thesis, Arizona State University, 1994.

Grossman, G. M., and Krueger, A. B., Environmental impacts of a North American free trade agreement, in *The U.S.–Mexico Free Trade Agreement*, Garber, P., Ed., MIT Press, Cambridge, 1993.

Grossman, G. M., and Krueger, A. B., Economic growth and the environment, *Q. J. of Econ.*, (May), 353, 1995.

Grossman, G. M., Pollution and growth: what do we know? in *The Economics of Sustainable Development*, Goldin, I., and Winters, L. A., Eds., University of Cambridge Press, 1995.

Helpman, E., and Krugman, P. R., *Market Structure and Foreign Trade*, MIT Press, Cambridge, 1985.

Krugman, P. R., and Obstfeld, M., *International Economics*, 2nd edition, Harper Collins, New York, 1991.

Leamer, E. E., *Sources of International Comparative Advantage: Theory and Evidence*, MIT Press, Cambridge, 1984.

Pastor, R. A., NAFTA as the center of an integration process: the nontrade issues, in *North American Free Trade: Assessing the Impact*, Lustig, N., Bosworth, B. P., and Lawrence, R. Z., Eds., Brookings Institute, Washington, D.C., 1992.

Thomas, V. and Nash, J., Reform of trade policy: recent evidence from theory and practice, *The World Bank Res. Observer*, 6, 219, 1991.

World Almanac and Book of Facts: 1982, Pharos Books, New York, 1982.

World Bank, *World Development Report*, IBRD, Washington, D.C., 1981, 1983, 1991, 1992.

World Health Organization, *Air Quality in Selected Urban Areas, 1979–1980*, WHO, Geneva, 1983.

Index

Index   **331**